sipri

Stockholm International Peace Research Institute

SIPRI is an independent international institute for research into problems of peace and conflict, especially those of arms control and disarmament. It was established in 1966 to commemorate Sweden's 150 years of unbroken peace.

The Institute is financed mainly by the Swedish Parliament. The staff and the Governing Board are international. The Institute also has an Advisory Committee as an international consultative body.

The Governing Board is not responsible for the views expressed in the publications of the Institute.

Stockholm International Peace Research Institute
Frösunda, S-169 70 Solna, Sweden
Cable: SIPRI
Telephone: 46 8/655 97 00
Telefax: 46 8/655 97 33
Email: sipri@sipri.se
Internet URL: http://www.sipri.se

Cover design: *The scorpion is an arachnid with a long, slender, jointed tail ending in a curved poisonous sting. As an animal that is said to even sting and poison itself, it is analogous to mankind's use of chemical and biological weapons, which easily contaminate the users themselves.*

The Challenge of Old Chemical Munitions and Toxic Armament Wastes

The Challenge of Old Chemical Munitions and Toxic Armament Wastes

Edited by
Thomas Stock and Karlheinz Lohs

sipri
Stockholm International Peace Research Institute

OXFORD UNIVERSITY PRESS

This book has been printed digitally and produced in a standard specification in order to ensure its continuing availability

OXFORD
UNIVERSITY PRESS

Great Clarendon Street, Oxford OX2 6DP

Oxford University Press is a department of the University of Oxford.
It furthers the University's objective of excellence in research, scholarship,
and education by publishing world-wide in

Oxford New York

Auckland Bangkok Buenos Aires Cape Town Chennai
Dar es Salaam Delhi Hong Kong Istanbul Karachi Kolkata
Kuala Lumpur Madrid Melbourne Mexico City Mumbai Nairobi
São Paulo Shanghai Taipei Tokyo Toronto

Oxford is a registered trade mark of Oxford University Press
in the UK and in certain other countries

Published in the United States
by Oxford University Press Inc., New York

Reprinted 2004

ISBN 0-19-829190-6

Antony Rowe Ltd., Eastbourne

Contents

Part III. Coping with old munitions and toxic armament wastes: technical experience

Part IV. Chemical weapons: past production activities

Part VIII. Final discussion and recommendations

Preface

The problems related to old chemical munitions and toxic armament wastes have long been ignored. Although there are earlier studies, the topic began to receive significant attention only in the late 1980s and the early 1990s. Old chemical munitions and toxic armament wastes pose considerable technical and administrative problems. World War I and World War II, as well as other military conflicts, led to the accumulation of large quantities of military hardware and the production of waste materials, some of which are toxic. Munitions are among the most dangerous weapons. They are not easy to produce, but destroying or rendering them harmless is difficult and expensive. Many munitions have been buried or dumped in bodies of water. Although hidden, they remain dangerous to people and the environment.

Old chemical weapons (CW) constitute one type of toxic waste. The Chemical Weapons Convention (CWC), which opened for signature in January 1993 and entered into force on 29 April 1997, defines old chemical weapons as those produced before 1 January 1946. They are addressed to a limited extent by the CWC, as are buried chemical weapons and CW dumped at sea.

In 1992, when the negotiations on the CWC concluded, it was clear that the Convention would not solve all of the problems related to old chemical weapons. However, the end of the cold war and growing awareness about these problems created a situation in which solutions could be debated and political decisions regarding old chemical weapons and toxic armament wastes could more easily be taken.

This study was conceptualized in 1992, and its primary focus is on the following: *(a)* the characteristics of chemical warfare agents, munitions and toxic armament wastes, including their deterioration, detoxification and long-term behaviour in soil and water; *(b)* the situation in Europe after World War I and World War II as regards CW; *(c)* the location of CW dumping sites; *(d)* the technical means for handling and destroying old chemical weapons; *(e)* the methods and means of treating contaminated areas; and *(f)* the national and international legal aspects of old chemical weapons and toxic armament wastes.

In October 1993, 18 experts from 10 countries met in Munster, Germany, at the NBC [Nuclear, Biological and Chemical] Defense Research and Development Center of the Federal Armed Forces (now the German Armed Forces Scientific Institute for Protection Technologies: NBC Protection) to discuss old chemical munitions and toxic armament wastes, focusing particularly on Europe. The experts included chemists, toxicologists, munition specialists, technicians, arms control experts, legal experts and journalists, all participating in their private capacities. Their commitment of time and effort was crucial to completion of the study.

The papers presented at the conference form the basis of this volume. Detailed data on the economic and financial aspects of disposing of old chemical munitions and toxic armament wastes are not presented, but it is clear that such costs are extremely large. The historical aspects of the use of toxic agents as weapons also are not discussed since these have been dealt with in scientific and technical literature in the past 30 years.

This study presents not only general information which relates to old chemical munitions and toxic armament wastes but also specific case studies to illustrate the approaches taken by various countries to a problem for which no one solution can be found. Each instance involving old chemical munitions or toxic armament wastes is unique, and in each case the approach to the problem must be tailored to the chemicals and munitions present and the degree of danger which they pose. The book also assesses the provisions of the CWC which are intended to deal with problems related to old chemical munitions and toxic armament wastes.

Two research assistants, Filip Tangen and Irina Schoulgin, worked diligently on the project and assisted in preparing the conference. Both provided significant help in the creation of the database of relevant literature on various aspects of the topic of old chemical munitions and toxic armament wastes which was established and which is now available at SIPRI. Thanks are also due Christina Barkstedt and Cathy Söderquist who assisted in preparing the project conference in 1993 and in many other aspects of this study.

Jetta Gilligan Borg deserves special thanks for her excellent editorial work and preparation of camera-ready copy of the text.

This volume, the 1993 conference in Munster and research that led to both were made possible by a generous grant from the Volkswagen Foundation, without whose support the study could not have taken place.

***

Professor Karlheinz Lohs, the co-editor of this volume, died on 26 June 1996. Unfortunately, he did not live to see the final product of his work. His death was a true loss to the scientific and disarmament community. Professor Lohs, who served on the SIPRI Governing Board from 1980 to 1990 and thereafter on the SIPRI Advisory Committee until his death, was a well-known specialist on toxicology and chemical disarmament. The publication of this volume commemorates Professor Lohs and his work.

Dr Thomas Stock
SIPRI CBW Project Leader, 1992–96

Dr Adam Daniel Rotfeld
Director , SIPRI
April 1997

Contributors

Dr Bernd Appler
Wehrwissenschaftliches Institut für Schutztechnologien-ABC-Schutz
Postfach 1142
D – 29623 Munster
Germany

Dr Reiner Escher
TÜV Ostdeutschland Sicherheit und Umweltschutz GmbH–Mitglied der TÜV Rheinland Gruppe
Müggelseedamm 109-111
D – 12587 Berlin
Germany

Professor Frode Fonnum
Chief Scientist
Norwegian Defence Research Establishment
Division for Environmental Toxicology
PO Box 25
N – 2007 Kjeller
Norway

Dr Frédéric Guir
Ingénieur en Chef de l'Armement
Ministère de la Défense
Délégation aux Affaires Stratégiques
Sous-Direction Technologie et Armement
14, rue Saint-Dominique
F – 00450 ARMEES
France

Dr Hans-Joachim Heintze
Ruhr-Universität Bochum
Institut für Friedenssicherungsrecht und Humanitäres Völkerrecht
Gebäude NA 02/28
D – 44780 Bochum
Germany

Mr Thomas Kurzidem
Research Assistant
Institute for Public Law
Johann Wolfgang Goethe-Universität
Senckenberganlage 31
D – 60054 Frankfurt am Main
Germany

Mr Fredrik Laurin
Journalist
SVT-Swedish Television
S – 105 10 Stockholm
Sweden

Dr Karl Lehmann
Manager
TÜV Ostdeutschland Sicherheit und Umweltschutz GmbH–Mitglied der TÜV Rheinland Gruppe
Müggelseedamm 109-111
D – 12587 Berlin
Germany

Professor Dr Karlheinz Lohs
(deceased)
formerly, Lobbe GmbH
Entsorgung und Sanierung
Iserlohn, Germany

Mr Ron Manley
Organisation for the Prohibition of Chemical Weapons
Technical Secretariat
Laan van Meerdervoort 51A
2517 AE The Hague
The Netherlands

Dr Hermann Martens
Wehrwissenschaftliches Institut für Schutztechnologien-ABC-Schutz
Postfach 1142
D – 29623 Munster
Germany

Professor Jirí Matousek
Director
Institute for Chemistry and Technology of Environmental Protection
Faculty of Chemistry
Technical University Brno
Veslarská 230
CZ – 637 00 Brno
Czech Republic

Ms Judith Perera
Science Correspondent-IPS
10 Bridgewater Road
Berkhamsted
Herts HP4 1HN
United Kingdom

Professor Dr Wolfgang P. W. Spyra
Professor for Chemical Engineering and Hazardous Waste
Brandenburgische Technische Universität Cottbus
Postfach 10 13 44
D – 03013 Cottbus
Germany

Dr Thomas Stock
Lobbe Xenex GmbH & Co.
Entsorgung, Sanierung, Umwelttechnik
Stenglingser Weg 4–12
D – 58642 Iserlohn
Germany

Dipl. Ing. Josef Stolz
Privy Council (ret.)
Arsenal Objekt 12/2/60
A – 1030 Vienna
Austria

Professor Ronald Sutherland
Head
Department of Chemistry
University of Saskatchewan
Saskatoon S7N 0WO
Canada

Colonel Kazimierz Szarski (ret.)
ul. Konski Jar 2/36
02785 Warsaw
Poland

Professor Dr Zygfryd Witkiewicz
Institute of Chemistry
Military University of Technology
01489 Warsaw
Poland

Dr Jean Pascal Zanders
Chemical and Biological Warfare Project Leader
Stockholm International Peace Research Institute
Frösunda
S – 169 70 Solna
Sweden

Acronyms and abbreviations

Acronyms not defined in this list are defined in the chapters of this volume.

ACAMS	Automated continuous agent monitoring system
ACW	Abandoned chemical weapons
ADS	Agent Destruction System (USA)
BATNEEC	Best Available Techniques Not Entailing Excessive Cost
BIOS	British Intelligence Objectives Subcommittee
BPEO	Best Practical Environmental Option
BW	Biological weapon
CAIS	CW agent identification sets
CAM	Chemical agent monitor
CAMDS	Chemical Agent Munitions Disposal System (USA)
CBDE	Chemical & Biological Defence Establishment (UK)
CCR	Commission Centrale de la Récupération (Belgium)
CD	Conference on Disarmament
CDE	Chemical Defence Establishment (UK)
CEB	Centre d'Etudes du Bouchet (France)
CIOS	Combined Intelligence Objectives Subcommittee
CON	Control operating system
CPSU	Communist Party of the Soviet Union
CW	Chemical weapon
CWC	Chemical Weapons Convention
DAP	Destruction Advisory Panel
DATS	Drill and transfer system
DFS	Deactivation furnace system
DRES	Defence Research Establishment Suffield (Canada)
DUN	Dunnage incinerator
EC	European Community
EEZ	Exclusive economic zones
EOD	Explosive ordnance disposal
EPA	Environmental Protection Act (UK), Agency (USA)
EPP	Environmental Protection Plan (Canada)
EU	European Union
GCMS	Gas chromatography–mass spectroscopy
GT	Frontier Troops (German Democratic Republic)
HELCOM	Helsinki Commission
HELCOM CHEMU	Ad Hoc Working Group on Dumped Chemical Munition of the Baltic Marine Environment Protection Commission of the Helsinki Commission
IAEA	International Atomic Energy Agency
ILC	International Law Commission

JACADS	Johnston Atoll Chemical Agent Disposal System
LIC	Liquid incinerator
MPF	Metal parts furnace
MOD	Ministry of Defence
NAS	National Academy of Sciences (USA)
NATO	North Atlantic Treaty Organization
NATO SOFA	NATO Status of Forces Agreement
NBC	Nuclear, biological and chemical
NDRETOX	Division for Environmental Toxicology of the Norwegian Defence Research Establishment
NSCM	Non-Stockpile Chemical Materiel (USA)
NVA	National People's Army (German Democratic Republic)
OCW	Old chemical weapons
OMGUS	Office Military Government United States
OPCW	Organisation for the Prohibition of Chemical Weapons
OTA	Office of Technology Assessment (USA)
OVT	Operational verification test
PAS	Pollution abatement system
PE	Polyethylene
ppm	Parts per million
PrepCom	Preparatory Commission
PTS	Provisional Technical Secretariat
R&D	Research and development
RCRA	Resource Conservation and Recovery Act (USA)
RMA	Rocky Mountain Arsenal (USA)
ROV	Remote-operated vehicle
SDM	Service de Destruction des Munitions (Belgium)
SMAD	Soviet Military Administration in Germany
tpm	Tonnes per month
UNCED	United Nations Conference on Environment and Development
UNSCOM	United Nations Special Commission on Iraq
WGT	Western Group of the former Soviet Army
WHO	World Health Organization

Conventions in tables

. .	Data not available or not applicable.
–	Nil or a negligible figure.

Part I. Introduction to the problem

1. Introduction

THOMAS STOCK and KARLHEINZ LOHS

I. The scope of the problem

The efforts at the Geneva Conference on Disarmament to achieve a convention banning chemical weapons (CW) were successfully completed in September 1992.[1] The Convention on the Prohibition of the Development, Production, Stockpiling and Use of Chemical Weapons and on their Destruction (CWC) was opened for signature in Paris on 13 January 1993.[2]

The CWC abolishes the development, production, stockpiling and transfer of chemical weapons and outlaws the use of such weapons (which is banned by the 1925 Geneva Protocol[3]). These prohibitions will be verified and monitored by the Organisation for the Prohibition of Chemical Weapons (OPCW).[4] The parties to the CWC are also obliged to destroy within 10 years any CW stockpiles which they possess. (In exceptional cases, the time period can be extended to 15 years.) The Convention entered into force on 29 April 1997 (as a consequence of the deposit of the 65th ratification on 30 October 1996), and it may be possible to destroy nearly all the chemical weapons by early in the next century.

It is often assumed that entry into force of the CWC will solve all the problems related to chemical weapons, including those associated with old and abandoned chemical weapons. However, the provisions of the CWC which address the destruction of chemical weapons make it clear that the problems related to old and abandoned chemical weapons will not be dealt with so easily. In addition, past production activities during World War II and earlier may have caused environmental damage.

There are three primary reasons why chemical weapons remain a serious problem. First, the time allowed for the destruction of stockpiled chemical weapons is 10–15 years. Second, states that are not parties to the Convention are not obliged to destroy their CW stockpiles. Third, many old chemical munitions or warfare agents which were produced before 1946 have not yet been discovered, and these pose a great risk to humans and the environment. The parties to the CWC are obliged to declare and subsequently destroy any old chemical weapons of which they have knowledge.

The history of modern chemical warfare began on 22 April 1915 when German forces attacked French and Canadian troops with chlorine gas at the Battle

[1] CW is used throughout this volume as an acronym for chemical weapons.

[2] Perry Robinson, J. P., Stock, T. and Sutherland, R. G., 'The Chemical Weapons Convention: the success of chemical disarmament negotiations', *SIPRI Yearbook 1993: World Armaments and Disarmament* (Oxford University Press: Oxford, 1993), chapter 14, pp. 705–56, including the text of the Convention. The text is also reproduced in Conference on Disarmament document CD/1173, 3 Sep. 1992. CWC is used throughout this volume as an acronym for the Chemical Weapons Convention.

[3] The text of the Protocol for the Prohibition of the Use in War of Asphyxiating, Poisonous or Other Gases, and of Bacteriological Methods of Warfare is reproduced in Goldblat, J., SIPRI, *Agreements for Arms Control: A Critical Survey* (Taylor & Francis: London, 1982), pp. 135–36.

[4] OPCW is used throughout this volume as an acronym for the Organisation for the Prohibition of Chemical Weapons.

of Ypres.[5] During World War I a wide range of new agents was introduced. The discovery of a new class of chemical warfare agents—the organophosphorus agents, including sarin, soman and tabun—in the late 1930s and 1940s added a new dimension to chemical armament.

Chemical weapons were not used in World War II. However, Germany, Japan, the Soviet Union, the UK and the USA stockpiled large amounts of these weapons in an arms race that was fuelled by the perception that other countries possessed chemical weapons and intended to use them. At the end of World War II Germany had a large CW stockpile consisting mainly of mustard gas, nitrogen mustard, soman and tabun and the World War I agents phosgene, chloroacetophenone, adamsite, hydrogen cyanide and Clark. In 1945, after the war, the Soviet Union, the UK and the USA set up the Allied Control Commission to deal with the problems associated with the discovery, dismantling and disposal of German CW stocks.[6] In addition, there was a need to dispose of the chemical weapons and related equipment which had belonged to the UK and the former German allies, such as Hungary and Italy. Dumping at sea was the method of disposal which was chosen, and the environmental consequences of these activities were essentially ignored. The complexity of the issue was not given adequate consideration, and the knowledge obtained from earlier disposal activities in Europe and elsewhere was ignored. The effect on the environment was not considered important at the time.

The changed political climate in Europe at the end of the 1980s and the current interest of most European states in environmental policy has led to the realization that there is a need for greater cooperation on environmental matters. There is also increased openness and readiness to discuss the problems related to the destruction of specific weapons and armaments in general. Nevertheless, the destruction of chemical weapons does not appear to be a top priority for most European states. The parties to the CWC are obliged to destroy existing CW stockpiles and old chemical weapons, some of which may not yet have been discovered. In Europe, facilities which were formerly used for the production of chemical and conventional weapons also pose technical challenges.

The environmental and technical problems which have occurred since the end of World War II have demonstrated the risks that old chemical munitions pose to people and the environment. Many countries must cope constantly with newly discovered chemical weapons and conventional munitions from World War I (Belgium, France, Germany and the UK are particularly affected). In many European countries old CW munitions in the soil and water, and the contamination created by them, are 'chemical time bombs'.

Former military facilities, especially munition production sites, constitute an additional problem. In Germany alone there are more than 100 known World War II sites which were used for the production of gunpowder and explosives (including facilities where chemical warfare agents were produced), and few of these sites have been decontaminated and redeveloped. The thousands of tonnes of chemically contaminated soil which remain to be decontaminated are also a problem. Little attention was given during World War II to the environmental and safety standards of facilities which produced and handled chemical and

[5] SIPRI, *The Problem of Chemical and Biological Warfare*, vol. 1, *The Rise of CB Weapons* (Almqvist & Wiksell: Stockholm, 1971), p. 30; and chapter 15 and appendix 15A in this volume.

[6] See chapters 19 and 20 in this volume.

conventional munitions. Currently, there is far greater awareness and knowledge of the environmental damage which such activities can cause.

The outcome of World War II meant that in many cases those who had been responsible for CW production activities were no longer in charge after the war. Destruction and decontamination activities were often inadequate; the techniques employed were unprofessional; and potential damage to the environment was largely ignored. Circumstances after World War II and in the early 1950s created a situation in which some former production sites remained as they were when the war ended. Decontamination and redevelopment were never carried out, and external factors such as weather, erosion, partial decomposition, and the like sometimes increased contamination.

The scientific studies which have been conducted thus far indicate the limitations of current knowledge about the effects on the environment and marine life of the chemicals present at such sites. In recent years there have been numerous reports of incidents in which individuals have come in contact with old CW munitions or chemical warfare agents. There have also been many accidents involving fishermen and others, some of whose injuries are permanent. However, there is also evidence that, in some instances, the problems were exaggerated because of limited knowledge of historical facts and lack of understanding of the impact such chemicals have on people and the environment.

There is a substantial amount of information on the location, handling, transport, storage and disposal of old or obsolete CW munitions, although more scientific and technical study is needed. Environmental and worker safety as well as the long-term behaviour and toxicology of old CW agents and contaminated wastes need further study. Old chemical munitions and toxic military wastes burden the environment and are a potential hazard to the people living close to places where they are located.

Smoke munitions and obscurants (including tactical and training smoke munitions and agents) were used extensively in World War I and World War II. Some of these munitions are probably stored or buried in former combat areas. The basic components of these weapons were often dual-use chemicals, that is, for both civilian and military use. These chemicals do not pose the same degree of risk as chemical warfare agents because much is known about them.

Old chemical munitions and toxic armament wastes can be disposed of by burial, dumping at sea, open burning, chemical degradation, incineration and chemical transformation followed by burial. However, not all of these methods can be used today. Dumping at sea and open burning are environmentally unsafe and prohibited by laws, agreements and conventions. The experience which has been gained in recent decades from the destruction of chemical weapons and toxic armament wastes has led to the consideration of new technologies. Some of the so-called 'alternative' or 'emerging' destruction technologies under investigation may simplify and encourage the destruction of old and obsolete chemical weapons.

The technical and practical matters which must be taken into account when coping with old chemical weapons differ from those which need to be considered when dealing with CW stockpiles. It may be difficult to determine whether a munition is chemical or conventional, and the origin of old chemical weapons may also be hard to identify. In some instances the explosive charges will still be active and unable to be removed. In addition, old chemical weapons often contain a complex mixture of various agents that have already undergone decomposition. Planning and destruction are complicated by the fact that old CW munitions are discovered randomly. The following steps can be used to structure the process of finding and destroying old chemical munitions: (*a*) reconnaissance and location, (*b*) unearthing and identification, (*c*) removal and transport, (*d*) intermediate storage, (*e*) preparation of the munitions for demilitarization, and (*f*) demilitarization.

The findings of the United Nations Special Commission on Iraq (UNSCOM) in the aftermath of the 1991 Persian Gulf War highlighted the problems which are involved in destroying former production facilities and chemical weapons, especially those which are leaking, in an efficient and environmentally safe manner.[7] The experience of UNSCOM in Iraq can provide valuable practical lessons about destruction.

Old chemical munitions and toxic armament wastes are also found in Africa, the Far East and the Middle East as a result of the numerous military conflicts which have occurred in these areas in the past 70 years. Little is known about the nature of these weapons, but they are probably a mix of conventional and chemical munitions. Some of the countries affected were occupied by colonial powers or foreign troops. Past military activities may not be well documented, or the country which formerly ruled may not wish to reveal its activities. Under the CWC if old or abandoned chemical weapons are discovered an obligation exists to ascertain the country which abandoned the weapons. It will be difficult to secure such sites and to destroy or store such weapons.

In addition to the decontamination problems at sites which were formerly used for the production of conventional and chemical munitions there are contaminated areas where troops were stationed during the cold war. Highly contaminated areas, which had been used by the military, were discovered in the former GDR and elsewhere after the withdrawal of Soviet troops. Damage to the environment created by the presence of troops and their various activities is not a new problem, but it is one that became more apparent when the cold war ended. The sites in question were used by the military for decades, and the contamination is extensive. Often dangerous wastes were dealt with inadequately or without regard to environmental safety so that various types of contamination may exist at a particular site. The decontamination and redevelopment of such sites are complex and require the participation of specialists with experience in areas such as chemical toxicology, analytical chemistry, geology, hydrography, and the like. The experience gained in the decontamination and redevelopment of former military production sites may also be applicable here.

[7] UNSCOM is used throughout this volume as an acronym for the United Nations Special Commission on Iraq. For a discussion of UNSCOM activities see Ekéus, R., 'The United Nations Special Commission on Iraq', *SIPRI Yearbook 1992: World Armaments and Disarmament* (Oxford University Press: Oxford, 1992), chapter 13, pp. 509–30; Ekéus, R., 'The United Nations Special Commission on Iraq: activities in 1992', *SIPRI Yearbook 1993* (note 2), chapter 13, pp. 691–703; Trevan, T., 'UNSCOM: activities in 1993', *SIPRI Yearbook 1994* (Oxford University Press: Oxford, 1994), chapter 19, pp. 739–58; Stock, T., Geissler, E. and Trevan, T., 'Chemical and biological arms control', *SIPRI Yearbook 1995: Armaments, Disarmament and International Security* (Oxford University Press: Oxford, 1995), chapter 19, pp. 749–60; Stock, T., Haug, M. and Radler, P., 'Chemical and biological weapon developments and arms control', *SIPRI Yearbook 1996: Armaments, Disarmament and International Security* (Oxford University Press: Oxford, 1996), chapter 15, pp. 691–99; and chapter 17 in this volume.

II. Definitions

In Article II of the Chemical Weapons Convention chemical weapons are defined as:

> (*a*) Toxic chemicals and their precursors, except where intended for purposes not prohibited under this Convention, as long as the types and quantities are consistent with such purposes;
> (*b*) Munitions and devices, specifically designed to cause death or other harm through the toxic properties of those toxic chemicals specified in subparagraph (*a*), which would be released as a result of the employment of such munitions and devices;
> (*c*) Any equipment specifically designed for use directly in connection with the employment of munitions and devices specified in subparagraph (*b*).

This definition reflects the 'general purpose criteria' of the CWC, the way the Convention distinguishes between prohibited and non-prohibited activities.

In the CWC chemical warfare agents are called 'toxic chemicals' and defined in Article II as: 'Any chemical which through its chemical action on life processes can cause death, temporary incapacitation or permanent harm to humans or animals. This includes all such chemicals, regardless of their origin or of their method of production, and regardless of whether they are produced in facilities, in munitions or elsewhere'. Article II also defines 'precursor chemicals', chemicals which are essential for the synthesis of toxic chemicals, as: 'Any chemical reactant which takes part at any stage in the production by whatever method of a toxic chemical. This includes any key component of a binary or multicomponent chemical system'.

These definitions reflect the debate in the late 1980s on a comprehensive definition of chemical weapons. It was argued that such a definition ought to include recent technical developments, such as binary chemical weapons, and that the definition had to be flexible enough to cover new chemicals.

The CWC does not provide a definition of chemical wastes, but they are generally regarded as substances and mixtures of substances which have industrial or municipal origins. These substances are no longer usable because of past use or deterioration in storage and must be destroyed by chemical, physical or biological treatment or by disposal at an appropriate facility.

Civilian and municipal wastes are classified as industrial waste. Military wastes include explosives, fuels, incendiaries, smoke agents and munitions, chemical warfare agents and munitions, and soil or water contaminated by these materials. Armament wastes, a type of military waste, comprise various chemicals and other materials that have been used in past production processes but which are no longer usable. Such wastes may be the remnants of past production activities or materials remaining from the dismantling of production units. Military wastes are materials which have been used; armament wastes are materials which are no longer needed because they are obsolete or too old to be used. Military wastes can also be defined as follows:

> Military wastes are the products of former military conflicts as well as the result of normal military activities in peacetime. In the latter case the wastes are related to the training of soldiers, maintenance of military equipment and manoeuvres. Military wastes also include old ammunition, chemical or conventional, which has to be

replaced after a certain period because it is no longer usable due to ageing, changed military–strategic perceptions, changes in weapons calibre or because it has been banned as a result of international agreements.[8]

Chemical wastes can include chemicals used for military purposes, but military waste is a broader term for materials including scrap, wastes from normal training, service activities, and the like.[9] Soil contaminated by pollutants from military activities can be regarded either as a kind of military waste or merely as soil contaminated by military pollutants.

Disposal measures can be categorized as systematic or disordered. Systematic disposal is the disposal of wastes at a waste disposal site in a controlled manner. Disordered disposal occurs when industrial or military wastes are destroyed without record keeping or supervision by the appropriate authorities.

It is important to remember that:

1. The CWC defines old chemical weapons or warfare agents as those which were produced before 1946.
2. Some former production sites were destroyed during wartime fighting.
3. Many munitions were disposed of by burial or dumping at sea.
4. After World War II the production facilities of the defeated powers were destroyed in an environmentally unsafe manner.

III. The focus of the study

This study investigates various aspects of the disposal of old chemical munitions and toxic armament wastes, including the environmental, toxicological, technological and legal ramifications. Wastes from former production sites and from military bases are also discussed. Most of the activities which created these wastes occurred in Europe before or during World War I and World War II, but the problem of dealing with them is also of relevance to other countries such as Canada and the USA.

This study does not focus on the historical aspects of the use of toxic agents for weaponry since the scientific and technical literature of the past 30 years has dealt extensively with this topic. However, a review of the literature reveals that the problems of old military wastes and chemical wastes (including old chemical warfare agents) have long been ignored or have received little attention. Although there are some national studies of the decontamination and redevelopment of former production or storage sites, there are no comprehensive evaluations of the various technical aspects. In part this is because of the unwillingness of states to admit or accept the fact that production and military activities have often had a negative effect on the environment.

Today these problems cannot be ignored. In Germany, for example, the environmental impact of military activities and toxic armament wastes is the focus

[8] The definitions and examples are given in Spyra, W., 'Rüstungsaltlasten in der Bundesrepublik Deutschland' [Old armament wastes in the Federal Republic of Germany], eds W. Spyra *et al.*, *Untersuchung von Rüstungsaltlasten* [Examination of sites contaminated by old armament wastes] (EF Verlag für Energie-und Umwelttechnik: Berlin, 1991), pp. 1–15; Martinetz, D., *Sanierung von Industrie- und Rüstungsaltlasten* [Cleaning up old industrial and armament-related wastes] (Verlag Harri Deutsch: Thun and Frankfurt am Main, 1994); and Pfaff-Schley, H. (ed.), *Militärische Altlasten: Militärisch genutzte Flächen 1993* [Military wastes: areas used by the military in 1993] (Erich Schmidt Verlag: Berlin, 1993).

[9] Müller, K. R., *Chemical Waste: Handling and Treatment* (Springer Verlag: Berlin, 1986).

of numerous studies which consider not only the technological, toxicological and environmental aspects, but also the legal aspects of the problem.[10] There is also increased public interest in the disposal and dumping operations of the late 1940s and 1950s in the Baltic Sea and elsewhere.[11]

Although the issue of old and abandoned chemical munitions was not one of the top priorities in the negotiation of the Chemical Weapons Convention, it was considered a significant issue. It is a sensitive topic which can create tension between states since identification of the country or countries responsible for abandoning or dumping munitions also leads to the assigning of blame.

The entry into force of the CWC creates a new security environment. The SIPRI–Volkswagen Project on old chemical munitions and toxic armament wastes analyses the available experience in dealing with the problems outlined above and reviews the relevant literature. The project focuses primarily on Europe, but relevant Canadian and US experiences are also reviewed.[12] The project studies and evaluates the following areas:

(*a*) the situation after World War I and World War II with respect to chemical weapons and chemical warfare agents, including information on former munition production plants—what was produced, stored, dumped, buried and abandoned (and the location of these activities);

(*b*) the location of the primary CW dumping sites in Europe, both on land and at sea;

(*c*) the technical evaluation of the long-term behaviour of CW munitions in soil and water, including the associated chemical, analytical and toxicological problems;

(*d*) the technical investigation of the status of recovered CW munitions which were buried or dumped at sea;

(*e*) the technical means of recovering, dismantling and destroying old CW munitions;

(*f*) the technical means of restoring and redeveloping contaminated areas at military bases, including discussion of applicable procedures and techniques;

(*g*) the experience of individual countries in coping with toxic wastes; and

(*h*) the national and international legislation, treaties and conventions which address old chemical weapons and toxic armament wastes.

The findings of the study provide information which may be of assistance to those who have to make decisions about the destruction of old chemical munitions and toxic armament wastes.

The study is divided into eight parts. Following this introduction, the second part of the study analyses the relevant chemical warfare agents and CW muni-

[10] Kiefer, K.-W., Pfaff-Schley, H. and Schimmelpfeng, L., *Rüstungsaltlasten '91: Untersuchungsmethoden Sanierungsmöglichkeiten-Verhinderung militärischer 'Neu'–Lasten* [Old armament wastes '91: methods of examination and possible means for cleaning up/preventing new military wastes] (Erich Schmidt Verlag: Berlin, 1991); Spyra *et al.*, *Untersuchung von Rüstungsaltlasten* (note 8); Spyra, W. *et al.*, *Verfahren zur Sanierung von Rüstungsaltlasten* [Clean-up methods for old armament wastes] (EF Verlag für Energie und Umwelttechnik: Berlin, 1992); Thomé-Kozmiensky, K. J. *et al.*, *Management zur Sanierung von Rüstungsaltlasten* [Managing the clean-up of old armament wastes] (EF Verlag für Energie und Umwelttechnik: Berlin, 1992); and Martinetz (note 8).

[11] Lundin, S. J., Stock, T. and Geissler, E., 'Chemical and biological warfare and arms control developments in 1991', *SIPRI Yearbook 1992* (note 7), chapter 6, pp. 171–73; and Stock, T., 'Chemical and biological weapons: developments and proliferation', *SIPRI Yearbook 1993* (note 2), chapter 7, pp. 282–83.

[12] See chapter 10 in this volume.

tions from World War I and World War II. In addition, there is discussion of toxic armament wastes and some smoke agents, toxicology, analytical investigations, decontamination, degradation in soil and water, and deterioration as a result of environmental factors.

The third part of the study focuses on contamination at former Soviet military bases in Germany. It includes reports of on-site investigations and risk assessments and of the methods used. The practical, technical and organizational problems of decontaminating the former small-scale German research and testing facility at Spandau Citadel are evaluated as are the various analytical techniques that were employed. Lessons are drawn about future redevelopment efforts, and the measures are presented which must be taken to ensure the safety of the people involved in the decontamination activities. Military production and troop training activities will continue in many countries, and the experience gained in the redevelopment of areas which are contaminated by wastes from past military activities is valuable information.

The fourth part of the study contains contributions by researchers from the Czech Republic, Germany and Poland on the history and current status of old chemical munitions in those countries. In addition, there is a contribution by a British author assessing the CW programme of the former Soviet Union (one of the two largest CW possessors) and its production, storage and dumping activities. Questions are raised about environmental policy in Russia, and evidence is presented that unless Russia receives foreign aid it will be unable to solve the problems related to chemical weapons which it now faces.

Contributions from Austria, Belgium, Canada, France, Germany and the UK are presented in the fifth part of the study, which focuses on the practical experience gained in locating, dismantling and destroying old chemical munitions. An evaluation of the destruction of the former Iraqi CW arsenal is presented which assesses the state of those stockpiles, including whether some weapons were obsolete or had deteriorated partially. Information about CW destruction in various countries is reviewed, and alternative destruction techniques are evaluated.

The sixth part of the study reviews the past dumping of chemical weapons in the Baltic Sea and the North Sea. Previously reported data are compared with new information. Dumping activities off the coast of Norway are given particular attention, and recent exploration of the dumping sites is discussed as is the effect of dumping on marine life. Technical and analytical activities are outlined, and the legal aspects of dumping are also examined.

The problems of dealing with partially disposed of or buried old munitions and toxic armament wastes within the confines of current national legal frameworks are discussed in part seven, which includes a German case study. In order to comply with the CWC countries must review their national legislation on toxic wastes and amend or pass new measures if needed. The ability of the CWC to deal with old or abandoned chemical munitions, the role of the OPCW and the responsibilities of individual parties to the CWC are discussed.

It is not easy to achieve a balance between general issues and specific cases when discussing old chemical weapons and toxic armament wastes since the problems involved tend to overlap. However, it is hoped that the material presented here provides the reader with both a general overview and specific examples which, together, create a comprehensive picture of the current situation. Those who study the topic of old chemical weapons and toxic armament wastes soon realize the need to explore its history. However, there are many aspects of this topic which have not yet been fully examined, and no comprehensive study exists. This volume attempts to begin to fill that gap.

Part eight of the study answers the following questions based on the material presented earlier in the volume:

1. Are old chemical munitions and toxic military wastes primarily a challenge for European countries?
2. Does appropriate and efficient technology exist to dismantle and destroy old munitions in an environmentally safe manner?
3. What is needed to address the problem of old chemical munitions and toxic armament wastes; what framework exists to deal with the problem; and what is the responsibility of each individual country?
4. How can states cooperate under the CWC to deal with the unresolved issues related to old chemical weapons and toxic military wastes?

This study, which was made possible by the generous support of the Volkswagen Foundation, Hanover, Germany, analyses various aspects of old chemical munitions and toxic armament wastes, an issue which is currently receiving significant attention. These problems affect not only Europe, but also other countries which possess old chemical munitions and toxic armament wastes that have resulted from past military activities.

The contributions to this study were made by 20 experts from various areas, including technicians, disarmament and arms control experts, chemists, toxicologists, technical engineers, legal experts and journalists. They represent 10 countries but write here as private individuals, and the views presented are their own. This study contributes to discussion of the issue of old chemical munitions and toxic armament wastes by presenting new findings, analysis and proposals.

The study does not focus on the economic and financial aspects of decontamination, but rather on its technical feasibility. Experience in Germany has shown that enormous financial resources are needed merely to assess the areas contaminated by military and toxic armament wastes.[13] Cleaning and redevelopment of such sites, the next step, can only be carried out at a few sites owing to financial constraints.

The editors of this volume and its contributors hope that this study will help focus attention on the problem of old chemical munitions and toxic armament wastes and encourage states to take action. The study is intended to provide information to political and scientific authorities to assist in making decisions about disarmament, decontamination and redevelopment.

[13] Thieme, J. *et al.*, *Bestandsaufnahme von Rüstungsaltlastverdachtsstandorten in der Bundesrepublik Deutschland* [Inventory of suspected old armament waste sites in the Federal Republic of Germany], report no. UBA-FB 93029 (Umweltbundesamt: Berlin, Mar. 1993).

Part II. Chemical warfare agents and munition and toxic armament wastes

2. Characteristics of chemical warfare agents and toxic armament wastes

KARLHEINZ LOHS and THOMAS STOCK

I. Introduction

Chemical weapons have a long history of use.[1] In modern times they were first employed extensively in World War I, beginning with their use on the Western front by Germany at Ypres on 22 April 1915. Chemical weapons were intended to create a tactical advantage and to demoralize the enemy. The initial success of CW in World War I and subsequent developments in the chemical industry led countries such as Canada, France, Germany, Japan, the former Soviet Union, the UK, the USA and others to conduct extensive research and development (R&D) and testing of new and potentially more effective chemical weapons and agents.

Despite the development of more effective chemical weapons in the period between World War I and World War II political and military leaders were uncertain about their value in combat. Nevertheless, various types of chemical munitions was stockpiled in the arsenals of the major powers. The chemical warfare agents which were used in these munitions can be categorized according to their effect and include the following:

1. Blister agents were designed to inflict casualties, restrict the use of terrain and slow down troop movement. These agents affect the eyes and lungs and blister the skin; they include mustard gas, arsenicals, nitrogen mustards and mixtures of mustard and lewisite.
2. Blood agents are absorbed primarily by inhalation. They prevent normal utilization of oxygen by body cells and rapidly damage the body. Cyanogen chloride and hydrogen cyanide are examples of such agents.
3. Choking agents, such as phosgene and diphosgene, damage the respiratory tract (nose, throat and lungs) and cause death because of lack of oxygen.
4. The nerve agents (sarin, soman and tabun) cause an increase in acetylcholine in the body by blocking the enzyme cholinesterase. Their effect on the skeletal muscles, parasympathetic end organs[2] and the central nervous system leads to death.
5. Riot control agents or tear gases (e.g., bromobenzyl cyanide and chloroacetophenone) cause discomfort, induce tears and irritate the skin.
6. Vomiting agents—such as diphenylchloroarsine, diphenylcyanoarsine and adamsite—irritate the eyes, nose and throat, leading to vomiting.[3]

[1] Lewin, L., *Gifte in der Weltgeschichte* [Poison in history] (Georg Silke Verlag: Berlin, 1926); and Hanslin, R., *Der chemische Krieg* [The chemical war] (Mittler & Sohn: Berlin, 1937).

[2] An end organ is an encapsulated sensory nerve ending.

[3] Table 3.1 in chapter 3 lists chemical warfare agents which were discovered and introduced into military arsenals in World War I and later. Only agents which were considered militarily useful at the time are included. Chemical warfare agents such as VX (O-ethyl S-2-diisopropylaminoethyl methylphosphonothiolate) and BZ (3-quinuclidinyl benzilate), which were discovered and stockpiled in the 1950s, are not included, although two significant chemicals from 1952 and 1962 are listed.

II. Chemical and toxicological properties of chemical warfare agents

This section describes the chemical and toxicological action and reaction of some chemical warfare agents, including their detoxification products and precursors. Information is presented on the toxicity of the chemical warfare agents and on the products of detoxification and their toxicity. It is an overview but may be of value to those involved in the dismantling and destruction of old chemical munitions.

Sulphur mustard

Mustard gas is a powerful chemical weapon that was first used in World War I by Germany at the Battle of Flanders near Ypres in mid-July 1917. Mustard gas is the US name for bis(2-chloroethyl)sulphide. In German it was called Schwefel-Lost (S-Lost) or Gelbkreuz (yellow cross). France and the USSR referred to it as yperite. HS was the designation used by the Allies for a type of mustard gas in which the sulphur mustard was produced by the Levinstein process (resulting in 20–30 per cent impurities of which most are sulphur). HD, another type of mustard gas, is distilled or nearly pure sulphur mustard.

César Despretz discovered mustard gas in 1822 while studying the reaction of sulphur chloride and ethylene.[4] The German chemists W. Lommel and Wilhelm Steinkopf[5] developed the production process for S-Lost[6] in 1916–17. Their process was used for large-scale industrial production of mustard gas for warfare purposes in Germany, and later in France, the UK and the USSR.

In the mid-1930s there was extensive research on new mustard compounds, which led to the development of another class of chemical warfare agents—the nitrogen mustards. The most well-known of these are: HN-1, bis(2-chloroethyl)ethylamine; HN-2, bis(2-chloroethyl)methylamine; and HN-3, tris-(2-chloroethyl)amine. Only Germany maintained stocks of nitrogen mustards in World War II.

Sulphur mustard is a clear, colourless, oily liquid which boils at 215–217°C and freezes at 14.4°C. Its specific gravity is 1.278. Mustard gas has a characteristic garlic or onion odour, and both the liquid and vapour forms are soluble in oils, fats and organic solvents. Mustard gas is soluble in alcohol, ether, gasoline, kerosene and carbon tetrachloride, but it has low solubility in water. In low concentration in water, pure mustard gas hydrolyses rapidly (its half-life is 5 minutes (min) at 37°C) producing hydrochloric acid and thiodiglycol. Increasing the temperature speeds up the reaction, and at 90°C its half-life is 2.3 seconds. Under normal conditions hydrolysis takes much longer.[7] The dissolution of mustard gas in water is the limiting step for hydrolysis.

In preparation for World War II Germany, in particular, produced and stockpiled agents that were modifications of sulphur mustard. Any substance which dissolves in sulphur mustard will lower its freezing point, and thickeners were

[4] Despretz, C. M., *Annales de Chimie et de Physique* [Annals of chemistry and physics], vol. 21, no. 2 (1822), p. 438.

[5] Lommel and Steinkopf's names were later combined in the acronym LOST.

[6] S-Lost is a designation for sulphur mustard or sulphur Lost.

[7] See also chapter 3 in this volume.

used to increase the viscosity of the sulphur mustard and reduce its degree of breakup when dispersed from a munition.[8] Thickeners also cause droplets to adhere to surfaces, thus increasing the difficulty of decontamination. Germany used polystyrene or bleached montan wax as thickeners in a mixture of sulphur mustard and arsine oil (arsinöl)[9] to produce Zählost.[10] The thickener makes the agent resistant to water and impedes hydrolysis, which can be of importance as regards old chemical weapons.

Germany also produced Winterlost: sulphur mustard mixed with various additives, such as chlorobenzene or arsine oil, to lower the melting point.[11] Britain used alloprene, a chlorinated rubber, as a thickener in the chemical agent runcol.[12] Japan employed thickeners that were mixtures of polyvinyl chloride and methyl methacrylate, and the US Army used acrylate, methacrylate and other polymers as thickeners.[13]

The name Winterlost was also used for mixtures of sulphur mustard and lewisite which were used by the former Soviet Army among others. The freezing point for the sulphur mustard and lewisite mixtures ranged from – 24.5°C to – 42°C, depending on the ratio of the mixture. Britain used another mixture, 13 per cent sulphur mustard and 87 per cent diisopropyl phosphorofluoridate, which remained solid even at – 36°C.

Sulphur mustard is the most significant of the mustard agents that are found in old chemical weapons. Problems are created in handling and chemical detoxification by its presence in many stockpiles, its insolubility, the inclusion of thickeners and the formulation of the gel and solid phases which occur in munitions that have been stored for long periods.

The toxic effects of sulphur mustards include, among others, inhibition of mitosis and cell cycle phases, depletion of nicotinamide adenine dinucleotide and decreased tissue respiration. The organs that are affected by exposure to sulphur mustard are the skin, eyes, lungs and respiratory tract. If sulphur mustard comes in contact with the skin itching and erythema occur, followed by the formation of blisters. A concentration of 1000–2000 $mg \cdot min/m^3$ of sulphur mustard is required to cause skin blisters. Sulphur mustard is also a potent carcinogen. Most of its delayed toxic effects result from the alkylation of DNA.[14] Figure 2.1 presents an overview of some possible metabolic pathways for sulphur mustard.

The toxic effects of sulphur mustard on the macromolecular level have been studied in detail[15] and include its effects on the skin, lungs, enzymes and depletion of glutathione. The acute and delayed toxic effects of inhalation of sulphur mustard have been known since the late 1930s, and they have been described

[8] Perry Robinson, J. P. and Trapp, R., 'Production and chemistry of mustard gas', ed. S. J. Lundin, *Verification of Dual-use Chemicals under the Chemical Weapons Convention: The Case of Thiodiglycol*, SIPRI Chemical & Biological Warfare Studies no. 13 (Oxford University Press: Oxford, 1991), p. 8.

[9] Arsinöl is a mixture of mainly phenyldichloroarsine, triphenyl arsine and diphenylchloroarsine.

[10] Brauch, H. G. and Müller, R.-D., *Chemische Kriegsführung, Chemische Abrüstung* [Chemical warfare, chemical disarmament] (Arno Spitz: Berlin, 1985), p. 161.

[11] Brauch and Müller (note 10).

[12] Toler, I. J., 'War gases in the Second World War', *British Army Review*, Aug. 1990, pp. 67–72.

[13] Perry Robinson and Trapp (note 8).

[14] For a summary of mustard gas and lewisite see Pechura, C. M. and Rall, D. P. (eds). *Veterans at Risk: The Health Effects of Mustard Gas and Lewisite* (National Academy Press: Washington, DC, 1993).

[15] Somani, S. M., 'Toxicokinetics and toxicodynamics of mustard', ed. S. M. Somani, *Chemical Warfare Agents* (Academic Press: San Diego, Calif., 1992), p. 26.

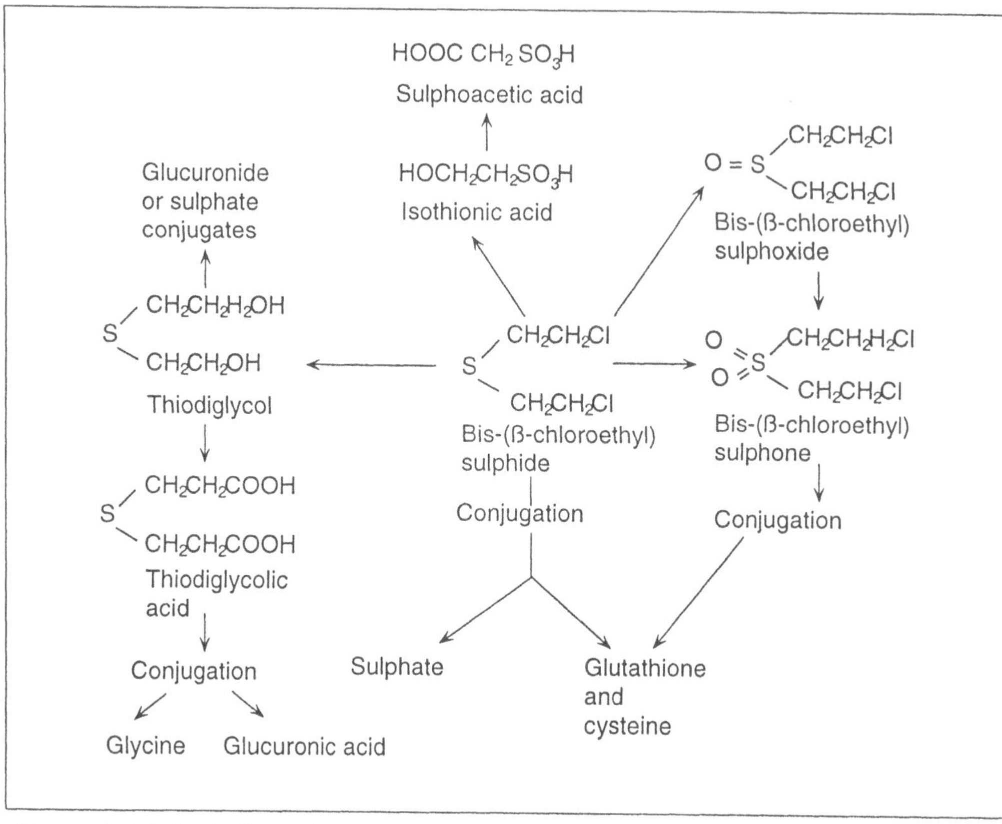

Figure 2.1. Metabolic pathways of sulphur mustard

Sources: Somani, S. M., 'Toxicokinetics and toxicodynamics of mustard', ed. S. M. Somani, *Chemical Warfare Agents* (Academic Press: San Diego, Calif., 1992), p. 24; and Pechura, C. M. and Rall, D. P. (eds), *Veterans at Risk: The Health Effects of Mustard Gas and Lewisite* (National Academy Press: Washington, DC, 1993).

and investigated extensively since then.[16] In World War I approximately 80 per cent of the chemical casualties were caused by sulphur mustard. Fewer than 5 per cent of those who received medical treatment died as a result of being injured by sulphur mustard, and fewer than 3 per cent of those hospitalized as sulphur mustard casualties died. Although mustard gas poisoning has been the subject of numerous studies medical data on human mustard gas exposure remains inadequate.

In the 1980s the World Health Organization published information obtained from investigations of Iranian patients exposed to mustard gas in the 1980–88 Iraq–Iran War. It is estimated that approximately 45 000 people were exposed to mustard gas in that war.[17] The effects of inhalation of mustard gas and the associated delayed effects have both been investigated in detail.[18]

[16] Muntsch, O., *Leitfaden der Pathologie und Therapie der Kampfstofferkrankungen* [Textbook of pathology and treatment of chemical warfare agent diseases], 6th edn (G. Thieme: Leipzig, 1941); Spielberg, U., 'Monographie: Psychopathologisch-neurologische Spätschäden nach Einwirkung synthetischer Gifte' [Monograph on psychopathological and neurological long-term effects after the action of synthetic poisons] (Wehr und Wissen Verlagsgesellschaft: Stuttgart, 1963); and Thomas, A., *Effects of Chemical Warfare: A selective review and bibliography of British state papers*, SIPRI Chemical & Biological Warfare Studies no. 1 (Taylor & Francis: London, 1985).

[17] Carus, W. S., *Chemical Weapons in the Middle East* (Washington Institute for Near East Policy: Washington, DC, 1988).

[18] Somani (note 15); World Health Organization, Report of the specialists appointed by the Secretary-General to investigate allegations by the Islamic Republic of Iran concerning the use of chemical

Exposure to sulphur mustard causes numerous clinical symptoms.[19] The pattern is related to the organs that are most exposed to the gas, which include the skin, eyes, respiratory tract and, in some cases, the gastrointestinal tract. Neuropsychiatric symptoms may also occur. In general, the duration and severity of the symptoms depend on the concentration of the gas and the period of exposure. The effects of mustard gas intoxication are not immediately apparent. The first symptoms often do not appear until hours after exposure, and there is no effective preventive treatment for sulphur mustard poisoning. Because of the alkylating properties of sulphur mustard changes in nucleic acids, cell membranes and proteins occur, and no antidote can counteract the acute or non-specific alkylating properties of the agent.[20]

Organoarsenic chemical warfare agents

Organoarsenic chemical warfare agents have been stockpiled since World War I. These agents were considered militarily useful because early gas masks relied on filters to absorb chlorine or phosgene, and some organoarsenic compounds could penetrate the filters and cause soldiers to remove their protective equipment. The first organoarsenic chemical warfare agent used in World War I was diphenylchloroarsine, a 'vomiting agent', which was introduced by Germany in 1917 and known as Blue Cross.[21] It was also known as Maskenbrecher ('mask-breaker'). Germany also referred to diphenylchloroarsine as Clark I, while its US code-name was DA. Diphenylcyanoarsine or Clark II (US code-name DC), another organoarsenic chemical warfare agent, was also a vomiting agent. Diphenylcyanoarsine is more toxic than diphenylchloroarsine but acts in a similar fashion.

Germany introduced the harassing agent phenyldichloroarsine, also known as Pfiffikus (US code-name PD), in 1917 and ethyldichloroarsine (US code-name ED) in 1918. Ethyldichloroarsine was a blister agent of moderate persistence, and phenyldichloroarsine was both a vesicant and a vomiting agent.

The USA developed adamsite[22] (10-chloro-5,10-dihydrophenarsazine, US code-name DM), a vomiting agent, and lewisite[23] (phenyldichloroarsine, US code-name L), a potent blister agent. Adamsite is irritating to the nose, throat and respiratory tract. It can be synthesized by condensation of arsenic trichloride or by the reaction of arsenic trioxide with diphenylamine. Its effect is simi-

weapons, United Nations Security Council document S/16433, 26 Mar, 1984; and World Health Organization, Report of the mission dispatched by the Secretary-General to investigate allegations of the use of chemical weapons in the conflict between the Islamic Republic of Iran and Iraq, United Nations Security Council document S/19711, 12 Mar. 1986. See also Pechura and Rall (note 14)

[19] For a comprehensive overview of the toxicology of sulphur and nitrogen mustard poisoning see Whitfield, D., *A Literature Review upon the Toxicology, Mechanism of Action and Treatment of Sulphur and Nitrogen Mustard Poisoning*, Technical Note no. 840 (Chemical Defence Establishment: Porton Down, Salisbury Wilts, Mar. 1987).

[20] Lohs, Kh., 'Schwefel-Lost (2,2'-Dichlordiethylsufid): noch immer toxikologisch aktuell' [S-Lost (2,2'-dichlorodiethylsulphide): still toxicologically topical], *Zeitschrift ärztliche Fortbildung*, vol. 87 (1993), pp 659–64.

[21] Blue Cross received its name from the German markings on the shell casings.

[22] Adamsite was discovered in 1918 by a team working under Major Roger Adams at the University of Illinois. A British team did simultaneous research on the substance. In Germany a patent was filed in 1914. SIPRI, *The Problem of Chemical and Biological Warfare*, vol. 1, *The Rise of CB Weapons* (Almqvist & Wiksell: Stockholm, 1971), p. 41.

[23] Lewisite was also called 'dew of death'.

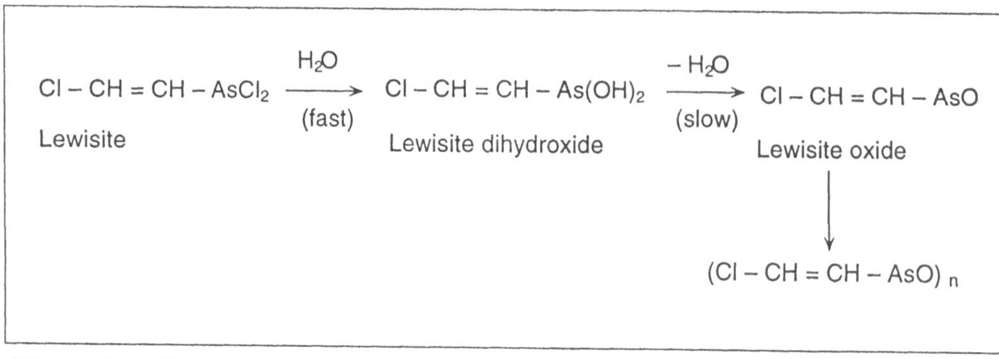

Figure 2.2. The hydrolysis of Lewisite

Source: Trammell, G. L., 'Toxicodynamics of organoarsenic chemical warfare agents', ed. S. M. Somani, *Chemical Warfare Agents* (Academic Press: San Diego, Calif., 1992), p. 261.

lar to that of diphenylchloroarsine, but it develops more slowly. Lewisite, a blister agent, is normally produced in a process in which arsenic trichloride and acetylene react in the presence of aluminium trichloride, a catalyst. The final product is a mixture of Lewisite 1, 2-chlorovinyldichloroarsine (α-lewisite); Lewisite 2, bis(2-chlorovinyl)chloroarsine; and Lewisite 3, tris(2-chlorovinyl)arsine.[24] It is readily absorbed through the skin and lungs and causes blistering of the skin and eyes.

Organoarsenic compounds are toxic because of their high affinity for the sulphydryl groups present in certain essential enzymes, where they act as enzyme inhibitors. Methyldichloroarsine (US code-name MD) and ethyldichloroarsine irritate the eyes, skin and respiratory tract and are absorbed through the skin. Lewisite, a typical example of such agents, hydrolyses as shown in figure 2.2. The most important representatives of the organoarsenic chemical warfare agents and their properties are complied in table 2.1.

After World War I arsenic warfare agents were stockpiled by both the former Soviet Union and the USA. Russia assumed control of the Soviet CW stockpile and now possesses the world's largest stockpile of lewisite (approximately 6800 tonnes), which is stored at Kambarka, Kizner (Republic of Urdmurtia) and Gorny (Saratov oblast).[25]

There is a broad spectrum of organoarsenic warfare agents, and they are often found in old chemical munitions. The final products of the hydrolytic degradation of adamsite, Clark I, Clark II and other organoarsenic compounds contain oxide and hydroxide derivatives as well as arsenic and arsenic oxide. These toxic compounds must be disposed of in an environmentally safe manner.

[24] A comprehensive overview of lewisite is presented in Goldman, M. and Dacre, J. C., 'Lewisite: its chemistry, toxicology, and biological effects', *Reviews of Environmental Contamination and Toxicology*, vol. 110 (Springer Verlag: Berlin, 1989), pp. 75–115; and Pechura and Rall (note 14).

[25] Stock, T., 'Chemical and biological weapons: developments and proliferation', *SIPRI Yearbook 1993: World Armaments and Disarmament* (Oxford University Press: Oxford, 1993), chapter 7, pp. 277–80.

Table 2.1. Properties of organoarsenic chemical warfare agents

Agent, US Army code-name	Freezing point (°C)	Boiling point (°C)	Volatility mg/m^3 at 20°C	Vapour density compared to air	Vapour pressure	LCt_{50}[a] for inhalation	LCt_{50} for skin	ICt_{50}[a]
Lewisite, L[b]	– 18	190	4480	7.1	0.394	1200–1500	100 000	300 (eye) 1500 (skin)
Mustard/ lewisite[b]	– 25.4	190	2730	6.5	0.248	1500	100 000	200 (eye) 1500–2000 (skin)
Phenyl-dichloro-arsine, PD[c]	– 20	252–255	390 (at 25°C)	7.7	0.033 (at 25°C)	2600	..	1800 (skin) 16 (vomiting)
Ethyl-dichloro-arsine, ED[b]	– 65	156	20 000	6.0	2.09	3000–5000	100 000	5–6 (inhalation)
Methyl-dichloro-arsine, MD[b]	– 55	133	74 900	5.5	7.76	..		25 (inhalation)
Diphenyl-chloro-arsine, DA[d]	41–44.5	333 (dec.)	7	–	0.0036 (45°C, calc.)	15 000	..	12 (inhalation)
Adamsite, DM[d]	195	410	–	–	..	15 000	..	22 (inhalation, 1 min) 86 (60 min)
Diphenyl-cyano-arsine, DC[d]	31.5–35	2.8	3	0.0002	10 000	..	..	30 (inhalation, 30 sec. exposure) 20 (5 min exposure)

LCt_{50} = Medium lethal dosage; ICt_{50} = Medium incapacitating dosage; dec. = Decomposing; calc. = Calculated.

[a] Mg · min/m^3.
[b] Blister agent.
[c] Blister and vomiting agent.
[d] Vomiting agent.

Sources: Based on Departments of the Army and Air Force, 'Military chemistry and chemical compounds', FM 3-9/AFR 355-7 (Department of the Army: Washington, DC, 1975), quoted in Trammell, G. L., 'Toxicodynamics of organoarsenic chemical warfare agents', ed. S. M. Somani, *Chemical Warfare Agents* (Academic Press: San Diego, Calif., 1992), p. 258; Goldman, M. and Dacre, J. C., 'Lewisite: its chemistry, toxicology, and biological effects', *Reviews of Environmental Contamination and Toxicology*, vol. 110 (Springer Verlag: Berlin, 1989), pp. 75–115; and Pechura, C. M. and Rall, D. P. (eds), *Veterans at Risk: The Health Effects of Mustard Gas and Lewisite* (National Academy Press: Washington, DC, 1993).

Choking chemical warfare agents

Choking agents such as chlorine, chloropicrin, diphosgene and phosgene injure unprotected persons by damaging the respiratory tract.[26] Phosgene was first used by Germany against British troops in Ypres on 19 December 1915. The UK began to produce phosgene at Porton Down at approximately that time, and it was first used by British forces in June 1916. US production of phosgene began at Edgewood Arsenal in 1918. There are some estimates that phosgene caused more than 80 per cent of all chemical warfare agent fatalities in World War I. It was manufactured as a chemical warfare agent by many countries in the 1920s and has remained in stockpiles (e.g., in Russia).

Diphosgene was produced after World War I as an improved form of phosgene. The gas mask filters which had been used in World War I were able to screen out phosgene but were destroyed by chloroform. Diphosgene decomposes first into chloroform and phosgene, and then into hydrochloric acid and carbon dioxide. It can destroy gas mask filters, creating the more lethal agent phosgene. Table 2.2 presents a compilation of the most significant data on phosgene and diphosgene.

The effects of phosgene exposure have been studied extensively.[27] Phosgene irritates the eyes initially, and its toxic effects are attributed to its hydrolysis and conversion to hydrogen chloride in the aqueous medium. Additional toxic effects are the result of alkylation reactions. Phosgene is highly reactive and combines with functional groups, such as the thiol (–SH), amino ($-NH_2$) and hydroxyl (–OH) groups, of the biological macromolecules. The human body cannot detoxify phosgene, and it therefore accumulates. First, the capillaries of the lungs are attacked, followed by the membranes of the lung sacs. Later, there is a direct exchange of phosgene into the blood, and the lungs become flooded with fluid. The effect of lethal phosgene exposure may not appear until approximately four hours after exposure.

Diphosgene converts into phosgene, chloroform and water, and its toxicological effects are similar to those of phosgene. It differs from phosgene in that it is slightly irritating to the nose and eyes because of its high chlorine content.

Chloropicrin (trichloronitromethane, US code-name PS), another choking chemical warfare agent, was first used by Russia in August 1916 and later by all of the belligerents in World War I. It is a powerful lachrymator and was able to penetrate the gas masks then in use. Chloropicrin was called Klop in Germany, Aquinite in France and PS in Britain. It was used in Tbilisi by the former Soviet forces as recently as April 1989.[28]

[26] Chlorine and phosgene are primarily lung poisons.

[27] Mehlman, M. A., 'Heath effects and toxicity of phosgene: scientific review', *Defence Science Journal*, vol. 37, no. 2 (Apr. 1987), pp. 269–79.

[28] Lundin, S. J., 'Chemical and biological warfare: developments in 1989', *SIPRI Yearbook 1990: World Armaments and Disarmament* (Oxford University Press: Oxford, 1990), chapter 4, pp. 109–10; Dörte, H., 'Einsatz chemischer Waffen in Tbilissi/Georgien: gesundheitliche Folgen einer Massenkatastrophe am 9. April 1989' [The use of chemical weapons in Tbilisi/Georgia: health consequences of a mass disaster on 9 April 1989], doctoral dissertation (medicine), University of Witten Herdecke, 1993; and Schmidt-Troschke, S., 'Massenpsychogene Erkrankung und Posttraumatische Belastungsstörung in Tbilissi/Georgien: gesundheitliche Folgen einer Massenkatastrophe am 9. April 1989' [Mass psychogenic illness and post-traumatic stress disorder in Tbilisi/Georgia: health consequences of a mass disaster on 9 April 1989], doctoral dissertation (medicine), University of Witten Herdecke, 1993.

Table 2.2. Comparative data on phosgene and diphosgene

Type of data	Phosgene	Diphosgene
NATO agreement code	CG	DP
Chemical name	Carbonyl chloride	Trichloromethyl choloroformate
Common name	Phosgene, collognite, D-Stoff	Diphosgene
Chemical formula	$COCl_2$	$ClCOOCCl_3$
Molecular weight	98.92	197.85
Vapour density (air = 1)	3.4	6.8
Vapour pressure at 20°C (mm Hg)	1.173	4.2
Volatility (kg/m^3)	4.3 at 7.6°C	0.045 at 20°C
Latent heat of vaporization (cal/g)	59	57
Freezing point	– 128°C	. .
Melting point	. .	– 57°C
Boiling point	– 7.6°C	127°C
Decomposition	800°C	300°C
Flash point	None	None
Rate of hydrolysis	Rapid	Slow
Hydrolysis products	HCl, CO_2	$COCl_2HCl$, CO_2
Stability in storage	Stable in steel containers	Unstable in metal containers
Action with metals	None in pure form	Reacts to form phosgene
Skin toxicity	None	None
Eye toxicity	Irritating	Lachrymatory
Rate of action	Immediate and cumulative	Immediate and cumulative
Protection required	Mask, respirator	Mask, respirator

Sources: Marthur, B. and Krishna, G., 'Toxicodynamics of phosgene', ed. S. M. Somani, *Chemical Warfare Agents* (Academic Press: San Diego, Calif., 1992), p. 239; and also compiled from Compton, J. A. F., *Military Chemical and Biological Agents: Chemical and Toxicological Properties* (Telford Press: Caldwell, N.J., 1987).

Blood gases

The most common blood gases are hydrogen cyanide (US code-name AC) and cyanogen chloride (US code-name CK). Both were first used in World War I in 1916. After World War I hydrogen cyanide was stockpiled by Germany, Japan, the USA and the USSR. Cyanogen chloride was stockpiled by Germany, the USA and the USSR. Hydrogen cyanide was also used in the Iraq–Iran War.[29]

Hydrogen cyanide is a non-persistent agent that acts by being either inhaled or absorbed through breaks in the skin. It is irritating to the eyes and upper respiratory tract. It kills rapidly, dissipates quickly and leaves no toxic residue. The mechanism of toxic damage by cyanide has been studied extensively. Hydrogen cyanide inactivates certain enzyme systems, and as the dose increases it causes loss of consciousness, cessation of respiration and death. Exposure to a lethal dose causes violent contraction of the blood vessels.

[29] Lundin, S. J., 'Chemical and biological warfare: developments in 1988', *SIPRI Yearbook 1989: World Armaments and Disarmament* (Oxford University Press: Oxford, 1989), chapter 4, pp. 100–103; Lundin, S. J., Perry Robinson, J. P. and Trapp, R., 'Chemical and biological warfare: developments in 1987', *SIPRI Yearbook 1988: World Armaments and Disarmament* (Oxford University Press: Oxford, 1988), chapter 5, pp. 106–107.

Table 2.3. Properties of hydrogen cyanide and cyanogen chloride

Agent	Rate of action	State at 20°C	Odour	Volatility (mg/m^3)	LCt_{50}, ICt_{50} ($mg \cdot min/m^3$)	Vapour density (x air)	Vapour pressure (mm Hg)
Hydrogen cyanide	Rapid	Colourless gas or liquid	Bitter almonds	1 080 000 at 25°C	LCt_{50} = 2000 to 4500[a], ICt_{50} varies[a]	0.93	742 at 25°C
Cyanogen chloride	Rapid	Colourless gas	Sharp or pepper	2 600 000 at 12.8°C	LCt_{50} = 11 000 ICt_{50} = 7000	2.1	1000 at 25°C

LCt_{50} = Medium lethal dosage; ICt_{50} = Medium incapacitating dosage.

[a] Depending on concentration.

Source: Compiled from Compton, J. A. F., *Military Chemical and Biological Agents: Chemical and Toxicological Properties* (Telford Press: Caldwell, N.J., 1987).

Cyanogen chloride is less lethal than hydrogen cyanide. It is easier to deploy because it is not explosive, and it is very effective when delivered as an aerosol. Its usefulness as a chemical warfare agent was re-examined in the 1930s, and the USA developed stabilizers to inhibit its polymerization. Cyanogen chloride, as a gas or a liquid, is either inhaled or absorbed through breaks in the skin. Once in the body it is converted into hydrogen cyanide, and inactivation of enzymes results. Its effect on the body is similar to that of hydrogen cyanide. Table 2.3 illustrates the properties of hydrogen cyanide and cyanogen chloride.

Incapacitating agents: tear gases and sensory irritants

Tear gases are also referred to as lachrymators and harassing, incapacitating and riot control agents. The following discussion concentrates on the most significant tear gas agents prior to 1945: bromobenzylcyanide (CA); choroacetophenone (CN); a mixture of chloroacetophenone, carbon tetrachloride and benzene (CNB); and a mixture of chloroacetophenone and chloroform (CNC).

The first riot control agent was ethyl bromoacetate (EBA). It was not used extensively, but was employed by France in World War I. During World War I Germany developed and produced lachrymators, and the first of these, a mixture of xylyl and xylene bromide isomers and/or benzyl bromide (T-Stoff), was used in January 1915 against Russian troops. Germany also used an irritant known as K-Stoff which was a mixture of chloromethyl and dichloromethyl chloroformates. By the end of the war in 1918 France had also developed CA, and CN had been developed by the UK and the USA.

Riot control agents can cause intense sensory irritation, and sweating amplifies their effect. If riot control agents are used in closed spaces lethal concentrations may result, but their effect under field conditions is difficult to predict. If the dosage is high the skin, eyes and respiratory tract may experience acute toxicity. Delayed toxicity may cause cancer and mutagenic or teratogenic effects, but there has been limited research in this area.[30] The properties of some riot control agents are outlined in table 2.4.

[30] For a discussion of toxicology and delayed and long-term effects see National Research Council, *Possible Long-term Health Effects on Short-term Exposure to Chemical Agents*, vol. 2, *Cholinesterase*

Table 2.4. Properties of selected riot control agents

Property	CN	CA	CNB	CNC	CNS
Physical state	Solid	Solid	Liquid	Liquid	Liquid
Colour	White	White	Brown	Clear	Clear
Odour	Apple blossoms	Rotten fruit	Benzene	Chloroform	Flypaper
Melting point (°C)	59	18.5 - 25.5	. .	. .	. .
Boiling point (°C)	. .	. .	75–247	60–247	60–247
Vapour pressure at 20°C	5.4×10^{-3}	1.1×10^{-3}	62	127	78
ICt_{50} (mg·min/m^3)	80	30	80	80	60

CN = Chloroacetophenone, CA = Bromobenzylcyanide, CNB = Mixture of chloroacetophenone, carbon tetrachloride and benzene, CNC = Mixture of chloroacetophenone and chloroform, CNS = Mixture of chloroacetophenone, chloropicrin and chloroform, ICt_{50} = Medium incapacitating dosage.

Sources: Hu, H., 'Toxicodynamics of riot-control agents (lacrimators)', ed. S. M. Somani, *Chemical Warfare Agents* (Academic Press: San Diego, Calif., 1992), p. 274; and Malhotra, R. C. and Kumar, P., 'Chemistry and toxicity of tear gases', *Defence Science Journal*, vol. 37, no. 2 (Apr. 1987), pp. 281–96.

Nerve agents

Sarin and tabun are the organophosphorus compounds that can be found in old chemical weapons. They were discovered, produced and stockpiled prior to the end of World War II by Germany. Phosphorous organic chemical agents of this type are 'super toxic lethal substances' that are lethal after only minutes of exposure. Table 2.5 presents some of the main characteristics of sarin and tabun. Data on soman are also listed, but soman was not stockpiled in World War II. Battlefield use of nerve agents was limited, but they were stockpiled in large quantities by the USA and the Soviet Union after World War II.

Nerve agents can be classified as G- and V-agents or as non-persistent and persistent agents. The G-agents[31]—sarin (O-isopropyl methylphosphonofluoridate), soman (O-pinacolyl methylphosphonofluoridate) and tabun (O-ethyl N,N-dimethylphosphoramidocyanidate)—are non-persistent and generally cause death by inhalation. The main representative of the V-agents is VX (O-ethyl S-2-diisopropylaminoethyl methylphosphonothiolate). V-agents are persistent and kill by inhalation and absorption through the skin. Thickened soman also belongs to this group. (Non-persistent nerve agents can be made persistent by adding thickeners.) The following discussion focuses on G-agents since the discovery and stockpiling of VX took place after World War II. Table 2.5 lists some of the physicochemical properties of the G-agents.

Reactivators, Psychochemicals, and Irritants and Vesicants (National Academy Press: Washington, DC, 1984).

[31] Sarin, soman and tabun are the German code-names. The NATO and US codes are GA, GB and GD for tabun, sarin and soman, respectively. The German production code-names for tabun were Gelan I, Trilon 83, T 83 and Stoff 100. For sarin they were Trilon 46, T 144 and Gelan III.

Table 2.5. Physicochemical properties of nerve agents

Agent, code-name, molecular weight	Dissemination and likely odour	Saturated vapour pressure at 20°C (torr)	Volatility at 25°C (mg/m^3)	Stability in water	Main chemical reactions
Tabun,[a] GA 162.30	Vapour/ liquid, fishy	. .	576	At pH 7.0, 20°C $t_{1/2}$ = 8.5 hours	. .
Sarin,[b] GB, 140.10	Vapour/ liquid, fruity	1.48	16 400	Rate of hydrolysis depends on pH; at pH 7.0, 25°C $t_{1/2}$ = 5.4 hours	Ready reactions with nucleophiles, hypochlorite anions, peroxide anions, phenols, oximes, hydroxamic acids, amines, alcoholates
Soman,[c] GD, 182.18	Vapour/ liquid, camphor	0.50	3060	Comparable to GB (pH dependence) At pH 7.0, 20°C $t_{1/2}$ = 82 hours	As for sarin

[a] Synthesized by Gerhard Schrader in Germany in 1937.
[b] Synthesized by Gerhard Schrader in Germany in 1938.
[c] Synthesized by Richard Kuhn in Germany in 1944.

Sources: Somani, S. M., Solana, R. P. and Dube, S. N., 'Toxicodynamics of nerve agents', ed. S. M. Somani, *Chemical Warfare Agents* (Academic Press: San Diego, Calif., 1992), p. 71; and Department of the Army, Program Executive Officer, Program Manager for Chemical Demilitarization, *Chemical Stockpile Disposal Program: Final Programmatic Environmental Impact Statement*, vol. 3 (Department of the Army: Aberdeen Proving Ground, Edgewood, Md., Jan. 1988), appendix N.

Nerve agents are stable and easy to disperse and can kill within minutes at low doses (see table 2.6). These agents cause a reaction which increases the level of acetylcholine in the body. The parasympathetic nervous system releases acetylcholine, a neurotransmitter, which plays a significant role in the control of many structures in the central nervous system, including the skeletal muscles. Normally, the body produces an enzyme, acetylcholinesterase, which prevents the accumulation of acetylcholine after its release by breaking it down into choline and acetic acid. Nerve agents, which are also called cholinesterase inhibitors, permit acetylcholine to persist at the ends of the neurons. (The symptoms of nerve gas poisoning are listed in table 2.7.) The behaviour of sarin illustrates the general biochemical mechanisms of the organophosphorus agents: at the molecular level acetylcholinesterase has two active sites, the anionic and the esteratic sites, which act as reactive sites.[32]

The pharmacological effects and toxicity of the organophosphorus agents depend on their stability, rate of absorption, distribution, ability to cross the blood barrier, rate of reaction with acetylcholinesterase, selectivity for acetylcholinesterase and cholinesterase enzymes at specific foci and their stability if

[32] Somani, S. M., Solana, R. P. and Dube, S. N., 'Toxicodynamics of nerve agents', ed. Somani (note 15), p. 77.

Table 2.6. Estimates of toxicity of nerve agents to humans

Agent	LD_{50} (skin) (mg)	LD_{50} (orally) (mg)	LCt_{50} (inhalation) (mg · min/m^3)
Tabun	200–1000	25–50	150–400
Sarin	100–500 (1700)[a]	5–20	50–100
Soman	50–300	5–20	25–70

[a] Number in parentheses is given in Franke, Koehler and Zaddach (see table sources).

Sources: Somani, S. M., Solana, R. P. and Dube, S. N., 'Toxicodynamics of nerve agents', ed. S. M. Somani, *Chemical Warfare Agents* (Academic Press: San Diego, Calif., 1992), p. 76; and Franke, S., Koehler, K. F. and Zaddach, H., *Chemie der Kampfstoffe* [Chemistry of chemical warfare agents], part 1 (K. F. Koehler GmbH: Munster, 1994).

Table 2.7. Clinical effects of nerve agents in humans

Body part	Effect of nerve agent
Eye	Unilateral or bilateral miosis, conjunctival injection; pain in or around eye; dim or blurred vision
Nose	Rhinorrhea
Mouth	Salivation
Pulmonary tract	Bronchoconstriction and secretions, cough; 'tight chest', shortness of breath; wheezing rales, rhonchi
Gastrointestinal tract	Increase in secretions and motility; nausea, vomiting, diarrhoea; abdominal cramps, pain
Skin and sweat glands	Sweating
Muscles	Fasciculations ('rippling'), local or generalized; twitching of muscle groups, flaccid paralysis; twitching, weakness
Cardiovascular system	Decrease or increase in heart rate; usually increase in blood pressure
Central nervous system	
Acute effects of severe exposure	Loss of consciousness, convulsions (or seizures after muscular paralysis); depression of respiratory centre to produce apnoea, coma
Acute effects of limited exposure or lingering effects (days to weeks)	Forgetfulness, irritability, impaired judgement, decreased comprehension; feeling of 'tenseness' or 'uneasiness'; depression, insomnia, nightmares, difficulties with expression

Source: Sidell, F. R., 'Clinical considerations in nerve agent intoxication', ed. S. M. Somani, *Chemical Warfare Agents* (Academic Press: San Diego, Calif., 1992), p. 162.

attached at the enzyme active sites.[33] Figure 2.3 illustrates the neurotoxic effect on the neurons and the sites where the organophosphorus agents act.

Estimates of the toxicity of tabun, sarin and soman vary, but the LCt_{50} for tabun is estimated at 400 mg · min/m^3, for sarin at 100 mg · min/m^3 and for soman at 50 mg · min/m^3. The percutaneous LD_{50} for tabun is estimated at 1000 mg, for sarin at 1700 mg and for soman at 100 mg. After exposure as a vapour the inhibition of acetylcholinesterase occurs within minutes or even seconds. Percutaneous exposure will lead to inhibition in 1–30 minutes. If exposure is limited, the effect depends on other factors such as temperature and

[33] Somani, Solana and Dube (note 32), p. 76.

Nerve gas exposure
↓
Inhibition of acteylcholinesterase
↓
Accumulation of acetylcholine at the synapse
↓
Enhanced and continuous stimulation of cholinergic neurons
↓
Opening of Ca^{2+} channels and enhanced uptake of extracellular Ca^{2+}
↓
Increase in intracellular free Ca^{2+}
↓
Enhanced release of excitatory amino acid neurotransmitters, such as glutamate and aspartate
↓
Convulsions
↓
Cell death

Figure 2.3. Neurotoxic effects of nerve gas exposure

Source: *Medical Chemical Defense: A Bulletin on Research and Development*, US Army MRICD, vols 2, 3–4 (1988), quoted in Somani, S. M., Solana, R. P. and Dube, S. N., 'Toxicodynamics of nerve agents', ed. S. M. Somani, *Chemical Warfare Agents* (Academic Press: San Diego, Calif., 1992), p. 85.

the part of the body which comes in contact with the agent. Immediately after percutaneous exposure the effects are insignificant and commonly not recognized. The effects of vapour and dermal exposure have also been studied.[34]

III. Military smoke agents

Man-made smoke has been used for military purposes for thousands of years with varying success,[35] but it has been of military significance only in the 20th century. Special military units have been trained for this type of warfare (e.g., the German Nebeltruppe—'smoke troops'). The Nebeltruppe ostensibly were trained to use smoke as camouflage and to disorient the enemy, but they also engaged in the preparation and use of chemical weapons.[36] By the end of World War II there were substantial supplies of smoke agents.

Large amounts of smoke agents were used in both World War I and World War II, but the demilitarization (dismantling of factories and storage sites) of these weapons was not a major problem because most of the chemicals were dual-use chemicals that could be used in the civilian chemical industry. In addition, these agents were not controlled by the Allies after World War II, as were chemical warfare agents. This is one of the reasons why no accurate figures about production of these agents exist.

[34] Sidell, F. R., 'Clinical considerations in nerve agent intoxication', ed. Somani (note 15), pp. 173, 175.

[35] Worral, P. R., *Smoke Tactics* (Gale and Polden: London, 1919).

[36] Rielau, H., *Geschichte der Nebeltruppe, ABC- und Selbstschutzschule* [History of the smoke troops, NBC and self-protection school] (Bundesministerium der Verteidigung: Cologne, 1966).

Military smoke agents continue to be regarded as potential weapons.[37] In addition, new developments have taken place.[38] The following sections provide a review of smoke-producing chemicals from World War I and World War II. There are three basic types of military smoke agents and obscurants: toxic, screening and signal smoke.

Toxic smoke

Toxic smoke is a multi-component system. Such agents are generally 'smoke carriers' that transport toxic substances, especially chemical warfare agents. They are known as 'warfare smoke' (Kampfnebel in German). They employ a variety of chemical warfare agents which are thermally stable and do not decompose by evaporation or sublimation. They can be transported and evenly dispersed on a battlefield.

Toxic smoke was used extensively in World War I, especially in the early part of the war when the relatively primitive gas masks were incapable of filtering most suspended matter. Finely dispersed substances, such as the nose and throat irritants Clark I and Clark II, as well as adamsite, were used as 'mask breakers'. These substances could penetrate the mask filters to irritate the upper respiratory tract and mucous membranes of the eyelid. Troops panicked, ripped off their masks and died following exposure to chemical warfare agents.

Nose and throat irritants and lung and skin poison were added, and phosgene and mustard gas were released to raise the death toll. The German Army used smoke candles (Rauchkerzen) which were filled with iron filings and sulphur powder in a distillate that created a smoke screen. The Allies employed mixtures of 80 per cent nitrochloroform (chloropicrin, CCl_3NO_2) and 20 per cent stannic chloride (tin tetrachloride) or silicic tetrachloride. Later, chloroacetophenone (CN-Stoff), a powerful eye irritant, was also used.

Camouflage smoke

There is no commonly accepted term for camouflage smoke, which functions as a smoke screen for the purpose of concealment. Man-made smoke for military camouflage or screening serves various purposes. Sensitive areas such as production sites, supply depots and troop assembly areas can be concealed, and the movement of troops and the deployment of weapons can be hidden. In both world wars convoys of ships that sailed between the USA and England were often veiled in man-made fog to prevent detection and air attacks. In the early part of World War I, when submarine detection capabilities were primitive, this fogging technique was used to conceal ships from submarines. In battle, warships used smoke screens and concealment fogs. On land, harmless smoke screens were often used after a gas attack with chlorine or phosgene. Troops that were concealed by smoke could attack enemy troops who were still wearing gas masks that limited their fighting ability.

[37] Klose, K. and Hartmann, H., *Der Einsatz von Nebelmitteln* [The use of smoke agents] (Deutscher Militärverlag: Berlin, 1964).

[38] 'Spezialisten für Nebel, Brand, IR- und Radarabwehr' [Specialists for smoke, fire, IR and radar defence], *Wehrtechnik*, no. 4 (1977), pp. 92–97.

Signal smoke

Signal smoke is coloured smoke which is fired from flare pistols, illuminating rockets or smoke mines. The colour of the smoke may serve to signal troops. Smoke signals are used in rescue missions to locate ships in danger or downed aircraft and to mark landing zones for paratroopers or emergency landing sites. The following sections describe various kinds of smoke, focusing on their chemical characteristics and the direct and indirect effects these substances have on the environment and humans.

Oil smoke

It is common military practice to generate smoke by burning oil (crude or used oil or other petroleum distillates). Navies often use this method to conceal ships at sea since smoke generators are easily installed on almost any vessel. Oil smoke only becomes an environmental contaminant if ships dispose of the oil improperly by dumping it on land or at sea. If soil becomes contaminated it must be cleaned by microbial processes to remove hydrocarbons.

Phosphorus smoke

Elemental phosphorus (technical-quality phosphorus is referred to as red, white or yellow phosphorus) is extremely suitable for use as a smoke generator. When burned it emits thick white clouds of phosphorus pentoxide (phosphoric anhydride)—an oxide with hygroscope properties. Orthophosphoric acid produces the clouds. One part phosphorus produces 3.16 parts cloud-forming phosphoric acid. In order to prevent spontaneous combustion in air white phosphorus must be stored under water.

In World War II phosphorus was the basic substance used for phosphorous organic chemical warfare agents of the tabun–sarin group. In Germany the supply was limited, and phosphorus was restricted to military use (e.g., in incendiary bombs). In World War II Germany was able to produce white phosphorus at a rate of 1300 tonnes per month.

Sulphur acid smoke

Two kinds of smoke can be made from sulphuric acid: sulphuric trioxide (dissolved in sulphuric acid it becomes 'acid smoke' or 'smoking sulphuric acid', also known as oleum) and chlorosulphonate acid.

Sulphur trioxide smoke

Sulphur trioxide is similar in effect to elemental phosphorus. It is an inexpensive and effective smoke substance that is ideal for smoke shells and smoke mines. It has been used for military purposes since World War I, but after World War II sulphur acid smoke largely replaced sulphur trioxide. In its pure form sulphur trioxide is a crystalline substance, but even slight humidity will turn it into sulphuric acid. Sulphur trioxide is dissolved in concentrated sulphuric acid to increase ease of handling. The smoke which is generated can contain up to 40 per cent free sulphur trioxide as well as sulphuric acid and is extremely corrosive.

There are various devices for generating sulphur trioxide smoke on land, at sea and in the air. These devices do not harm the environment since only small amounts of agent residue remain after their use. Caution must be exercised when sulphur trioxide is used as a smoke generator and during its disposal. In World War II Germany was able to produce 8000 tonnes per month of sulphuric acid for smoke purposes.

Chlorosulphone acid smoke

Chlorosulphone acid, HSO_3CI, has chemical characteristics similar to those of sulphur trioxide acid. It can easily be used to generate smoke for military purposes. Chlorosulphone acid is a fluid that boils at 158°C. On contact with humid air it is immediately transformed into hydrochloric acid and sulphuric acid, and huge quantities of smoke are released. Like sulphur trioxide it is usually mixed with concentrated sulphuric acid, and the mixture is classified as acid smoke. It is often confused with acid smoke that contains no chlorine.

Sulphuryl chloride smoke

The intense irritant sulphuryl chloride, SO_2CI_2, is less important as a chemical warfare agent. In World War I it was tested in experimental trials as a chemical warfare agent. It was also used as a solvent for other CW agents such as phosgene and chloropicrin. The disposal and detoxification of weapons believed to contain sulphuryl chloride require the utmost caution because such old munitions may contain both the smoke agent and CW agents.

Acid chloride smoke

Acid chlorides are still used to generate smoke in some countries although smoke generation using phosphorus and sulphur is cheaper and simpler.

Silicon tetrachloride smoke

In moist air silicon tetrachloride turns into a smoke of finely dispersed silicic acid and hydrochloric acid. It is a liquid with a boiling point of 58°C. The amount of smoke and its ability to conceal can be enhanced by adding ammonia, which causes the evaporation of ammonium chloride smoke. The concealment efficiency of silicon tetrachloride is twice that of phosphorus smoke.

Titanium chloride smoke

In World War I only the USA had access to the raw material needed to produce titanium chloride. It is expensive to manufacture, and at its boiling point of 136°C it is cracked into titanium hydroxide and hydrochloric acid if water is added. As with silicon tetrachloride, adding ammonia enhances the concealment properties of the generated smoke. In World War II titanium chloride was produced at Leverkusen, Germany, where the Bayer company manufactured approximately 50 tonnes per month.

Stannic chloride smoke

Stannic chloride (tin tetrachloride) boils at 114°C. On contact with humid air it produces smoke composed of hydrochloric acid and tin oxide. The addition of ammonia optimizes its use for concealment, but stannic chloride smoke is less effective than other smoke. It was used during and after World War I and was a basic ingredient in smoke-generating hand grenades. The British Army used a mixture of 20 per cent stannic chloride and 80 per cent chloropicrin. The Allies also used stannic chloride in chemical weapons as a solvent for hydrogen cyanide and phosgene. Phosgene residue may still be present in buried military wastes.

Zinc smoke

Zinc smoke is excellent for concealment but difficult to use. The most common zinc smoke is Berger mixture, named after the French chemist who developed it. Zinc smoke is made from zinc dust and an aliphatic organochlorine compound (carbon tetrachloride or crystalline hexachloroethane) that reacts violently when heated, forming a mixture of hygroscopic zinc chloride particles and finely diffused carbon.

Chemically stable smoke flares are produced by combining Berger mixture and an absorbent carrier, such as zinc oxide or diatomaceous earth, with sawdust or pulverized plastic waste. Iron powder and potassium permanganate mixtures are used as igniters. Metal canisters (called fog or smoke pots) are filled with the mixture and used in military manoeuvres.

Ammonium smoke

Ammonium smoke, also known as salmiac or sal ammoniac smoke, is produced when gaseous ammoniac (NH_3) reacts with a strong acid or acid chloride. Hydrochloric or sulphuric acids or their derivatives are commonly used to cut costs.

Plastic smoke

Plastic smoke was in the development stage at the end of World War II, but experiments had been conducted with phenolic resin. Later, epoxy resin and polyurethane smoke substances were used.

Other types of smoke

The military showed interest in the use of smoke for protection against bacterial or chemical warfare agents only near the end of World War II. However, technology was inadequate at that time, and protective smoke was not developed. After World War II changes and improvements were made.[39] Odorous sub-

[39] Lohs, Kh., 'Chemie, Toxikologie, Umwelttoxikologie und Beseitigung von Nebelstoffen' [Chemistry, toxicology, ecotoxicology and disposal of smoke agents], eds G. Rippen and D. Martinetz, *Handbuch-Rüstungsaltlasten* [Old armament wastes handbook] (Ecomed Verlagsgesellschaft: Landsberg, 1996), pp. 423–42.

stances were added to training smoke to simulate battlefield conditions during manoeuvres.

Summary

Smoke-generating substances continue to be significant weapons. The majority of the agents used are dual-use chemicals that are widely employed in the chemical industry. Many of the chemicals which have military application can also be used in civilian manufacturing activities.

IV. Toxic armament wastes

Toxic armament wastes are currently receiving attention because of the environmental problems created by past military activities. There is awareness that disarmament by simple destruction may be harmful to the environment and expensive, particularly as regards chemical weapons. Toxic armament wastes are the remnants of past military activities.[40] They include the chemical substances and precursors that were used in the production of munitions such as gunpowder, high explosives and incendiary agents like napalm. In addition, service activities produced wastes such as spills of oil, petrol, flammable special fuels and other chemicals. The dismantling of weapons after military conflicts has also generated waste, and some dismantling methods have harmed the environment.[41]

Toxic wastes from past military activities are found at former production sites for conventional munitions and chemical warfare agents, and at intermediate and final munition storage areas, destruction and dismantling sites, military training grounds, shooting-ranges, blasting and other test sites and at filling stations. The wastes include the chemical warfare agents discussed in this chapter and the raw material, solvents and precursor chemicals used to manufacture them. Chemical syntheses often produce a variety of chemical by-products that pose a threat to the environment.

Identification of the toxic armament wastes present at a particular site can be facilitated by studying data on past production activities. At some sites the soil and water are contaminated, but the pattern and extent of contamination is dependent on many factors such as the length of production, the chemicals

[40] In a 1990 answer by the German Government to a request by the Green Party the following definition of sites contaminated by armament activities (Rüstungsaltlasten) was given: 'is not only limited to areas where chemical armament production and munition storage were conducted. [The German Government] has an understanding that in a more comprehensive way all the soil-, water- and air contamination by chemicals from conventional and chemical warfare agents are covered'. See also Pfaff-Schley, H., 'Rüstungsaltlasten: mehr als ein aktuelles Umweltthema' [Armament wastes: more than a current environmental topic], eds K.-W. Kiefer, H. Pfaff-Schley and L. Schimmelpfeng, *Rüstungsaltlasten '91: Untersuchungsmethoden Sanierungsmöglichkeiten-Verhinderung militärischer 'Neu'–Lasten* [Old armament wastes '91: methods of examination and possible means for cleaning up/preventing new military wastes] (Erich Schmidt Verlag: Berlin, 1991), pp. 11–15.

[41] In an answer by the Government to the Bundestag the following substances were named: chemical warfare agents, explosives, incendiary, smoke, fog and propellant agents, tactical additives to chemical warfare agents, precursors and waste products from the production of chemical warfare agents, wastes from the destruction of conventional munition and chemical warfare agents. Drucksache 11/6972 'Gefährdung von Mensch und Umwelt durch kontaminierte Standorte der chemischen Rüstungsproduktion, Rüstungsaltlasten' des Deutschen Bundestages, 11. Wahlperiode, Sachgebiet 2129, 26. Apr. 1990 [Printed Matter 11/6972, Bundestag, 'Endangering mankind and the environment by sites contaminated by chemical armament production, old armament wastes', 11th election period, subject 2129, 26 Apr. 1990].

used, the storage conditions for raw material and final products and the treatment of wastes. The structure of the soil and the location of the surface water also affect the pattern of contamination. It can be difficult to assess current contamination because many of the activities which caused it occurred more than 50 years ago. Some chemical warfare agents remain dangerous, and adequate protection must be provided for the individuals involved in redevelopment activities at contaminated sites.

Explosives are an important group of toxic armament wastes. They are the basis of all conventional munitions and have been and are still produced in large quantities. The most common explosive is TNT (2,4,6 trinitrotoluene),[42] which has been used as an explosive since the early 1900s. TNT is a threat to the environment because of its persistency and its ability to enter easily into ground water. Pure TNT is often found at former production sites because of the type of storage, destruction and disposal techniques that were employed.

The industrial production of TNT involves the nitration of toluene with sulphuric and nitric acid, and various partially nitrated products are produced as by-products. A major source of environmental contamination of the soil at former production sites is related to 'washing' the TNT, after which the water contains a high concentration of aromatic nitrogen compounds.[43] In the past, the acidic water was neutralized, and following sedimentation it was disposed of in surface water. The mud used in the neutralization process was disposed of by burying.[44] These methods contaminated both the ground and surface water.

The process which produced TNT and some of the aromatic nitrogen compounds as by-products of production or decomposition is harmful to people and the environment, and several recent studies have addressed this problem.[45] TNT undergoes abiotic decomposition (photolysis and sorption) and biological transformation; more than 30 compounds have been identified in water containing TNT.[46]

The spectrum of toxic armament wastes is broad, and the contamination caused by them affects the soil, water and air. Toxic armament wastes are regional environmental problems and, in some instances, could pose global threats. Many of the technical problems related to military wastes are similar to waste problems that exist in chemical industry, especially those related to toxic residues from pesticides. Cost is a significant element in the redevelopment of contaminated areas, and it is often the deciding factor when decisions about redevelopment are made.[47]

[42] The final TNT product contains 2,3,4 trinitroluene and 2,3,6 trinitroluene.

[43] In order to wash 1 kg of 'raw' TNT 1.6 litres of water are needed. Afterwards, the water will contain a broad spectrum of aromatic nitrogen compounds.

[44] Emmrich, M. and Rüden, H., 'Umwelthygienische Aspekte von Rüstungsaltlasten: Am Beispiel des TNT' [Environmental hygiene aspects of armament wastes: the example of TNT], eds K. J. Thomé-Kozmiensky *et al.*, *Management zur Sanierung von Rüstungsaltlasten* [Managing the clean-up of old armament wastes] (EF Verlag für Energie und Umwelttechnik: Berlin, 1992), pp. 343–52.

[45] Layton, D. *et al.*, *Conventional Weapons Demilitarization: A Health Effect and Environmental Effect Data Base Assessment, Explosives and Their Co-Contaminants, Final Report Phase II* (US Army Medical Research and Development Command, Fort Detrick: Frederick, Md., 1987), quoted in Emmrich and Rüden, (note 44), pp. 343–52.

[46] Schneider, M. and Thomé-Kozmiensky, K. J., 'Biologische Behandlung von Explosivstoffen und explosivstoffbehafteten Böden' [Biological treatment of explosive materials and soil contaminated with explosives], eds Spyra *et al.*, *Verfahren zur Sanierung von Rüstungsaltlasten* [Redevelopment methods for old armament wastes] (EF Verlag für Energie und Umwelttechnik: Berlin, 1992), pp. 177–228.

[47] Martinetz, D., *Sanierung von Industrie- und Rüstungsaltlasten* [Cleaning up old industrial and armament-related wastes] (Verlag Harri Deutsch: Thun and Frankfurt am Main, 1994).

3. Old chemical munitions and warfare agents: detoxification and degradation

THOMAS STOCK and KARLHEINZ LOHS

I. Introduction

Knowledge about old chemical munitions and toxic armament wastes that remain from activities in the first half of the 20th century can be valuable to those involved in the investigation of contaminated areas and old munition finds and in the redevelopment of sites previously used for the production, filling or storage of such munitions and agents. This chapter discusses the problems and risks of dealing with old munitions and toxic armament wastes. In addition, the chapter provides information on the detoxification of water and soil contaminated by these materials.

Old chemical munitions and chemical warfare agents are defined in this chapter as those produced and possibly stockpiled before 1946, the definition used in the CWC. Table 3.1 provides an overview of the chemical warfare agents which were used and stockpiled in World War I and later.

A chemical weapon is composed of its chemical warfare agent ('toxic chemicals and/or precursors'[1]) and the surrounding shell. Together, they are designed to cause death or injury when used. The toxic agent acts by inhalation, penetration, absorption or percutaneous reception. The CWC definition also includes 'any equipment specifically designed for use directly in connection with the employment of munitions and devices'. Until the end of World War II the most common means of delivery were grenades, bombs, artillery shells, mines and spray tanks.[2] Many areas are contaminated by old chemical munitions or toxic wastes as are vehicles (e.g., tanks and vans), storage areas (e.g., bunkers) and old CW production sites and training grounds.

II. Deterioration of old chemical munitions

Discoveries of old chemical munitions present identification problems. Military strategy dictated that weapons which did not explode were designed to confuse the enemy. Coding and camouflage paint were commonly used on chemical munitions to mask their origin and contents or to make identification difficult.[3]

[1] This is the definition in the CWC. The text of the Convention on the Prohibition of the Development, Production, Stockpiling and Use of Chemical Weapons and on their Destruction is reproduced in *SIPRI Yearbook 1993: World Armaments and Disarmament* (Oxford University Press: Oxford, 1993), appendix 14A, pp. 705–56, and in Conference on Disarmament document CD/1173, 3 Sep. 1992.

[2] In World War II Germany introduced a portable spray can, the Sprühbüchse, which was filled with sulphur mustard.

[3] The inspections conducted by the United Nations Special Commission on Iraq (UNSCOM) revealed that the Iraqi CW programme marked ammunition shells in a confusing and illogical manner. Munitions with various markings were used for different CW agents apparently based only on the availability of empty shells.

Table 3.1. Chemical warfare agents used in World War I and later

Year of first use	Type of use	Agent
Harassing incapacitating agents: tear gases and other irritants		
1914	C	Ethyl bromoacetate
1914	C	O-dianisidine chlorosulphonate
1914	C	Chloroacetone
1915	B	Xylyl bromide
1915	C	Xylylene bromide
1915	B	Benzyl bromide
1915	C	Capsaicin
1915	C	Methyl chlorosulphonate
1915	C	Ethyl chlorosulphonate
1915	C	Chloromethyl chloroformate
1915	C	Dichloromethyl chloroformate
1915	B	Bromoacetone
1915	C	Bromomethylethyl ketone
1915	C	Iodoacetone
1915	C	Dimethyl sulphate
1915	A, B	Ethyl iodoacetate
1915	C	Benzyl iodide
1915	C	O-nitrobenzyl chloride
1915	C	Benzyl chloride
1916	C	Acrolein
1917	A, B	Diphenylchloroarsine (DA)
1918	C	Thiophosgene
1918	A, B	Diphenylcyanoarsine (DC)
1918	C	N-ethylcarbazole
1918	A, B	1-bromo-1-tolunitrile (CA)
1918	C	Phenyldibromoarsine
1918	A	10-chloro-5,10-dihydrophenarsazine (adamsite, DM)
1919	A	1-chloroacetophenone (CN)
Lethal agents: choking gases		
1915	B	Chlorine
1915	C	Bromine
1915	C	Perchloromethylmercaptan
1915	A, B	Phosgene
1916	A, B	Trichloromethyl chloroformate (diphosgene)
1916	A, B	Chloropicrin
1917	C	Phenylcarbylamine chloride
1918	C	Bis(chloromethyl) ether
1918	C	Bis(bromomethyl) ether
Lethal agents: blood gases		
1916	A, B	Hydrogen cyanide
1916	C	Hydrogen sulphide
1916	C	Cyanogen bromide
1916	A, B	Cyanogen chloride
1918	C	Cyanoformate esters
Lethal agents: blister gases		
1917	A, B	Bis(2-chloroethyl) sulphide (mustard gas)
1917	A, B	Phenyldichloroarsine
1918	C	Ethyldichloroarsine
1918	C	Methyldichloroarsine
1918	C	Ethyldibromoarsine

Year of first use	Type of use	Agent
1918	A	2-chlorovinyldichloroarsine (lewisite)
1938	A	Bis(2-chloroethylthioethyl) ether
1940	A	Tris(2-chloroethyl) amine
Lethal agents: nerve gases		
1942	A	Ethyl N,N-dimethylphosphoramidocyanidate (tabun)
1952	A	O-isopropyl methylphosphonofluoridate (sarin)
(1962)?	. .	O-pinacolyl methylphosphonofluoridate (soman)

A = Stockpiled in World War II, B = Heavy World War I use, C = Minor World War I use.

Source: Robinson, J. P. P., 'The chemical industry and chemical-warfare disarmament: categorizing chemicals for the purposes of the projected Chemical Weapons Convention', *The Chemical Industry and the Projected Chemical Weapons Convention: Proceedings of a SIPRI/ Pugwash Conference*, SIPRI Chemical & Biological Warfare Studies no. 4 (Oxford University Press: Oxford, 1986), p. 86.

Many of the old chemical munitions that are discovered today were disposed of more than 50 years ago. Their markings are often no longer legible, making identification difficult or impossible. In addition, the condition of the old munitions may have been altered by physical damage, weather conditions, humidity, erosion and degradation in soil or water. Strict safety precautions must be used when handling old munitions. Each munition is unique, and the highest safety standards must be applied until there is sufficient information to determine the munition type and the agent with which it is filled.

There are standard techniques and procedures for taking samples. An old chemical munition must first be penetrated by drilling holes in the shell to reach the chemical agent. Special safety procedures must be used for such shells and for leaking munitions when they are stored, transported or destroyed.

Table 3.2 lists the shell designations and code- or camouflage names of the most commonly produced and stockpiled World War I chemical munitions. The table presents an overview of these munitions. The quantity which was stockpiled is the primary criterion for inclusion in the list. (There are other compilations which provide comprehensive lists of these munitions.[4]) A list of the acronyms and code-names which were used for some chemical warfare agents in the World War II period is presented in table 3.3.

The following discussion addresses some of the differences between toxic armament wastes and old chemical munitions. The most distinct difference is that in old munitions the chemical warfare agents may have deteriorated.

[4] Lohs, K. and Spyra, W., *Chemische Kampfstoffe als Rüstungsaltlasten* [Chemical warfare agents as old armament wastes] (EF Verlag für Energie und Umwelttechnik: Berlin, 1992); Franke, S., *Lehrbuch der Militärchemie,* Band 1, *Chemische Kampfstoffe* [Military chemistry handbook, vol. 1, chemical warfare agents] (Deutscher Militärverlag: Berlin, 1967); Stöhr, R., *Chemische Kampfstoffe und der Schutz vor chemischen Kampfstoffen* [Chemical warfare agents and protection against chemical warfare agents] (Militärverlag der DDR: Berlin, 1985); and Somani, S. M. (ed.), *Chemical Warfare Agents* (Academic Press: San Diego, Calif., 1992). In addition, the SIPRI 6-volume series on *The Problem of Chemical and Biological Warfare*, and the 15-volume SIPRI Chemical & Biological Warfare Studies provide useful information.

Table 3.2. Shell designation and code- or camouflage names of World War I chemical munitions

Shell designation	Filling designation	Composition of shell filling	Composition of shell filling (%)
Harassing agents munitions			
German weapons			
Ni-Shrapnell	Niespulver	O-dianisidine chlorosulphonate	. .
T-Granate	T-Stoff	Xylyl and xylylene bromide	. .
		isomers, and/or benzyl bromide	. .
T-Granate, grün		T-Stoff,	50
		bromoacetone or	50
		xylyl bromide,	90
		bromomethyl ethyl ketone	10
K-Granate	K-Stoff	Chloromethyl chloroformate,	90
		dichloromethyl chloroformate	10
Blaukreuz		Diphenylchloroarsine,	50–100
		N-ethylcarbazole with or without phenyldichloroarsine as solvent	0–50
Blaukreuz 1		Diphenylcyanoarsine, with or without phenyldichloroarsine as solvent	. .
French weapons			
No. 6	Palite	Chloromethyl chloroformate,	75
		stannic chloride	25
No. 8	Papite	Acrolein,	75
		stannic or titanium chloride	25
No. 9	Martonite	Bromoacetone,	60
		chloroacetone,	20
		stannic chloride	20
No. 9B	Homomartonite	Bromomethylethyl ketone,	60
		chloromethylethyl ketone,	20
		stannic chloride	20
No. 10	Bretonite	Iodoacetone,	75
		stannic chloride	25
No. 11	Cederite	Benzyl bromide,	75
		stannic chloride	25
No. 12	Fraissite	Benzyl iodide,	60
		benzyl chloride,	20
		stannic chloride	20
No. 13	Sulvinite	Ethyl chlorosulphonate,	75
		stannic chloride	25
No. 14	Cyclite	Benzyl bromide,	80
		titanium chloride	20
No. 15	Lacrymite	Thiophosgene,	75
		stannic chloride	25
No. 16	Rationite	Dimethyl sulphate,	75
		chlorosulphonic acid or	25
		methyl chlorosulphonate	25
No. 21	Camite	α-bromobenzyl cyanide	100

Shell designation	Filling designation	Composition of shell filling	Composition of shell filling (%)
British weapons			
	SK	Ethyl iodoacetate	100
	KSK	Ethyl iodoacetate,	70
		ethanol and ethyl acetate	30
Irritants and choking agent munitions			
German weapons			
Grünkreuz	. .	Trichloromethyl chloroformate, with or without a lesser proportion of chloropicrin or bromomethylethyl ketone	. .
Grünkreuz 1	. .	Phenylcarbylamine chloride and halogenated ketones	. .
Grünkreuz 2	. .	Phosgene,	60
		trichloromethyl chloroformate,	30
		diphenylchloroarsine	10
Grünkreuz 3	. .	Gelbkreuz (see below)	. .
French weapons			
. .	Manganite	Hydrogen cyanide,	50
		arsenic trichloride	50
No. 4	Vincennite	Hydrogen cyanide,	50
		stannic chloride,	15
		arsenic trichloride,	30
		chloroform	5
No. 4B	Vitrite	Cyanogen chloride,	. .
		arsenic trichloride	. .
No. 5	Collongite	Phosgene,	75
		stannic chloride	25
No. 7	Aquinite	Chloropicrin,	75
		stannic chloride	25
British weapons			
	CG	Phosgene	. .
	CBR	Phosgene,	50
		arsenic trichloride	50
	PS	Chloropicrin	. .
	PG	Phosgene,	50
		chloropicrin	50
	NC	Chloropicrin,	80
		stannic chloride	20
	JL	Hydrogen cyanide,	50
		chloroform	50
	VN	See French Vincennite above	
	JBR	Hydrogen cyanide,	50
		arsenic trichloride,	25
		chloroform	25
	CB	Cyanogen bromide	. .
Mustard agent munitions			
Germany			
Gelbkreuz	. .	Bis(2-chloroethyl) sulphide	70–90
		and chlorobenzene, nitrobenzene, carbon tetrachloride and/or bis(chloromethyl) ether	30–10

Shell designation	Filling designation	Composition of shell filling	Composition of shell filling (%)
Gelbkreuz 1	. .	Ethyldichloroarsine,	5–50
		bis(chloromethyl) ether,	50–95
		or ethyldichloroarsine,	40
		ethyldibromoarsine,	40
		bis(chloromethyl) ether	20
France			
No. 20	Yperite	Bis(2-chloroethyl)sulphide	80
		chlorobenzene or carbon tetrachloride	20
Great Britain			
. .	BB	Bis(2-chloroethyl)sulphide	. .

Source: SIPRI, *The Problem of Chemical and Biological Warfare*, vol. 1, *The Rise of CB Weapons* (Almqvist & Wiksell: Stockholm, 1971), pp. 44–45, 48–49 and 52.

Chemical warfare agents and their precursors are generally reactive compounds that tend to be unstable when stockpiled. In munitions their composition may change as a result of various chemical reactions. This process is dependent on: (*a*) the degree of purity of the agent; (*b*) the amount of by-product in the agent (although large-scale production of chemical warfare agents was technically advanced, by-products were often present in the final product); (*c*) the amount of stabilizing chemical additives in the agent (used to slow detoxification and added to the final product before the shells or bulk containers were filled, such as the German process of stabilizing tabun with chlorobenzene); (*d*) the percentage of chemical additives used to customize warfare agents for use in particular climatic or other conditions (e.g., Germany produced Zählost, sulphur mustard to which plastics or elastomer polymers were added to increase persistence, and Winter-Lost, mustard with chemical additives to decrease the freezing point and make it more usable in winter or cold climate[5]); and (*e*) the way climate affects the storage of warfare agents and filled munitions (e.g., the need for heat and humidity control measures in tropical weather conditions).

The toxicological properties of the agents are affected by the chemicals that are added to them to achieve military tactical advantage and improve their stability in storage. These properties and the biological effect on humans and the environment of both the agents and the chemical additives alter when they are mixed. This must be considered when decontamination is conducted at former production sites or when old chemical munitions are handled.

Chemical warfare agents undergo natural degradation in air, soil and water, which can lead to their detoxification. The degradation process can be physical (transport, immobilization, dilution), biological (action of micro-organisms) or chemical (oxidation, reduction, hydrolysis, transformation). The following sections discuss the impact of water, light and temperature on chemical agents and munitions.

[5] The former Soviet Union also produced a chemical called Winterlost, but it was a mixture of sulphur mustard and lewisite.

Hydrolytic detoxification

When chemical warfare agents are exposed to water hydrolytic degradation begins, and agent toxicity diminishes until the agent is detoxified. The time needed for hydrolytic detoxification varies depending on the agent (e.g., a few minutes for phosgene, approximately one hour for sarin and several hours for sulphur mustard). Hydrolytic detoxification requires large amounts of water and thorough mixing. Catalytic additives can be used to accelerate hydrolysis of the chemical warfare agent, but detoxification will be limited if the additives and agent are not well mixed.

If sulphur mustard or nitrogen mustard are inadequately mixed in water, 'preservation' of the agent can occur. Hydrolysis will take place at the interface between the agent and the water, and at the point of contact the surface will harden to form a 'resin' (i.e., a solid casing). This resin will protect the agent, preserving its toxicological effect. Data about hydrolytic detoxification of agents which are based on laboratory investigation thus may be inaccurate for agents that are located in bodies of water.

Degradation in soil

The behaviour of chemical warfare agents in soil is highly dependent on factors such as the degree of soil humidity, its mineralogical composition (sand, clay, humic acid substances, etc.), the micro-fauna and -flora, plants, bacteria overgrowth, seasonal climatic changes, impact of agricultural activities, and the like. The abiotic, biotic, organic and inorganic components affect detoxification to varying degrees. Physico-chemical processes are primarily responsible for agent distribution and subsequent detoxification. The time needed for detoxification in soil is affected by pH value, moisture and temperature.

Physical degradation

Physical factors such as temperature, light absorption, ultraviolet light, wind and snow will influence the detoxification of chemical warfare agents. Photodegradation, absorption and chemical reactions cause detoxification or conversion of the agents, and many laboratory procedures can be used to obtain valuable data.[6] Frequent changes of climate and extreme climatic conditions will affect the deterioration of munitions and particularly the agents with which they are filled.

[6] Trapp, R., *The Detoxification and Natural Degradation of Chemical Warfare Agents*, SIPRI Chemical & Biological Warfare Studies no. 3 (Taylor & Francis: London, 1985); and Norway, Royal Ministry of Foreign Affairs, *Verification of a Chemical Weapons Convention: Sampling and Analysis under Winter Conditions*, parts 1 and 2 (Royal Ministry of Foreign Affairs: Oslo, 1982, 1983).

Table 3.3. Acronyms and code-names used for some chemical warfare agents in World War II

Name	Chemical agent	Country
AC	Hydrogen cyanide	UK, USA
Adamsite	10-chloro-5,10-dihydrophenarsazine	UK
Alpha-Lewisit	Lewisite I, 2-chlorovinyldichloroarsine	Germany
Arsinöl	Diphenylchloroarsine (35%) + phenyldichloroarsine (50%) + triphenylarsine (5%) + arsenic (III) chloride (5%)	Germany
Azin	Adamsite	Germany
Beta-Lewisite	Lewisite II, bis(2-chlorovinyl)chlorarsine	UK
Blauring 1	10-chloro-5,10-dihydrophenarsazine, adamsite	Germany
Blauring 2	Arsinöl	Germany
Blauring 3	Diphenylcyanoarsine (Clark II)	Germany
Blau-Gelbring	Arsin-Öl	Germany
Blue No. 1	Phosgene	Japan
Brown No. 1	Hydrogen cyanide	Japan
C 1	Diphenylchloroarsine (Clark I)	Germany
C I	Diphenylchloroarsine (Clark I)	Germany
C II	Diphenylcyanoarsine (Clark II)	Germany
C2	Diphenylcyanoarsine (Clark II)	Germany
C 6-Base	Nitrogen mustard	Germany
C 6-Salz	Nitrogen mustard hydrochloride	Germany
CAF	Chloroacetophenone	UK
CAP	Chloroacetophenone	UK
CG	Phosgene	USA
CK	Cyanogen chloride	USA
Clark I	Diphenylchloroarsine	Germany
Clark II	Diphenylcyanoarsine	Germany
C-Lost	Mixture of various Losts with silicon oil	Germany
CN	Chloroacetophenone	USA
CNB	Chloroacetophenone + 45% tetrachloromethane + 45% benzene	USA
CNS	24% chloroacetophenone + 36% chloropicrin + 40% chloroform	USA
D-4	Precursor for tabun (dimethylaminophosphoryl dichloride)	Germany
D-7	Tabun	Germany
DA	Diphenylchloroarsine	USA
DC	Diphenylcyanoarsine	USA
D-Ester	2,2'-dichlorodiethylsulphide	Germany
DFP	Diisopropyl phosphorofluoridate	UK
Didi	Methylphosphonyl dichloride/fluoride mixture	Germany
D-Lost	Direkt-Lost from ethylene and sulphur dichloride	Germany
DM	Adamsite	USA
Doppellost	Sulphur mustard	Germany
EBA	HN-1	UK
Ethyl-S	HN-1	Germany
Excelsior	5-chloro-5,10 dihydroarsacridine	Germany
Fluorlost	2,2'-difluorodiethylsulphide	Germany
Fluortabun	Ethyl dimethylamidofluorophosphate	Germany
G	Tabun munition	Germany
GA	Tabun, with 20% chlorobenzene	Germany, USA
Gamma Lewisit	Tris(2-chlorovinyl)arsine	Germany

Name	Chemical agent	Country
Gelan	Tabun	Germany
Gelan I	Tabun	Germany
Gelan III	Sarin	Germany
Gelbring-1	S-Lost (sulphur mustard)	Germany
Gelbring-2	Zählost (2,2'-dichlorodiethylsulphide with thickeners)	Germany
Grün 1	Phosgene	Germany
Grün 3	Nerve agent, possibly soman	Germany
Grünring 1	N-Lost (HN-3)	Germany
Grünring 2	Phosgene	Germany
Grünring 3	Tabun	Germany
Grünring 4	Sarin	Germany
Grünring 5	Hydrogen cyanide	Germany
HQ	Mixture of 1,2-bis(beta-chloroethylthio)ethane and dichlorodiethyl sulphide	USA
HT	Mixture of bis-(2-chloroethyl)ether and dichlorodiethyl sulphide	USA
Isanol	Isopropanol for sarin	Germany
K2-Stoff	Diphosgene	Germany
KB 10	Ethyl-N-(2 chloroethyl)-N-nitrosocarbamate	USA, UK
KB 16	Methyl-N-(2 chloroethyl)-N-nitrosocarbamate	USA, UK
L I	Lewisite I, 2-chlorovinyldichloroarsine	UK
L II	Lewisite II, bis(2-chlorovinyl)chlorarsine	UK
L III	Lewisite I, tris(2-chlorovinyl)arsine	UK
Le 100	Tabun	Germany
Lewisit I	2-chlorovinyldichloroarsine	Germany
Lewisit II	Bis(2-chlorovinyl)chlorarsine	Germany
Lewisit III	Tris(2-chlorovinyl)arsine	Germany
LN	Herbicide	UK
LN 8	2,4 D	UK
LN 14	2,4,5 T	UK
M 30	Adamsite, mixture with cellulose nitrate, urea, diethyl phthalate and magnesium oxide	Germany
M 49	Adamsite, mixture with cellulose nitrate, urea, and diethyl phthalate	Germany
MBA	HN-2	UK
NC-Mixture	80% chloropicrin + 20% tin tetrachloride (two green rings)	UK, USA
NL	HN-3	Germany
N-Lost	HN-3	Germany
N-Stoff	Chlorotrifluoride	Germany
N-Yperit	HN-3	Germany
O	Oxollost	Germany
OA	Oxollost/Oxol mixture	Germany
OB	Winter-Lost	Germany
OCS	Chlorosulphonic acid and sulphuric acid	Germany
Oel-UP/9	N-Lost (HN-3)	Germany
OKM	Lost + dichlorodipropylsulphide (Winter-Lost)	Germany
Öl-F	Diphosgene	Germany
O-Lost	Bis(2-chloroethyl) ether	Germany
Omega-Salz	Chloroacetophenone	Germany
O-Salz	Chloroacetophenone	Germany
Oxol	Thiodiglycol	Germany
Oxol-Lost	2,2'-dichlorodiethylsulphide (Lost from the oxol process, chlorination of thiodiglycol)	Germany

Name	Chemical agent	Country
Produkt G	Tabun	Germany
Q	1,2-bis(beta-chloroethylthio)ethane	UK, USA
R-15	Adamsite	USSR
R-43	Lewisite	USSR
R-43A	Thickened lewisite	USSR
R-74	Sulphur mustard	USSR
Red no. 1	Diphenylcyanoarsine	Japan
RK-7	Mixture of lewisite and sulphur mustard	USSR
S	HN-2	UK
Sarin	O-isopropyl methylphosphonofluoridate	Germany
Sauerstoff Lost	Bis(2-chloroethyl)ether	Germany
Sauerstoffyperit	Bis(2-chloroethyl)ether	Germany
Soman	O-pinacolyl methylphosphonofluoridate	Germany
Stickstofflost	N-Lost	Germany
Stickstoffsenfgas	N-Lost	Germany
Stoff 100	Tabun	Germany
Stoff 146	Sarin	Germany
Stoff Oder	Tabun	Germany
Substanz 83	Tabun	Germany
S-Yperit	Sulphur mustard	Germany
T	Bis(2-chloroethyl)ether	USA
T-9	N-Lost (NH-3)	Germany
T-46	Sarin	Germany
T-83	Tabun	Germany
T-114	Sarin	Germany
T-144	Sarin	Germany
T-150	Chlorocyanogen	Germany
T-155	Hydrogen cyanide	Germany
T-300	BI IV 99 (chemical alloy of arsenic, magnesium and aluminium)	Germany
T-773	N-Lost (HN-3)	UK
T-1035	Dimethylfluorophosphate	UK
T-1036	Diethylfluorophosphate	UK
T-1703	DFP	UK
T-1835	Di-sec-butyl-fluorophosphate	UK
Tabun	O-ethyl N,N-dimethylphosphoramidocyanidate	Germany
Tabun A	95% tabun + 5% chlorobenzene	Germany
Tabun B	80% tabun + 20% chorobenzene	Germany
TBA	HN-3	UK
Trilon	Soman	Germany
Trilon 46	Sarin	Germany
Trilon 83	Tabun	Germany
Trilon 113	Ethylsarin	Germany
Trilon 144	Sarin	Germany
Trilon 300	Arsenic hydride	Germany
UP	N-Lost (HN-3)	Germany
VIR	Thickened mustard	USSR
Weißring	Chloroacetophenone	Germany
White No. 1	Arsenic (III) chloride	Japan
Winter-Lost	50% sulphur mustard and 50% phenyldichloroarsine or mixtures of sulphur mustard with various additives, such as arsinöl, chlorobenzene and anthracene	Germany
Y 3	60% Levinsteinlost, 40% oxygen mustard	UK

Name	Chemical agent	Country
Y 5	Mustard gas from sulphur dichloride and ethene in tetrachlorocarbon	UK
Y 5A	Y 5 in chlorobenzene	UK
Yellow no. 1	S-Lost, sulphur mustard	Japan
Yellow no. 2	Lewisite	Japan
Zählost	2,2'-dichlorodiethylsulphide with thickeners	Germany
Zähyperit	2,2'-dichlorodiethylsulphide with thickeners	Germany
ZL	Zählost	Germany
Zyklon	Hydrogen cyanide	Germany
Zyklon A	Hydrogen cyanide with 10% chloroformic acid ester, stabilized with kieselguhr	Germany
Zyklon B	Hydrogen cyanide stabilized with kieselguhr	Germany

Sources: Compiled by the authors, based on Bretschneider, W., 'Tarnbezeichnungen, Begriffe und Abkürzungen chemischer Kampfstoffe' [Code-names, definitions and abbreviations for chemical warfare agents], K. Lohs and W. Spyra, *Chemische Kampfstoffe als Rüstungsaltlasten* [Chemical warfare agents as old armament wastes] (EF Verlag für Energie und Umwelttechnik: Berlin, 1992); Kopecz, P., 'Verdachtsstandorte von Rüstungsaltlasten in Deutschland' [Suspected sites of old armament wastes in Germany], vol. 5, *Teilvorhaben Kampfstofflexikon* [Abridged warfare agents dictionary], Forschungsbericht 103 40 102, UBA-FB 93-029 (Umweltbundesamt: Berlin, 1993); and Santesson, J., *C-Ordbok-förkortningar, akronymer och trivialnamn* [CW dictionary: abbreviations, acronyms and common names], FOA report A 40048-C1 (FOA: Umeå, Sweden, June 1984).

III. Detoxification and degradation of chemical warfare agents

This section presents data on the degradation and detoxification of chemical warfare agents, focusing on their hydrolysis and conversion.

Behaviour in water

The behaviour of chemical warfare agents in water is dependent on several factors including temperature, pH value and salt content. The most important consideration for hydrolysis of these agents is their solubility in water. Table 3.4 presents data on the solubility of some chemical warfare agents.

Sulphur mustard hydrolyses slowly in water to form hydrochloric acid and thiodiglycol, both of which are non-toxic. Hydrolysis is dependent on temperature, density, viscosity, pH value and pressure. It differs depending on whether distilled or sea water is used. In distilled water the half-life of sulphur mustard is 8.5 minutes at 25°C; in sea water its half-life is 60 minutes at 25°C. The reported values differ. In sea water the half-life of sulphur mustard has also been estimated at 15 minutes at 25°C, 49 minutes at 15°C and 175 minutes at 5°C.[7] The rate of hydrolysis of sulphur mustard also depends on the salt content (cations and anions) of the solution.

[7] Epstein, J. *et al.*, *Summary Report on a Data Base for Predicting Consequences of Chemical Disposal Operations*, ESAP 1200-12 (Department of the Army, Edgewood Arsenal: Aberdeen Proving Ground, Edgewood, Md., 1973), quoted in *Chemical Stockpile Disposal Program: Final Programmatic Environmental Impact Statement,* vol. 3 (US Army: Aberdeen Proving Ground, Edgewood, Md., Jan. 1988), appendix N, pp. N-15–16.

Table 3.4. Solubility of chemical warfare agents

Chemical agent	Solubility in water [g · l^{-1}]
Chloroacetophenone	1
Diphenylchloroarsine (Clark I)	2
Diphenylcyanoarsine (Clark II)	2
Adamsite	Negligible
Phosgene	9
Sulphur mustard	0.8
HN-3	0.16
Lewisite	0.5
Tabun	120

Source: Compiled by the authors.

The solubility of sulphur mustard is demonstrated by the following example: in still water at 18°C, 15 days would be needed for half the mass of a drop of sulphur mustard with an initial diameter of 1 cm to dissolve. This would allow other processes to slow or halt further dissolution and hydrolysis (e.g., formation of oligomeric/polymeric layers).[8] Hydrolysis accelerates when the temperature is increased; 99 per cent hydrolysis of a saturated sulphur mustard solution at 20°C will occur in 110 minutes (at 50°C, in 4 minutes).[9] Sulphur mustard is soluble in water with a solubility of 0.8g · l^{-1} at 20°C. When dissolved, sulphur mustard hydrolyses more rapidly.

Sulphur mustard is more dense than water and sinks in it.[10] Small drops tend to dissolve, but larger drops collect in sediment and can survive for a long period. Sulphur mustard can resist degradation in water or soil for years.[11] The rate of hydrolysis of sulphur mustard on the seabed, where it may become oil-like, is affected by ocean currents and agitation of the sea. In the deep ocean a one-tonne block of sulphur mustard would take approximately five years to dissolve. (The dissolution rate is lower than the rate of hydrolysis and is therefore the key factor.[12]) The low temperature and high pressure in the deep ocean can also cause sulphur mustard to solidify.

Zählost, a modified form of sulphur mustard, is more stable in water. It only hydrolyses to a significant degree after the sulphur mustard has diffused; the remaining thickeners slow further diffusion and preserve any active agent for years. Such fragments are elastic and protected from mechanical destruction. Hydrolysis occurs where the agent and water interface, and that surface hardens to form a resin. Sulphur mustard that has undergone this process does not tend to dissolve. Products of the initial hydrolysis form oligomeric or polymeric

[8] Small, M. J., *Compounds Formed from the Chemical Decontamination of HD, GB, and VX and Their Environmental Fate*, technical report 8304 (US Army Medical Bioengineering Research and Development Laboratory, Medical Research and Development Command: Fort Detrick, Frederick, Md., 1984), quoted in *Chemical Stockpile Disposal Program* (note 7), p. N-16.

[9] Trapp (note 6), p. 17.

[10] Perry Robinson, J. P. and Trapp, R., 'Production and chemistry of mustard gas', ed. S. J. Lundin, *Verification of Dual-use Chemicals under the Chemical Weapons Convention: The Case of Thiodiglycol*, SIPRI Chemical & Biological Warfare Studies no. 13 (Oxford University Press: Oxford, 1991), p. 11.

[11] Small (note 8), p. N-16.

[12] Sage, G. W. and Howard, P. H., 'Environmental fate assessments of chemical agents VX & HD', CRDEC-eccr-043, Aug. 1989, ADB137325L, quoted in Donnelly, P., J., 'Environmental fate of chemical agents in deep seawater', *Literature Review*, 3 Aug. 1990.

structures that effectively slow hydrolytic degradation of the remaining sulphur mustard, and oligomeric and dimeric products form a protective surface.

Nitrogen mustards generally react more slowly than sulphur mustards, and the products of hydrolysis are non-toxic. Hydrolysis of nitrogen mustards has been studied extensively.[13] The solubility in water of HN-3 is $0.16 g \cdot l^{-1}$. Approximately three weeks in fresh water are needed for total decomposition.[14] (The solubility in water of the tertiary amine, HN-1, is higher.)

Hydrolysis of *diphenylchloroarsine* (DA, Clark I) creates diphenylarsenious acid and hydrochloric acid; when diphenylcyanoarsine (DC, Clark II) hydrolyses, hydrogen cyanide and diphenylarsenious oxide are created. Both acids are toxic but quickly detoxify in water. The organoarsenic compounds that result from hydrolysis may later decompose to toxic inorganic arsenic compounds. As for all arsenic compounds, accumulation of arsenic may occur in the biosphere.

Adamsite hydrolyses very slowly in water and is almost insoluble. Its reaction products are hydrochloric acid and diphenylarsenious acid.

Lewisite (α-lewisite or 2-chlorovinyldichloroarsine) is only slightly soluble in water;[15] when dissolved it hydrolyses rapidly into 2-chlorovinylarsine oxide, which is relatively soluble in sea water. In an alkaline solution a further reaction may occur that produces acetylene and arsenic acid. In large volumes of turbulent water lewisite may dissolve more rapidly.[16] Lewisite is composed of 36 per cent arsenic by weight, and a toxic element remains when it is dissolved.

Aqueous solubility is the primary determining factor for hydrolysis of nerve agents. *Sarin* is more soluble than *soman*, and extensive research has been done on the hydrolytic behaviour of sarin, which undergoes spontaneous hydrolysis in water. The first step of the hydrolytic reaction results in diisopropyl methylphosphonate and hydrofluoric acid; the production of hydrofluoric acid accelerates in the second step. The half-life of sarin in water at different temperatures and pH values has also been well studied.[17] Its half-life decreases as the temperature and pH value increase. At 30°C and pH 9.0 the half-life of sarin in water is 15 minutes.[18] Sarin hydrolyses faster in sea water than in fresh water, primarily because of the increased concentration of salts (e.g., magnesium ions) in sea water.[19] Table 3.5 presents data on the half-life of sarin in water.

Tabun is quite soluble in water. The rate of its hydrolysis is dependent on the pH value and the temperature of the water, but it is always rapid. The products

[13] Schindlbauer, H., *Analytik und Reaktionsverhalten von N-Lost aus alten Kampfstoffgranaten* [Analysis and reaction behaviour of N-Lost from old chemical agent shells] (Institut für Chemie und Technologie von Erdölprodukten, Technische Universität: Vienna, 27 June 1990).

[14] *Chemische Kampfstoffmunition in der südlichen und westlichen Ostsee: Bestandsaufnahme, Bewertung und Empfehlungen*, Bericht der Bund/Länder-Arbeitsgruppe Chemische Kampfstoffe in der Ostsee [Chemical munitions in the southern and western Baltic Sea: compilation, assessment and recommendations, Report by the Federal/Länder Government Working Group 'Chemical Munitions in the Baltic Sea'] (Federal Maritime and Hydrographic Agency: Hamburg, May 1993), p. 22.

[15] *Chemical Stockpile Disposal Program* (note 7), p. N-16.

[16] *Chemical Stockpile Disposal Program* (note 7), p. N-16.

[17] Epstein, J., 'Properties of GB in water: water technology and quality', *Journal of the American Waterworker Association*, vol. 66 (1974), pp. 31–37.

[18] Somani, S. M., Solana, R. P. and Dube, S. N., 'Toxicodynamics of nerve agents', ed. Somani (note 4), p. 72.

[19] Michels, H. O., Gordon, E. C. and Epstein, J., 'Detection and estimation of isopropyl methylphosphonofluoridate and O-ethyl S-diisopropylaminoethylmethylphosphonothiolate in seawater in parts-per-trillion level', *Environmental Science Technology*, vol. 11 (1973), pp. 1045–49, quoted in *Chemical Stockpile Disposal Program* (note 7), p. N-15.

Table 3.5. Half-life of sarin in water

Figures are in hours.

	pH				
Temperature (°C)	6.5	7.0	7.5	8.0	9.0
0	8 300	2 650	830	265	26.5
10	1 870	591	187	59	6.0
20	461	146	46	15	1.5
25	237	75	24	7.5	0.8
30	125	39	12.5	4	0.4

Source: Trapp, R., *The Detoxification and Natural Degradation of Chemical Warfare Agents*, SIPRI Chemical & Biological Warfare Studies no. 3 (Taylor & Francis: London, 1985), p. 13.

of hydrolysis are phosphonic acid and hydrogen cyanide. The half-life of tabun in sea water at 7°C is approximately 5 hours.[20] The half-life of phosphor-organic agents in fresh water differs from their half-life in sea water. The half-life of *soman* in water at 20°C (pH 7.0) is 4950 minutes.[21]

Phosgene is a highly reactive compound. In sea water it rapidly hydrolyses into carbon dioxide and hydrochloric acid. At 0°C a 1 per cent solution of phosgene hydrolyses in less then 20 seconds. Table 3.6 presents data on the half-life of several chemical warfare agents in water at various temperatures.

Behaviour of chemical warfare agents in groundwater[22]

A chemical warfare agent generally reaches the groundwater by leaching, and some of the agent is adsorbed by soil particles in the process. Decontamination occurs following adsorption. The amount of agent retained by the soil varies based on the properties of the soil and agent. An agent can reach the groundwater quickly if it moves through a geological fault, but this is uncommon. It is difficult to predict the rate at which the agent will reach the groundwater because of the effect of factors such as the type and porosity of the soil, rainfall, temperature, humidity, and the like.

Behaviour of chemical warfare agents in soil

Soil is a complex ecosystem of biotic and abiotic components, organic and inorganic material and water, and these affect the behaviour of chemical warfare agents in soil. The persistence and degree of natural degradation of an agent in soil depends on: (*a*) the chemical and physical properties of the agent; (*b*) the meteorological conditions (e.g., wind speed, temperature and humidity); (*c*) the properties, composition, temperature, pH value and moisture of the soil; and (*d*) the type and uniformity of vegetation.

Physico-chemical processes distribute the agent after its deposit in the soil. The most important detoxification step is adsorption, followed by degradation.

[20] *Chemische Kampfstoffmuntion in der südlichen und westlichen Ostsee* (note 14) p. 22.
[21] Trapp (note 6), p. 15.
[22] The following discussion is based on *Chemical Stockpile Disposal Program* (note 7), pp. N-25–26.

Table 3.6. Stability of chemical warfare agents in fresh water

Agent	Half–life	pH	Temperature (°C)
GA (tabun)	7 hours (hr)	4–5	20
	8.5 hr	7.0	20
GB (sarin)	$(5.4 \times 10^8) \times 10^{-pH}$	7.0–13.0	25
	5.4 hr	. .	. .
	837 hr	7.5	0
	453 hr	7.5	4
	46 hr	7.5	20
	24 hr	7.5	25
GD (soman)	82 hr	7.0	20
	996 hr	7.0	25
	428 hr	7.5	25
	13 years	7.6	4
	184 hr	8.0	25
HD (sulphur mustard)	110 minutes (min.) = 99% hydrolysis	. .	20
	8.5 min. (distilled H_2O)	. .	25
	7.4–15.8 min.	. .	20
	158 min.	. .	0.6
HN-3 (nitrogen mustard)	Hydrolyses slowly in water; pH-dependent quantitative hydrolysis occurs only when temperature is above 90°C.	. .	. .
L (lewisite)	Rapid for dissolved lewisite	. .	. .

Sources: Leffingwell, S. S., 'Public health aspects of chemical warfare agents', ed. S. M. Somani, *Chemical Warfare Agents* (Academic Press: San Diego, Calif., 1992), p. 329; Trapp, R., *The Detoxification and Natural Degradation of Chemical Warfare Agents*, SIPRI Chemical & Biological Warfare Studies no. 3 (Taylor & Francis: London, 1985); and Department of the Army, Program Executive Officer, Program Manager for Chemical Demilitarization, *Chemical Stockpile Disposal Program: Final Programmatic Environmental Impact Statement*, vol. 3 (Department of the Army: Aberdeen Proving Ground, Edgewood, Md., Jan. 1988), appendix N.

The humic substances that are part of the system of non-living organic material play a significant role. They contain aromatic structures and various chemical groups such as alcoholic hydroxy, carboxy, methoxy and phenolic groups, carbonyl functions and ether bridges. An increase in temperature accelerates desorption. Leaching, volatilization and surface run-off also affect the degree of detoxification.[23]

Chemical degradation reactions are influenced by the occurrence of cation-containing material in soil, which may initiate catalytic chemical reactions. Clay minerals such as kaolinite, montmorillonite and certain zeolites can be used to degrade organophosphorus compounds and mustards.[24]

Biological degradation is also important. Microbial metabolism in soil is complex and affects the rate and degree of detoxification of a chemical warfare agent. Available data on the microbial metabolism of chemical warfare agents are limited. Only a few micro-organisms can break the strong and stable

[23] For a more detailed discussion of physico-chemical processes see Trapp (note 6), pp. 25–30.
[24] Trapp (note 6), pp. 32–33.

carbon–phosphorus bond of the organophosphorus agents. Their activity is promoted by warm soil temperature, moisture and the presence of organic matter.

The rate of evaporation affects soil decontamination. For example, 1.8 years are required to remove 90 per cent of sulphur mustard from a 10-cm layer of soil via evaporation.[25] Sulphur mustard is the most persistent chemical warfare agent in soil. It and one of its breakdown products may still be present in soil at Fort McClellan, Alabama, after 9–29 years of environmental weathering.[26] Sulphur mustard that was dumped in 1941 at Edgewood Arsenal could still be detected 30 years later.[27] Thickened sulphur mustard (e.g., Zählost) also produces long-term soil contamination.[28]

The persistency of an agent in soil is a significant factor that is dependent on the boiling and freezing points, volatility and rate of hydrolysis of the agent. Table 3.7 lists some of the characteristics of various chemical warfare agents.

Munition deterioration

Munition deterioration is affected by external and internal factors. External factors include temperature, humidity, climate, physical storage conditions and shell casing material. Long-term storage may be the most significant parameter affecting the 'life expectancy' of a filled shell. Internal factors include the physical stability of the agent, its chemical purity, the composition of stabilizers or additives, the degree of decomposition and the pressure within the shell.

External factors contribute to the rate of corrosion, but it is impossible to estimate the rate of deterioration that results from prolonged exposure to air, water, sun and soil. The chemical and physical factors that promote corrosion (e.g., electrolytic composition, dissolved gases, pH value, electrical resistance, pressure and environmental change) interact with other factors that delay it (e.g., natural surface films and artificial protection such as paint and metal casings).

Carbon steel, aluminium and brass alloys are used for shell casings. Steel shells react differently in soil or sea water than other shells, and aluminium shells are less durable than steel shells. The thickness of the walls of a shell is also significant. (Thick-walled shells are more resistant to external deterioration.) It is difficult to predict how long filled shells will survive in normal storage conditions. The situation is further complicated if the munitions are located in water or soil.[29] Contact between a shell and bottom sediment or sludge will cause a steady rate of corrosion.

If the filling agent in a shell can be identified it is possible to determine how it will react with the shell casing. For example, hydrogen cyanide is very unstable even if stabilizers are added, and organophosphorus compounds are stable only if stabilizers are added. Most chemical warfare agents are produced

[25] *Chemical Stockpile Disposal Program* (note 7), p. N-24.

[26] Small, M. J., *Soil Detection Limits for Potential Chemical Warfare-Related Contaminants at Fort McClellan, Alabama*, technical report 8208 (US Army Medical Bioenginering Research and Development Laboratory: Fort Detrick, Frederick, Md., 1983), quoted in *Chemical Stockpile Disposal Program* (note 7), appendix O, p. O-27.

[27] Epstein *et al.* (note 7), appendix O, p. O-27.

[28] Perry Robinson and Trapp (note 10), p. 13.

[29] In the aftermath of the Gulf War it was discovered that Iraqi 122-mm chemical munitions burst a few months after filling.

Table 3.7. Persistency of chemical warfare agents

Chemical warfare agent	Characteristics
Lewisite	Heavily splashed liquid will persist up to one day, long duration in snow, vaporized agent hydrolyses in hot humid climate
HN-1	Mildly persistent, up to one day in moderate climate, less persistent in dry conditions, may last for approximately a week in snow
HN-2	Heavily splashed liquid may last up to one day in moderate weather, much longer in extreme cold
HN-3	Considerably more persistent than mustard, may last for several days in temperate weather, longer in extreme winter conditions
Mustard–lewisite mix	Lasts approximately 1–1 1/2 days in temperate weather, more than a week in winter conditions
Mustard	Heavily splashed liquid may last several days, may last up to a month in winter conditions
Hydrogen cyanide	Vapour persistency short, vapour density 0.93 times the weight of air
Cyanogen chloride	Vapour persistency short, vapour density 2.1 times the weight of air
Phosgene	Short duration, high volatility, vapour density 3.4 times the weight of air
Diphosgene	Short duration, more volatile then phosgene, vapour density twice as heavy as phosgene
Tabun	Splashed liquid may remain for several days in temperatures below 25°C, hydrolyses in snow
Sarin	Non-persistent, evaporates at approximately the same rate as water, high volatility
Soman	Mildly persistent, may last for one day or more in special circumstances
Diphenylchloroarsine (Clark I)	Short duration, dissemination as aerosol
Diphenylcyanoarsine (Clark II)	Short duration, dissemination as aerosol
Adamsite	Short duration, dissemination as aerosol
Bromobencylcyanide	Most persistent of the tear gases, heavily splashed liquid remains for up to two days, permeates most porous surfaces
Chloroacetophenone	Short, practical dissemination by thermal source or aerosol

Source: Based on data in Compton, J. A. F., *Military Chemical and Biological Agents: Chemical and Toxicological Properties* (Telford Press: Caldwell, N.J., 1987).

in industrial batches, and impurities in the final product may initiate decomposition of the agent in storage. Some decomposition reactions can increase pressure inside a shell, causing it to deteriorate or burst (e.g., organocyano compounds create high pressure when they decompose inside shells).[30]

Table 3.8 presents data on how chemical warfare agents react with metals and other materials. The reactions of these agents and their behaviour in storage are based on experience with pure agents. Impurities, decomposition or chemical additives can cause other reactions to occur.

[30] Lohs and Spyra (note 4), p. 27.

Table 3.8. Stability in storage and corrosive action of chemical warfare agents

Agent	Corrosive action/stability in storage
Mustard	When pure little corrosive action, stable in steel or aluminium
Lewisite	None on ferrous alloys, stable in steel or glass canisters
HN-1	Corrosive to ferrous alloys at 65°C, polymerizes slowly
HN-2	No significant corrosive action, high tendency to polymerize
HN-3	No significant corrosive action, slow polymerization, gradual formation of polymeric quaternary ammonium salts
Mustard–lewisite mixture	No significant corrosive action, no hazardous chemical reactions between the components
Hydrogen cyanide	No significant corrosive action when pure, impurities such as phosphoric acid or sulphur dioxide will contribute to instability
Cyanogen chloride	No significant corrosive action when pure, stable for less than one month in canister munitions, polymerization forms compounds which are explosive
Phosgene	No significant corrosive action when pure, when hydrolysed produces HCl and enhances corrosion
Diphosgene	Unstable in anything but glass, most metals catalyse diphosgene in phosgene, if hydrolysed HCl enhances corrosion
Tabun	No significant corrosive action when pure, stable in the absence of water
Sarin	Slightly corrosive to ferrous alloys, stable in steel canisters
Soman	Slightly corrosive to most metals, less stable than tabun and sarin
Diphenylchloroarsine	No significant corrosive action when dry, stable in pure form
Diphenylcyanoarsine	No significant corrosive action when dry, very stable in storage
Adamsite	Slightly oxidizing when dry, corrosive to ferrous alloys when wet
Bromobencylcyanide	Liquid highly corrosive to all common metals, reacts with pure iron, special storage required (highly corrosive), moderately stable in glass enamel-lined canisters
Chloroacetophenone	Slight oxidation of ferrous alloys, stable in steel and plastic containers

Source: Based on data in Compton, J. A. F., *Military Chemical and Biological Agents: Chemical and Toxicological Properties* (Telford Press: Caldwell, N.J., 1987).

IV. Conclusions

Chemical warfare agents and munitions that have been buried or dumped in water for a long period react based on the chemical and physico-chemical properties of the chemical substance and on external factors such as temperature, pH value, moisture, physical surroundings and so on. Whether an agent remains toxic is related to how it reacts in the environment where it is situated. If the agent is encased in a shell, a degree of self-detoxification may have occurred. This chapter summarizes some of the data on the long-term behaviour of chemical warfare agents and CW munitions. They remain highly dangerous, and each instance of discovered agents or munitions is unique. Before measures can be taken to neutralize the potential danger the toxicity, degree of detoxification and the condition of the munition must be determined. Past experience, laboratory investigations and the body of information currently available can only serve as a guide.

Part III. Coping with old munitions and toxic armament wastes: technical experience

4. Decontamination of former military bases in eastern Germany

KARL LEHMANN and REINER ESCHER

I. Introduction

In what is today the Federal Republic of Germany (FRG) an area of approximately 950 000 hectares (ha) is used or has been used for military purposes. Close to one-third of the federal state of Brandenburg was used by the military. Four groups—the German Army (Bundeswehr); the Allied forces and the North Atlantic Treaty Organization (NATO); the National People's Army (NVA) and the Frontier Troops (GT) of the former German Democratic Republic (GDR); and the Western Group of the former Soviet Army (WGT)—each used approximately one-fourth of the area.

The reunification of Germany in October 1990 led to a number of changes. The NVA and GT were disbanded and the size of the Bundeswehr was reduced. Both the WGT and Allied forces departed. This meant that a number of military areas were abandoned or put to new uses.

The sites in the five new federal states and in Berlin that were used by the WGT were freed for civilian use in 1991–94. The military relinquished more than 1000 sites of various size and with a total area of approximately 240 000 ha (see figure 4.1). Brandenburg had the greatest number of sites and the largest area, followed by the federal state of Sachsen-Anhalt.

At the end of 1990 approximately 380 000 soldiers and officers of the WGT remained in eastern Germany. Thirty per cent of these troops were withdrawn each year in 1991–93, and the remaining 10 per cent departed in 1994. By the end of 1992 more than one-half the troops had left eastern Germany together with their equipment and supplies. The states of Thüringen and Sachsen and the Lieberose Training Ground in Brandenburg were among the first to be totally free of WGT troops.

The more than 1000 sites that had been used by the WGT ranged in size from less then 0.01 ha to approximately 24 000 ha (see figure 4.2), and approximately one-third of these were as small as 1–10 ha. The majority were used for strictly military purposes such as troop training, military housing and storage areas for arms and other military equipment including ammunition and related materials.

In order to guarantee their independence the WGT bases were supplied with nearly all of the necessities to sustain military life in war or peace. There were the means to prepare food, including livestock, as well as medical facilities and shops. Table 4.1 presents examples of typical functions and services that were provided on WGT bases.

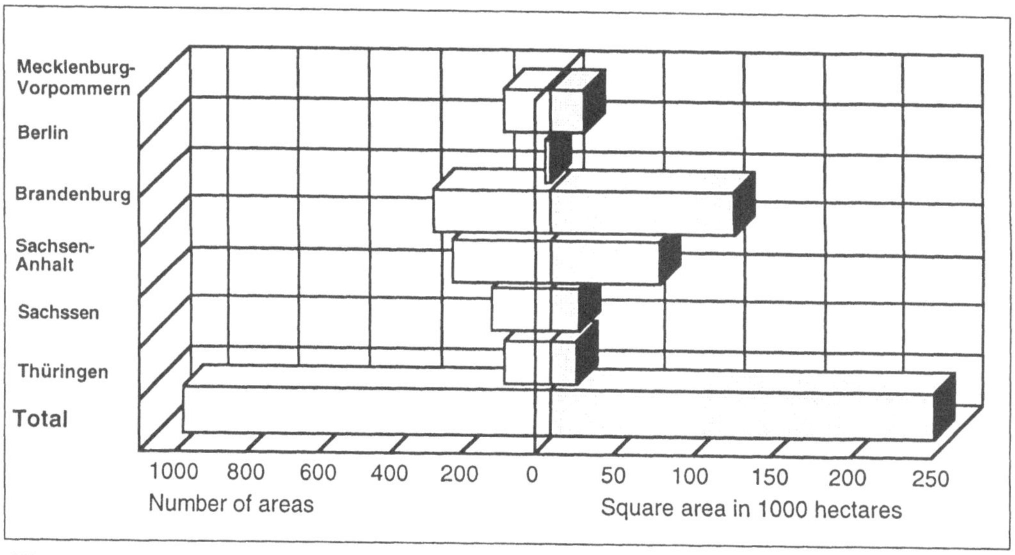

Figure 4.1. Distribution of former WGT bases in the new federal states and Berlin, as of July 1992

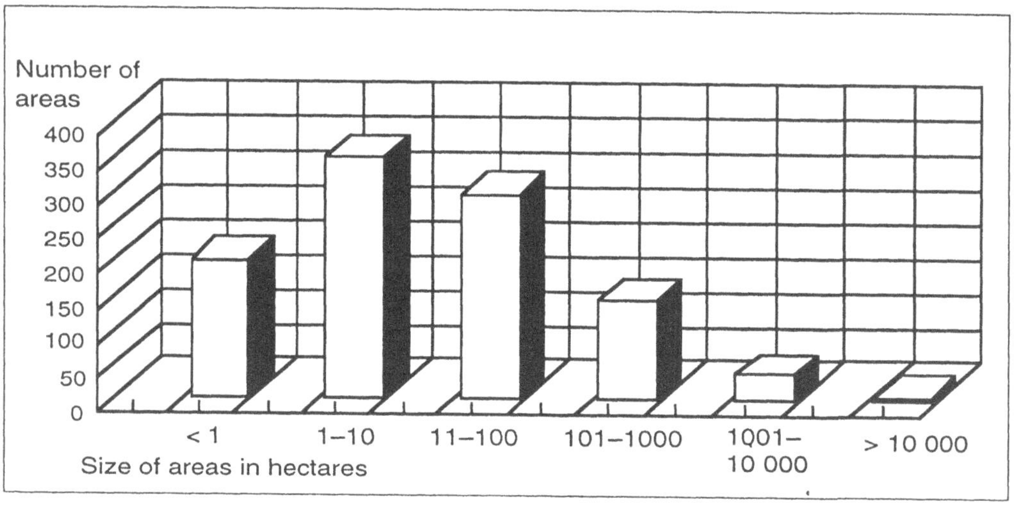

Figure 4.2. Distribution by size of the former WGT bases in eastern Germany, as of July 1990

Various types of activities were conducted at one area, and the typical WGT base was used for a variety of functions, including:

(*a*) training activities involving parade-grounds, playing-fields, steam baths and classrooms;

(*b*) military housing and storage areas for equipment and arms including barracks, equipment storage rooms, covered storage areas and weapon depots;

(*c*) military headquarters;

(*d*) storage of additional equipment and spare parts, of fuel and lubricants (including fuel tanks and filling stations), chemicals (e.g., solvents and detergents), solid fuel, building materials, food and feed for livestock;

(*e*) communication services such as radio stations;

Table 4.1. Typical functions and services available on WGT bases

Function or service	Examples of use
Training and parade-grounds	Training areas, bomb, shooting and artillery ranges
Headquarters	Supreme command, commandant and staff offices
Barracks for various units	Armoured corps, mobile land forces, artillery, anti-aircraft defence, air force, rocketry troops, para-troopers, engineers, intelligence service, supply and transport units
Air base	Aircraft (fighters, bombers, transport aircraft), four-engine and transport helicopters
Intelligence service	Radio and radar facilities
Storage	Ammunition, fuel and lubricants, chemicals (detergents, solvents, acids, lyes), lacquers and paints, engineering equipment
Maintenance and repair	Maintenance and repair facilities, battery-charging stations
Medical	Medical facilities (hospitals, sanatoriums)
Trade, supply, cleaning, waste disposal	Shops, bakeries, barns for livestock, sewage treatment plants, areas for waste products
Housing, recreation, and the like	Flats, schools, clubs, museums, memorial sites, guest houses, playing-fields, gymnasiums, cemeteries

Source: Compiled by the authors.

(*f*) maintenance and repair of equipment and arms and the areas where these activities are carried out, vehicle repair areas, cleaning facilities and battery-charging stations;

(*g*) medical facilities; and

(*h*) shops, supply areas and waste treatment facilities including storage areas, dining halls, deep wells, livestock barns (e.g., for pigs and sheep), heating plants and waste disposal facilities.

Since military facilities are put to so many uses they are potential sites of contamination, and there may be several types of hazardous waste and contamination in one area.

II. Areas of potential contamination

In general, any area where harmful substances have been stored, transported or processed in any way must be regarded as a potential hazardous waste site. The types of harmful substances present can often be determined if there is information about the way in which an area was used previously. Table 4.2 lists the types of hazardous wastes that are likely to be found at a military base. It is not comprehensive but provides an indication of possible sources of contamination. Together with table 4.1 it outlines a number of potential waste hazards.

Contamination can be present from military use that occurred many years in the past. In eastern Germany a number of the areas where contamination is present were used continuously by the military from the late 19th century.

Table 4.2. Types of hazardous waste likely to be found on a military base

Area	Type of hazardous waste
Battery-charging station	Potash lye, sulphuric acid, heavy metals, lubricants
Pipes and sewers	Mineral oil hydrocarbons, lubricants, aromatic hydrocarbons (BTEX), chlorinated hydrocarbons, heavy metals, tensides
Storage area for chemicals	Lyes, acids, pesticides
Storage area for lacquers and paints	Mineral oil hydrocarbons, BTEX, chlorinated hydrocarbons, heavy metals, nitro-compounds
Storage area for detergents and solvents	Mineral oil hydrocarbons, BTEX, chlorinated hydrocarbons, tensides
Storage area for fuel and lubricants	Mineral oil hydrocarbons, lubricants, BTEX, chlorinated hydrocarbons, heavy metals
Medical facilities	BTEX, chlorinated hydrocarbons, infected waste
Waste disposal facility	Dust, infected waste, heavy metals
Vehicle repair areas	Mineral oil hydrocarbons, lubricants, BTEX, chlorinated hydrocarbons, polycyclic BTEX, anti-freeze agents, heavy metals, tensides
Storage area for scrap materials	Mineral oil hydrocarbons, lubricants, BTEX, chlorinated hydrocarbons, anti-freeze agents, sulphuric acid, heavy metals, tensides
Motor pool	Mineral oil hydrocarbons, lubricants, BTEX, chlorinated hydrocarbons, polycyclic BTEX
Filling station	Mineral oil hydrocarbons, lubricants, BTEX, chlorinated hydrocarbons, polycyclic BTEX, heavy metals
Livestock facilities	Liquid manure
Transformer station	Chlorinated hydrocarbons, polychlorinated biphenyls, lubricants,
Munition storage	Lubricants
Laundry	Mineral oil hydrocarbons, BTEX, chlorinated hydrocarbons, tensides
Equipment cleaning facility	Mineral oil hydrocarbons, lubricants, BTEX, chlorinated hydrocarbons, heavy metals, tensides
Repair shop	Mineral oil hydrocarbons, lubricants, BTEX, chlorinated hydrocarbons, polycyclic BTEX, anti-freeze agents, sulphuric acid, heavy metals, tensides

Source: Compiled by the authors.

Many of the areas in question were battlefields or were used after wars for testing or destroying munitions. The presence or past use of conventional or chemical warfare agents must be considered when the potential hazard presented by these military sites is evaluated.

III. Treatment of hazardous waste sites

In Germany today, suspected hazardous waste sites are treated using an approach that is based on recommendations from experts and on practical studies. This includes recording information about the site, assessing the danger, research prior to restoration of the site, the actual restoration and associated protective measures, and monitoring.

All information of relevance about the site in question is recorded. This includes data about the size and shape of the site, its past use, its geological and hydrological properties and the types of hazardous wastes suspected to be present. Visual inspection of the site is made, followed by interpretation of all available information in order to arrive at an initial evaluation of the site, especially as regards possible danger to humans and the environment.

If hazardous waste is found the site is studied to localize possible hazards and to protect against them. This is followed by more extensive investigation to determine the kind, amount and distribution of the contamination and its effect on humans and the environment. At the end of the process of initial evaluation an assessment is made of the need for restoration, the extent to which the site is to be restored and the way this is to be accomplished.

Prior to actual restoration research is conducted to determine how best to restore the site, and the scientific, technical, organizational, legal and financial consequences of the restoration activities are considered. The next step is the actual restoration during which measures are taken to protect both humans and the environment. The final step involves monitoring the success of the restoration activities and the protective measures. If necessary, additional measures are taken to decontaminate the site.

In the first two and one-half years after withdrawal of the WGT forces from eastern Germany the steps involving recording information about the hazardous waste sites and initial evaluation of the sites were carried out. Next, information about contamination was gathered and detailed investigations were conducted at selected sites.

The authors of this chapter have investigated hazardous waste sites and been involved in the processes of recording information about sites and assessing danger.[1] These investigations focused on barracks, fuel depots and airports. The research was conducted according to a number of agreed steps.

First, detailed study was made of the former use and of the future use to which the site would be put, and a survey of the regional and local geological and hydrological conditions was undertaken.

This was followed by visual inspection and recording of information about the type and exact location of the suspected hazardous wastes. The next step was to determine what fieldwork was necessary (e.g., geophysical studies, drilling for groundwater and soil samples, sampling of the air, and physical and chemical measurements, samples and surveys). In addition, decisions were made about the chemical analyses that needed to be done in the laboratory. The actual fieldwork was then done, which included taking samples based on the results of the physical and chemical studies. Chemical analyses were then performed in the laboratory.

Finally, all of the results were evaluated and special consideration was given to the plans for future use of the site. If necessary, additional action was taken

[1] The authors of this chapter are employed by the German company TÜV Ostdeutschland Sicherheit and Umweltschutz GmbH–Mitglied der TÜV Rheinland Gruppe, which conducted these activities.

immediately or plans were made to conduct more research. Recommendations were made about the need, type and extent of restoration measures.

IV. Decontamination activities and findings

A number of conclusions can be drawn from the evaluations of the hazardous wastes which were left by the WGT forces.

1. *Each study of a hazardous waste site at a former WGT base is unique.* Even at similar sites there are peculiarities and problems that are specific to the particular site, in addition to the contamination problems that are found at all such sites. This is mainly the result of a differences in the type of activity that was conducted at the site, the length of its use, the geological and hydrological conditions and the types of harmful substances that were used.

2. *Research about the hazardous substances used at these sites is essential.* Such research must be extensive. These sites were used in various ways at different times and the type of contamination also varies. If there is information about what substances were used, the type of handling they received and any accidents that may have occurred, research activities can be tailored to the site and costs reduced. Often there is a lack of documentation about such activities and interviewing witnesses can be an important source of information.

The typical pattern for the production or depositing of hazardous wastes at former military facilities encompasses four phases: (*a*) use by the German Reichswehr or Wehrmacht prior to take-over by the Allied or WGT forces, (*b*) use by the Allied or WGT forces until their departure from the area, (*c*) the departure period, and (*d*) activities after the departure of military forces. In the first two periods contamination resulted from the use of hazardous materials in training and was also caused by equipment, storage facilities and filling stations. In addition, contamination can occur if there is improper handling of substances that are harmful to the environment and from leaks, accidents and various waste disposal activities or from waste storage.

When the WGT forces departed hazardous wastes remained deposited in the soil. These included harmful fluids and other materials that had not been disposed of properly. This led to damage to the environment not only on the military bases but also in adjacent areas. After the departure of the WGT forces damage was caused by demolition of the facilities, improper decontamination activities and vandalism. Visual inspection often revealed that the vegetation at almost the entire border of these bases was damaged or darkly coloured. Figure 4.3 shows how spot checks of the soil revealed concentrations of mineral oil hydrocarbons 10 to 100 times in excess of the maximum amounts allowed.

3. *A detailed inspection must be conducted.* Such an inspection should be carried out as part of the research into how the site was used and is useful in charting the location and type of suspected hazardous wastes. The inspection must closely scrutinize the entire base and all buildings. Past use must be reviewed and suspected sites where hazardous wastes may have been used must be documented. Note should be made of visible contamination, containers with dangerous substances, open tanks and shafts, suspicious locations where ammunition has been used or stored, and areas where there are signs that the soil

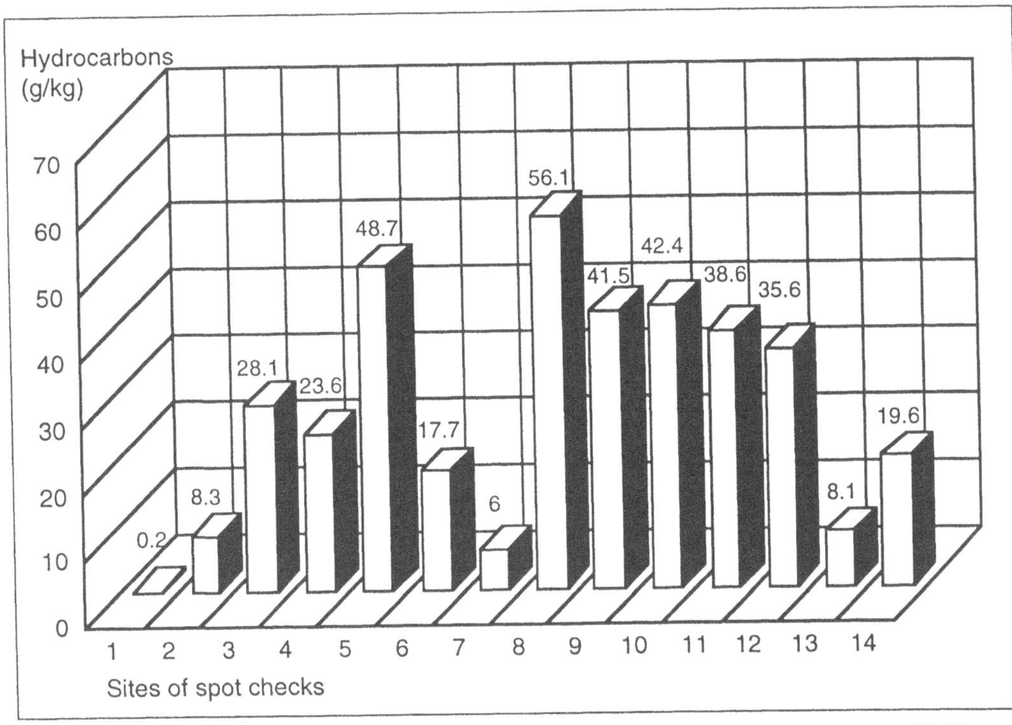

Figure 4.3. Contamination of the soil by mineral oil hydrocarbons on former WGT bases

has been disturbed recently or covered with sand or where there is no of vegetation. The scale of existing maps and charts must be checked to ensure their accuracy.

4. *At most former WGT bases the dominant type of contamination is similar to that in old industrial and manufacturing areas.* Fuel and lubricants are often found. In general, strictly military contamination such as ammunition, explosives and warfare agents is relatively rare and most common at training areas and storage depots. However, in order to maintain adequate safety all areas must be regarded as possible sites of such contamination.

5. *Fieldwork at former WGT bases must be done with attention to the hidden dangers which may be present.* In addition to normal measures for worker safety special attention must be given to hazards such as gas, water and power lines and to unexploded, buried munitions. Prior to drilling such dangers must be ruled out. The greatest danger of fire or explosion is presented by flammable fluids in unlocked tanks and containers which contain some fluid and by the remains of so-called 'imitation' agents as well as munitions and explosives. Acids, lyes and solvents which have been stored in an unsafe or improper manner can damage the skin or cause poisoning. Other dangers include structurally unsound buildings, unsafe or open shafts, oil separators and oil tanks.

6. *The facilities at these sites are old and may not function adequately.* This is especially true of the gas, water and power lines. Most former WGT bases had their own system for the treatment of harmful substances, and these pipes and sewers are often contaminated with various wastes including mineral oil hydrocarbons, halogenated hydrocarbons and the like and are in need of cleaning and decontamination. In addition, at places where surface water pooled there are often dangerous deposits.

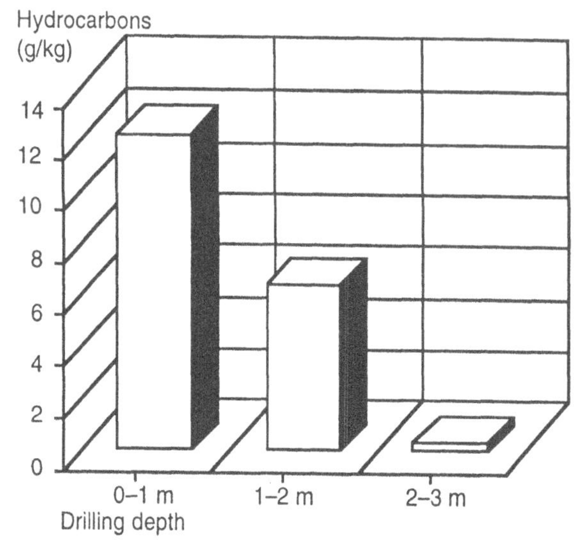

Figure 4.4. Contamination of the soil by mineral oil hydrocarbons at a filling station with a groundwater level of three metres

7. *The soil at most former WGT bases is porous or cracked.* Even where there is concrete its surface is generally cracked and porous owing to age and extensive use. The permeability of the soil has allowed harmful substances to penetrate it, especially in vehicle service areas (e.g., repair areas, cleaning facilities and filling stations), where both standard and specialized military vehicles are parked, and at storage areas for scrap materials. Often the groundwater is also contaminated.

When fuel or lubricants are pumped or poured into a vehicle there is often some 'filling loss' resulting in mineral oil hydrocarbon contamination; this is often found at base filling stations. Figure 4.4 presents the results of drilling to determine the degree of such contamination. The highest concentration of contaminants was at the surface and decreased at lower levels. The high level at the surface may have been caused by the layer of ash and slag from a former heating plant that covered the soil surface. This may have been the source of the heavy metals which had entered the soil and the groundwater.

8. *Areas where there were filling stations and fuel depots are contaminated.* In these areas there was continuous contamination of the soil and groundwater by fuel and lubricants, and monoaromatic hydrocarbons and lead deposits are present. The tanks that were used were generally situated in unprotected soil, and no provision was made to contain leaking fuel or lubricants. Figure 4.5 shows the results of soil analysis at such a fuel depot. The high concentration of mineral oil hydrocarbons raised suspicion that there had been damage to a fuel storage tank, and this later proved true. The leaking tank caused contamination of the groundwater.

9. *In order to reduce cost both detailed on-site investigation and chemical analysis in the laboratory are needed.* There are numerous suspected hazardous waste sites and many of them are large. Often the hazardous substances present are complex, and there is a lack of information about the past use of them. In

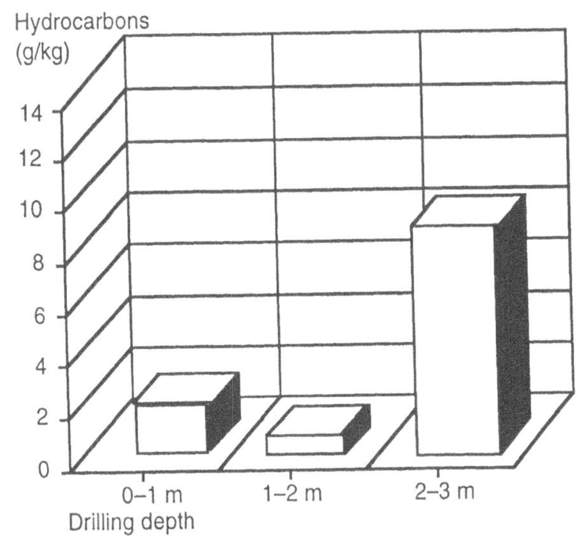

Figure 4.5. Contamination of the soil by mineral oil hydrocarbons at a fuel depot with a groundwater level of three metres

addition to the steps for the treatment of hazardous waste sites which are outlined above in subsection III, it is useful and economical to test the soil first using parameters that apply to groups of harmful substance. If the results are positive more tests can be conducted to identify the particular substances present. In general, research about past use and inspection of a site only reveal information about the groups of harmful substances present, and additional testing must be done.

10. *If investigations reveal that there is significant danger to humans and the environment immediate measures must be taken.* If the soil and groundwater are contaminated, safety measures must be implemented to prevent further spread of the contamination and to decrease the existing danger. At former WGT bases other immediate measures need to be taken, including: (*a*) testing the safety of gas, water and power lines and disconnecting or dismantling them if necessary to prevent accidents or the release of harmful substances into adjacent areas; (*b*) supporting roofs, buildings, open trenches and shafts in danger of collapse; (*c*) removal or disposal of containers filled with hazardous substances and of oil separators and elimination of pools for waste collection; (*d*) disposal of filled tanks; and (*e*) identification of widely scattered waste and scrap, followed by proper disposal, with special attention given to materials that endanger the environment.

V. Hazardous military waste and its conversion

Many former WGT bases are located in or near towns. Investment should centre on such areas since the infrastructure already exists to handle normal contamination and the ownership of the properties in these areas is known. In addition, there is high unemployment in eastern Germany, and the new jobs that could be created by development of these areas are much needed.

If the former WGT bases are to be converted to business and residential areas all dangerous substances must be removed and the total extent of contamination ascertained. The cost of the needed restoration measures must also be determined. The particular type of contamination at a site to be redeveloped must be considered and taken into account in planning for future use. Accurate assessment of the contamination problem is of importance when deciding where new buildings should be located and can help to minimize the cost of restoration and protection measures (e.g., excavation of contaminated soil or repair of damaged concrete and other surfaces). Special attention will need to be paid to the location of facilities where safety is of particular concern such as nurseries, schools, medical facilities and housing.

In general, the former WGT bases are most suited for commercial or governmental uses. The areas that were used to house troops or as headquarters are usually free of contamination, but the parade-grounds and the areas where technical activities were carried out are contaminated to varying degrees, depending on the way in which they were utilized. Naturally, each case will need to be evaluated separately.

It is not easy to judge the condition of the buildings, but in general those from the Reichswehr or Wehrmacht periods of use (often made of brick) and those that were built for special uses can continue to be used after renovation. Buildings built by the WGT are usually not worth saving and should be demolished. In most cases new gas, water and power lines must be installed.

The authors of this chapter were involved in the reconversion of a former WGT base in the Beeskow district of Brandenburg. Buildings which had been used to house troops and as headquarters were converted to schools and office buildings. A hospital and buildings that will be used for commercial purposes are still under construction.

In the future, new areas in Germany will need to be developed for various uses, and even areas that have been contaminated are likely to become places where residential or commercial centres will be located. In this context, the former WGT bases should be given greater attention.

Bibliography

Vertrag zwischen der BRD und der UdSSR über die Bedingungen des befristeten Aufenthaltes und die Modalitäten des planmäßigen Abzuges aus dem Gebiet der BRD vom Okt. 1990 [Treaty between the FRG and the USSR on the conditions of the limited stay and the modalities of organized retreat from the territory of the FRG of October 1990].

Kötter, L., Niklauß, M. and Toennes, A., Erfassung möglicher Bodenverunreinigungen auf Altstandorten [Registration of possible contamination of soil at old industrial sites] (Kommunalverband Ruhrgebiet [Municipal association of the Ruhr district]: Essen, 1989).

Ministerium für Umwelt, Raumordnung und Landwirtschaft des Landes Nordrhein Westfalen [Ministry of the environment, planning and agriculture of the federal state of North-Rhine Westphalia], Wegweiser für den Umgang mit Altlasten-Verdachtsflächen auf freiwerdenden militärisch genutzten Liegenschaften [Guide for the handling of suspected hazardous waste sites at former military sites] (Ministerium für Umwelt, Raumordnung und Landwirtschaft des Landes Nordrhein Westfalen: Düsseldorf, 1992).

Altlasten-Sondergutachten des Sachverständigenrates [Special report on old hazardous wastes by the council of experts for environmental matters] (Metzler Poeschel Verlag: Stuttgart, 1990).

Länderarbeitsgemeinschaft Abfall (LAGA) [Joint group on wastes of the federal states of the FRG], LAGA-Informationsschrift Altablagerungen und Altlasten [LAGA information publication on old waste deposits and old hazardous wastes] (Erich Schmidt Verlag: Berlin, 1991).

5. Decontamination of the Wehrmacht gas laboratory at Spandau Citadel, Berlin: technical and organizational experiences

WOLFGANG P. W. SPYRA

I. Introduction

Prior to World War II the German Army (Wehrmacht) established a specially trained unit—the smoke troops[1]—in preparation for the possible use of chemical warfare agents. The unit served as the chemical combat element of the army and was trained to: (*a*) cope with the use of chemical warfare agents by the enemy, and (*b*) use smoke and gas ammunition and high-grade, high-explosive munitions.

The key elements of the army's programme for combat preparation were a facility for testing chemical warfare agents at Raubkammer near Munsterlager and a gas laboratory at the historic Spandau Citadel in Berlin. Owing to their importance, both facilities were under the direct command of the Army Ordnance in Berlin.

The main tasks of the gas laboratory at Spandau Citadel were: (*a*) developing and testing protective equipment; (*b*) conducting detoxification experiments; (*c*) developing and testing chemical warfare agents; (*d*) studying captured chemical warfare agents and detoxicants; (*e*) carrying out experiments on animal and human subjects; (*f*) developing methods to produce blood and nerve agents such as P100 and sarin; and (*g*) finding substitute raw materials for products such as Tornesit (chlorine rubber), which was used to protect against the viscous blistering agent Lost.

When it was decided that Spandau Citadel, a significant monument of European culture and history, should be reconstructed it was obvious that great care would need to be exercised at all stages of the decontamination and reconstruction. In 1983 and 1984 this was confirmed when releases of chemicals occurred during the reconstruction work. A thorough investigation was conducted, which included studying laboratory records of past activities and scientific literature and interviewing witnesses. This revealed that decontamination of the site was essential. There was well-founded suspicion that the chemicals were nerve agents, and as a result some areas were closed off, the sources of the emissions were sealed and the site was monitored for further releases of toxic substances.

In 1985–87 work was still at a relatively early stage and some former employees of the gas laboratory were asked to assist in identifying the types of toxic chemicals likely to be present at the site. Together with an analytical chemical examination of soil samples from the site it became possible to ascertain the activities of the gas laboratory during World War II. The interviewing of witnesses turned out to be extremely difficult because the German Army had only informed its personnel on a 'need to know' basis. Information which was

[1] Rielau, H., *Geschichte der Nebeltruppe, ABC- und Selbstschutzschule* [History of the smoke troops, NBC and self-protection school] (Bundesministerium der Verteidigung: Cologne, 1966).

of value could only be obtained from former employees who had worked as assistant chemists or in other skilled capacities. The high-ranking officers who were still alive were generally able to provide only very specific information about work at the gas laboratory, and this was of little use in planning the decontamination and reconstruction work. Witness accounts of research activities at the gas laboratory were confirmed only to a limited extent by the examination and analysis of the material recovered from the site. Arsenic compounds had been developed and tested at the gas laboratory, and it appeared feasible that experiments could have been conducted to replace arsenic with antimony or selenium in order to achieve greater toxicity. One statement revealed that research had been conducted on combining chemical warfare agents with human peptides to achieve increased hypodermic absorption.

One witness claimed that even extreme ideas such as production of a 'mummifying warfare agent' had been entertained. While this information was not considered reliable, there is a spider in desert areas of South Africa which secretes a substance that causes the decomposition of cells with resulting skeletonization. It appeared likely that the gas laboratory at Spandau Citadel was involved not only in developing methods for producing chemical warfare agents but also in research on new agents.

The Berlin Senator for the Interior was made responsible for a group which prepared a plan to clean up Spandau Citadel.[2] Owing to the special political situation of Berlin the German Army was unable to carry out the decontamination and instead a select group of Berlin police officers was trained. Past experiences with decontamination were studied and included in their training. The operation was led by the head of the Berlin police laboratory and lasted from June 1988 until the end of 1989. The cost of decontamination was met by the State of Berlin. Figure 5.1 shows the location of the various deposits of chemical warfare agents, containers with unknown contents, and laboratory equipment and receptacles or damaged items.

II. Research

The opinions of the local community, government agencies in Berlin and commercial enterprises differed about the necessity and extent of the decontamination and reconstruction. However, all were united in their opposition to the approach chosen by the police.

For example, archaeologists feared that significant remains might be damaged or destroyed, and the Nature Conservation Department was concerned about damage to 400-year-old cultures of seeds and the resting-places of rare bats which are native to the area. While these arguments were taken into account, the safety of those involved in the decontamination and reconstruction was none the less considered paramount.

The numerous rules and regulations for the conduct of such activities added to the complexity of the situation. Allied Council Law no. 43, valid only in

[2] Senat von Berlin, Senatsinnenverwaltung, Projektgruppe Zitadelle Spandau, 'Handlungskonzept für die Ermittlung, Bergung und Beseitigung chemischer Kampfmittel im Bereich der Zitadelle Spandau' [The plan of action for the investigation, recovery and removal of chemical weapons in the vicinity of Spandau Citadel, Berlin], 1985.

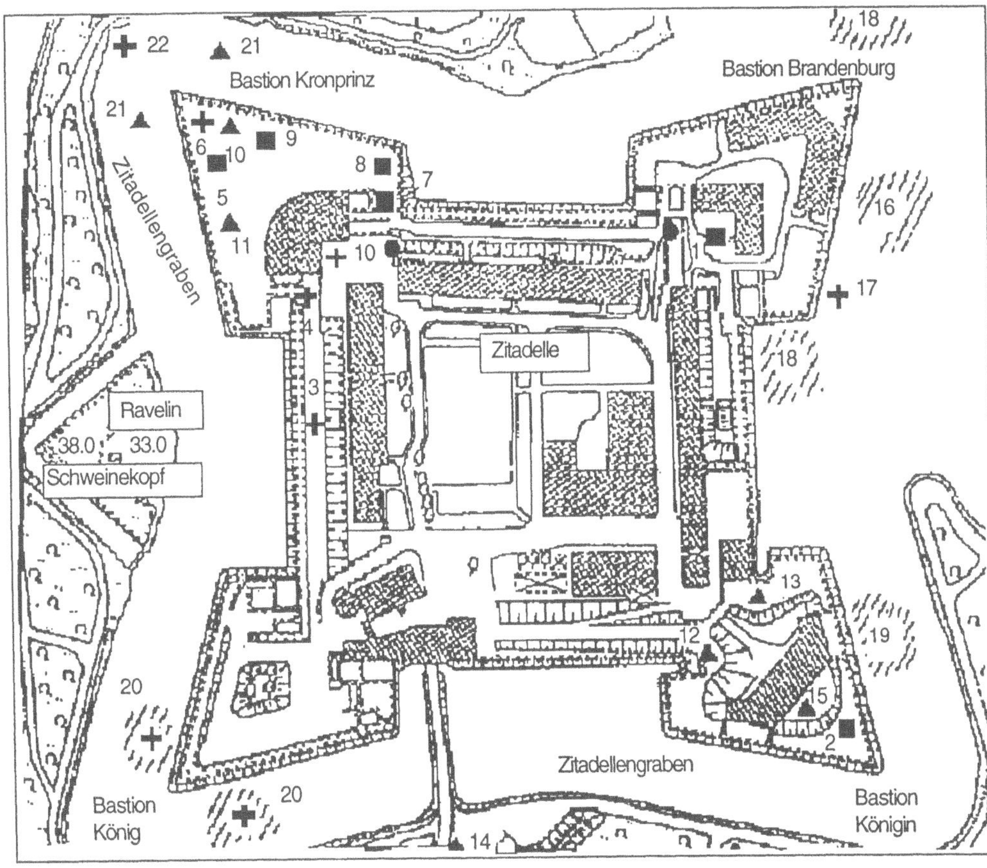

Figure 5.1. Chemical decontamination activities at Spandau Citadel

A figure together with a symbol indicates one examination procedure. + indicates chemical warfare agents; ▲ denotes containers with unknown contents; ■ indicates laboratory equipment and receptacles or damaged items. Numbers indicate the quantity of chemical warfare agents detected at that location.

Source: Prepared by the author.

Berlin and established by the Allied occupation powers in 1947, totally bans the production, import, export, transport, storage and handling of war materials. In order to comply with that law numerous other laws, rules and regulations were enacted, resulting in a complex legal structure.

In addition, it was difficult to obtain relevant literature on chemical warfare agents, especially those from World War II. The situation at Spandau Citadel was complicated by a 30 March 1945 order from the Chief of the Wehrmacht Supreme Command, Field Marshal General Wilhelm Keitel. It ordered that 'in no case chemical warfare agents must be marked as such; if there are any markings they are to be removed so that the enemy is not made aware of them'.

III. The decontamination strategy

The plan for the decontamination and reconstruction of Spandau Citadel developed a classification system based on evaluation of the various toxic materials and the risks that they were likely to pose. The system had three categories:

1. A 'red' classification was given to those areas where it was felt that there were chemical warfare agents. These areas were to be decontaminated under the direct supervision of the specially trained police.

2. A 'green' classification was applied to areas where, according to records from World War II, there was no contamination of the soil at a depth of 1 metre or more.

3. A 'white' classification was reserved for areas where there was no evidence of the presence of toxic chemicals. However, any future construction in these areas was required to be done under the supervision of the specially trained police.

Despite careful planning, considerable risk remained of danger to the health and safety of human beings and animals and possible damage to the environment in the event of accidents resulting from unexpected discoveries of toxic chemicals. In order to avoid this, plans were made to excavate all of Spandau Citadel except those areas where historical records could prove that there had been no changes since 1936 when the gas laboratory began its work.

IV. Hazards

A strategy was adopted to reduce risk by preventing the release of toxic substances and a zone was established to which only trained personnel were permitted access. The area outside the zone was regarded as safe, although the possibility remained that unknown hazards might exist there. The size of the zone of limited access was determined by several factors, including the degree of danger presented by the chemicals located there.

The following criteria were used in assessing the level of danger: (a) quantity of chemical warfare agent released; (*b*) toxicity and risk to human beings of the released chemical warfare agent; (*c*) type of release; (*d*) weather conditions; and (*e*) effectiveness of the available protective measures.

It is not possible to determine the degree of risk posed by deposited chemical warfare agents with total accuracy. In order to provide for maximum safety the assumption had to be made that any discovered compounds should be regarded as highly toxic chemical warfare agents. Experience has shown that for the decontamination or removal of chemical waste the zone of limited access should have an ellipsoidal shape. The shape of the zone is of greater importance than is often assumed owing to the fact that some weather conditions may create significant risk.

Several years of observation of climatic conditions at Spandau Citadel demonstrated that within several 100 metres of areas of possible hazard the special topography of the site caused winds which frequently changed direction. This, together with the general weather conditions near the site, made it impossible to determine the direction in which accidental emissions would be released.

One hundred thousand people live within 2.5 km of Spandau Citadel, and there is a junction for both public transportation and vehicular traffic 500 metres from the site.

V. Basic considerations

The group which investigated the situation at Spandau Citadel came to the conclusion that it was unlikely chemical warfare munitions would be discovered there, and planning therefore focused on the best methods to recover chemical warfare agents. Chemical warfare munitions consist of a chemical agent, a firing device and an explosive which discharges the chemical.

The main risk in the recovery[3] of chemical warfare agents is the release of highly toxic chemicals. All of the measures employed were therefore intended to ensure that: (*a*) no chemical warfare agents were released; (*b*) in the event of release, the source of the toxic substance could be sealed immediately; and (*c*) any released chemical compound would be bound by additives to prevent its release into the air. In order to be prepared for the accidental release of chemical warfare agents, their physical and chemical properties were studied and various recovery methods were investigated so that the most appropriate measures could be used when and if needed.

Recovery methods and the strategy for decontamination

The methods to be used for the recovery of the chemical warfare agents were worked out during the training of the special police who were to decontaminate Spandau Citadel. They ranged from mechanical procedures to special treatment of archaeological finds. It was difficult to provide appropriate and adequate safety measures for those working at the site since the higher the protection level, the lower the efficiency of the operation.

The person leading the operation had to be extremely knowledgeable about the chemical warfare agents present at the site. In addition, it was difficult to locate some chemicals because the chemicals normally used in laboratories are generally kept in glass or plastic containers which, when buried, cannot be located by standard detection methods. It was, however, possible to use the ferromagnetic sounding method for detection since statements by some witnesses revealed that for safety reasons toxic materials were always kept in metal containers. These containers had been buried at various locations at Spandau Citadel.

Physical properties

The physical properties of chemical warfare agents do not usually have a significant impact on worker safety. At Spandau Citadel the general assumption was that the substances likely to be recovered would not be limited to a few pure chemical warfare agents but would more likely be so-called 'tactical mixtures'[4] of various compounds. However, in planning for decontamination the state of

[3] 'Recovery' here means removal of chemical warfare agents from the soil of the site, which can be accomplished by digging up the material and subsequently processing it.

[4] In order to achieve a military advantage in certain special weather conditions, chemicals additives were mixed with the chemical warfare agents prior to the filling of shells. A tactical advantage was gained by increasing the persistence of the chemical agent, its stability and the rate of its decomposition on delivery. For additional information, see chapter 3 in this volume.

the chemical agents (solid, liquid or gas), the vapour pressure and the enthalpy of vaporization of the chemicals had to be considered in terms of worker safety.

Toxic chemicals can more easily be recovered at low temperatures than at high temperatures because these chemicals have lower vapour pressures at lower temperatures. Low temperatures also reduce the concentration of the toxic chemicals in the air. The minimum temperature which was acceptable for recovery work was the freezing point of water owing to the fact that flushing with large quantities of water was one of the methods to be used in the event of an accidental release of chemicals. The maximum acceptable temperature for decontamination work was related to the temperature at which workers wearing protective clothing could function efficiently. Workers wearing protective clothing cannot function for an extended period of time at a temperature higher than 38°C, approximately the normal body temperature. If work must be done in extreme weather conditions, including rising or falling temperatures, the period of time during which work can be conducted efficiently is substantially reduced. In such instances it is essential that the physical condition of the workers be monitored continuously.

Chemical properties

The chemical warfare agents buried at Spandau Citadel had chemical properties which were designed to meet specific military needs. They were supposed to be resistant to high temperatures and to the use of other chemicals as decontaminants. In addition, in combat they were intended to cause instantaneous and complete contamination of the air, to kill anyone exposed to them or to impair their ability to act normally. The chemical warfare agents were also designed to contaminate large areas for a limited period of time.

The reactivity of chemical warfare agents is of special importance not only as regards harm to humans, but also in terms of detection and decontamination efforts.

The decontamination activities led to some surprising discoveries. For example, the stability of hydrocyanic acid had to be re-evaluated after a 1989 discovery of several ampoules of liquid hydrocyanic acid. The ampoules were from 1945, and it was expected that almost complete polymerization would have occurred after such a long period. However, only a very limited polymerization was detected.[5]

VI. Safety precautions

At Spandau Citadel the safety precautions that were taken to protect the workers and the site itself were determined by the properties of the chemical warfare agents.

[5] Krammer, R. *et al.*, 'Lagerbeständigkeit von Blausäure: Untersuchung eines Altlastenfundes' [The storage stability of hydrogen cyanide: investigation of discovered military waste], *Abfallwirtschafts-Journal*, vol. 2, no. 11 (1990), p. 691.

Initial status of the site

Initially, the site was evaluated to determine its status and the precautions which would need to be taken during decontamination and reconstruction. The nature of the warfare materials and the environmental status of the site were considered important in determining how best to avoid further damage.

Protection of personnel

Protection of the personnel involved in the recovery activities focused on two main areas.

First, adequate protective clothing and respiratory devices were considered essential. The ability to function is directly related to the degree of physical stress. In general, greater use of protective equipment increases physical stress and decreases the ability to perform effectively. A wide range of protective clothing was therefore needed so that the workers would have adequate protection for the level of danger posed by a particular task without being unnecessarily encumbered by the protective clothing.

Second, a laboratory at the site enabled the immediate identification of any chemical agent. In the event of an accident, the physician was able to utilize a specific therapy tailored to the particular chemical agent which was causing the intoxication. Without the laboratory, only a general approach could have been taken to treating the worker who had been exposed to a toxic agent. In addition, the laboratory was used for the identification and classification of recovered material.

Making the site safe

It was important for both the personnel involved in the recovery activities and the general public that high standards of safety were maintained at Spandau Citadel.

The daily 'temporary danger zone'

In order to protect the general public an area had to be made 'off limits' each day that recovery work was carried out. The configuration of the 'temporary danger zone' varied from day to day depending on the work being done.

Factors which affected the determination of the zone included: the properties of the particular substances to be decontaminated, their toxicity, the amount of the substances and the weather conditions which had been forecast. Spandau Citadel is located in a highly populated area, and thus it was not possible to close off a sufficiently large area so that total safety could be guaranteed no matter what type of accident occurred. It was therefore necessary to plan for eventual accidents.

Preliminary investigations provided data on the toxic substances and the quantities which were likely to be found. This information was used to determine how to proceed with decontamination and reconstruction activities. A review of a two-year period of weather patterns revealed that for absolute safety an area of several hundred metres would need to be closed off. However, existing structures created limitations, and it was decided that a circular safety zone of 125 metres around the recovery area would be adequate. Considerable technical ingenuity was needed to find ways to limit the damage in the event of an accident. When recovery measures were not being carried out, the danger zone was made smaller.

Safety procedures used in the daily work

All of the measures used during recovery were intended to ensure that chemical warfare agents would not be released. This meant that qualified personnel and the proper equipment had to be immediately available at all times. Any chemical warfare agent that was accidentally released had to be prevented from escaping into the air and from spreading as vapour. Water walls (water screens) were used to avoid the spread of clouds of toxic substances.

In order to avoid their release into the air, the recovered materials (e.g., containers of chemicals, contaminated soil, etc.) were put into gas-tight receptacles (containers or drums) that were filled with adsorbents. Autoclaves were used to dispose of certain particularly dangerous chemicals.

Safety procedures for particularly dangerous situations

It was possible to recognize situations of particular danger and to take safety precautions to prevent accidents owing to an approach which consisted of three components. First, decontamination was carried out under the supervision of a bomb technician trained in chemical explosive decontamination. Second, both mobile and stationary analytic devices were available at the site. Third, chemical warfare agents and toxic materials were recovered only when weather conditions were favourable.

Procedures for coping with a serious accident

The risk of a serious chemical accident at Spandau Citadel was felt to be minimal owing to the recovery methods used and the strategy of careful decontamination. None the less, preparations were made to deal with such a possibility.

When a chemical accident occurs, the evacuation of a populated area should only be considered if it is possible to evacuate people without endangering them. If that is not feasible, it is best that they remain in their homes with their doors and windows closed and wait for further instructions from the police.

Medical protection and safety measures

A physician was always present when decontamination procedures were carried out. He was responsible for the psychological and medical well-being of the personnel, who were free to contact him if they suspected that they had suffered ill effects as a result of exposure to chemical warfare agents. A physician was also 'on call' at any time of the day. In turn, the physician could contact the police officer in charge of the operation at any time of the day and request the examination of daily specimens and samples if he felt this was required in order to ensure the safety of the decontamination personnel.

VII. Securing the site on a daily basis

Each day at the end of work the site had to be secured so that the area which had been declared a 'temporary danger zone' could be used. A 'permanent danger zone' remained closed at all times.

In addition to securing the site at the end of each working day, a monitoring system was established. In the event of an alteration of the normal composition of the air, an alarm would sound whether or not chemical warfare agents or other chemicals had been released. The command post immediately started a suction device at the site and informed police headquarters, which is staffed around the clock. Measures which had previously been worked out were then taken; these included: (*a*) summoning a bomb technician trained in chemical explosive decontamination, and (*b*) calling in a special police unit trained in recovery work.

VIII. Conclusions

The approach of providing the public with unrestricted information about the decontamination measures to be taken at Spandau Citadel and the risks involved led to many questions. Concerns were expressed about accidents, emergency procedures, medical care, measures for disaster control, and the like. Much time and patience was needed to adequately inform the public about the procedures that were to be carried out.

The chemical decontamination element of the reconstruction of Spandau Citadel was completed in 1 1/2 year instead of the 4 years that had been scheduled, and during this time normal activity at the Citadel was maintained to a limited extent. Approximately 70 tonnes of soil, partially or completely contaminated by chemical warfare agents (mostly arsenic organic compounds) were recovered and destroyed.

All types of chemical warfare agents except for nerve gas agents were found at Spandau Citadel. Unfortunately, it is currently not possible to dispose of the material which was found since there is only one German combustion plant for the destruction of chemical warfare agents. That plant is under the control of the Federal Ministry of Defence, and at the current time it is unable to meet the many demands on it. There are, however, plans to increase its capacity.

It would not have been possible to decontaminate Spandau Citadel without the generous support that was provided by the Ministry of Defence, in particular its NBC Defense Research and Development Center of the Federal Armed Forces at Munster[6] and the Command Staff of the Munster training facility.

[6] Now called the German Armed Forces Scientific Institute for Protection Technologies: NBC [nuclear, biological and chemical] Protection.

Part IV. Chemical weapons: past production activities

6. The production of chemical warfare agents by the Third Reich, 1933–45

BERND APPLER

I. German chemical warfare agent armament, 1919–45[1]

Soon after the end of World War I the German military (Reichswehr) became convinced that the 1919 Treaty of Versailles was so humiliating for Germany that another war was inevitable. A so-called 'struggle for liberation' was expected to start in the years just after the war. German military strength had been greatly weakened and it would not have been possible to increase Germany's conventional weapon capability rapidly. Instead chemical warfare agents were considered as a possible way to compensate for the lack of conventional capability. Even after the war the German chemical industry continued to hold a leading position both in terms of its production capacity and the scientific talent of its personnel. Those who favoured the development of chemical warfare agents, such as Colonel Max Bauer[2] and Professor Fritz Haber,[3] expressed the belief that these agents were 'natural' and 'superior' weapons as the use of them required both discipline and intelligence. Experiences in World War I had also demonstrated the effectiveness of chemical weapons against soldiers protected by sheltered positions in trenches.

Preparations for resumption of the production of chemical warfare agents began as early as 1923,[4] and by 1924 the Inspectorate for Armaments and Devices (Inspektion für Waffen und Gerät) was involved. The relevant documents from World War I were widely scattered. They were collected and scientists employed at universities and in industry were asked to cooperate in what was regarded as a project in the national interest. It is apparent from the minutes of the 25 April 1931 annual meeting of scientists conducting research on chemical warfare agents that many individuals were involved.[5]

[1] This chapter draws extensively on the following sources: Brauch, H. G. and Müller, R.-D., *Chemische Kriegsführung, Chemische Abrüstung* [Chemical warfare, chemical disarmament] (Arno Spitz: Berlin, 1985); Combined Intelligence Objectives Subcommittee (CIOS), G-2 Division, Supreme Headquarters Allied Expeditionary Force (Rear), *Investigation of Chemical Warfare Installations in the Munsterlager Area, including Raubkammer*, APO 413, 1945; Historical Division, Headquarters, European Command, *Chemical Warfare: A List of Reports of the Historical Division, Headquarters, European Command*, Frankfurt [no year] (manuscript); and an unpublished British report based on evaluations of CIOS and British Intelligence Objectives Subcommittee (BIOS) reports, investigations and interrogations and German material. In the chapter the following terminology is used for chemical warfare agents to some extent: for CN, chloroacetophenone (British/US term) or White-ring, Weißring (German term); for DA, diphenylchloroarsine (British/US) or Clark I (German); for DC, diphenylcyanoarsine (British/US) or Clark II (German); for Lost produced from ethylene and sulphur dichloride, direct mustard, D-mustard (British/US) or Direkt-Lost, D-Lost (German); and for DM, diphenylaminochloroarsine (British/US) or adamsite (German).

[2] See Brauch and Müller (note 1), p. 88.

[3] See Brauch and Müller (note 1), pp. 89–93.

[4] Evidence of this is provided by the minutes of Dr Hugo Stoltzenberg, a CW expert of the Army Ordnance Department (Heereswaffenamt or HWaA), quoted in Brauch and Müller (note 1), pp. 94–95.

[5] The minutes state: 'The head of the Heereswaffenamt Lieutenant-General von Bockelberg greets the attendees . . .' Among those attending were: Geheimer Regierungsrat Professor Dr Hantzsch, University of Leipzig; Geheimer Regierungsrat Professor Dr Hofmann, University of Charlottenburg; Professor Dr Flury, University of Würzburg; Professor Dr Steinkopf, Technical High School Dresden; Dr Zernik, University of Würzburg; Oberregierungsrat Professor Dr Hase, Biologische Reichsanstalt; Regierungsrat

The Treaty of Versailles prohibited Germany from testing chemical warfare agents, and thus all of the experiments which were carried out were classified as top secret. This made it impossible to conduct the necessary field tests in Germany, and in 1927 an agreement was reached with the Soviet Union for the common use of a proving ground near the Volga River. In addition, spray trials using aeroplanes were carried out at Polygon Field, located near Podosinski (renamed Uchtomskaja by the Soviets), south of Moscow. In 1927 a proving ground was opened near Orenburg, on the Ural River; it was not used, however, owing to its distance from Germany. Joint German–Soviet tests of live agents were conducted at a large field (approximately 400 square kilometres) near Vol'sk (also known as Tomka), on the Volga River. Owing to disagreements with the Soviet authorities the site was abandoned when the National Socialist (Nazi) Party assumed power in Germany.

A chronology of Germany's R&D activities related to chemical warfare agent production from 1924 to 1945 is presented in appendix 6A.

II. Production of chemical warfare agents in the Third Reich, 1933–45

Structure of the organization

Interest in the production of chemical warfare agents centred in the armed forces (Wehrmacht) and the Ministry of Armament and War Production (Reichsminister für Rüstung und Kriegsproduktion), under the leadership of Albert Speer. The Nazi special police force (Waffen-SS) was not interested in chemical warfare agents, for reasons that are not known, except for chlorotrifluoride (N-stoff), which was produced at Falkenhagen. It was thought initially that it could be used to ignite the charcoal in the filters of gas masks, but field tests later proved this untrue.

In the Wehrmacht the production of chemical warfare agents was the responsibility of the Supreme Command of the Army (Oberkommando des Heeres, OKH) and, in particular, of the Army Ordnance Department (Heereswaffenamt, HWaA). The Ordnance Department for Testing of Equipment (Wa Prüf 9) conducted research and development while the Ordnance Department for Munitions (Wa Mun 6) carried out production; both worked in cooperation with the Ministry of Armament and War Production.[6] The responsibility for the development of ammunition was divided: (*a*) the Ordnance Department for Testing of Equipment concentrated on the chemical aspects of ammunition development; (*b*) the Department for Munitions, Powders and Explosives (Wa Prüf 1) developed and constructed shells; and (*c*) the air force developed bombs. The maximum number of employees working at the various R&D, testing and production facilities was: Ordnance Department for Testing Activities, 200 employees; the army's gas laboratory at Spandau Citadel in Berlin,

Dr Phyl, Reichsgesundheitsamt; Professor Dr Jander, University of Göttingen; Professor Dr Schaarschmidt, Technical High School Charlottenburg; and Dr Ingenieur Stantien, Berlin-Dahlem. 'Niederschrift über Jahresbesprechung der wissenchaftlichen Mitarbeiter der Kampfstoff-Forschung am 25 April 1931' [Minutes from the annual meeting of the scientific collaborators in CW agent research, 25 Apr. 1931], in Brauch and Müller (note 1), p. 119.

[6] This is a slight simplification of the structural organization of the armed forces.

450 employees;[7] and the Army Proving Ground (Heeresversuchsstelle Raubkammer) at Munster, 800 employees.

The navy, the army and the air force had their own departments that were responsible for adapting chemical warfare to their needs. These included the Navy Gas Protection Inspection, the Navy Air Protection Inspection and a similar army department, Inspection 9. The air force maintained a testing facility at Munster, and, together with the navy, it made extensive use of the results of testing conducted by the army.

Organization for mass production of chemical warfare agents

During World War I existing civilian chemical plants were used for the production of chemical warfare agents, but during the Third Reich production was carried out in plants especially constructed and financed for that purpose. In addition, the Army Ordnance Department established a holding company (Montan Industries GmbH) in 1936 to produce gunpowder, explosives, chemical warfare agents, smoke munitions, and the like.

The Ministry of Armament and War Production directed the Army Ordnance Department to plan and construct the plants requested by the military. The finished plants were turned over to Montan Industries, which then leased them to private companies that were actually I. G. Farben facilities for the most part.

There were a number of consulting committees in the Ministry of Armament and War Production that provided support to the Office for Procurement of Raw Material. The most important of these was the Committee for Explosives, led by a man named Sarrazin, which had three subcommittees. Special Committee C, one of the subcommittees, was headed by the director of I. G. Farben, Otto Ambros, and it was responsible for chemical warfare agents.

History of production

In the 1920s the German military planned to produce chemical warfare agents in cooperation with Dr Hugo Stoltzenberg, a key figure in German preparation for chemical warfare,[8] in plants that appeared to be producing something else. The plan failed because the I. G. Farben company was unwilling to participate and was also reluctant to lose its monopoly on chlorine gas, which is needed to produce chemical warfare agents. This led to the failure of projects such as the planned mustard gas plant at Gräfenhainichen.

Prior to the assumption of power by the National Socialist Party in 1933 the quantity of chemical warfare agents produced in Germany was negligible. Initially, this did not change owing to the fact that the production of conventional armaments was given top priority. However, when the chemical industry became interested in increasing its production the situation altered. In March 1936 the stockpile of chemical weapon agents was only 300 tonnes of chloroacetophenone and 700 tonnes of thiodiglycol (Oxol), a precursor used in the production of mustard gas. An additional 300 tonnes of thiodiglycol were in

[7] See chapter 5 in this volume.

[8] See SIPRI, *The Problem of Chemical and Biological Warfare,* vol. 1, *The Rise of CB Weapons* (Almqvist & Wiksell: Stockholm, 1971), pp. 60, 257 and 293.

production, and the industry possessed the capability to manufacture the chemical warfare agents Clark I, Clark II and Perstoff.

By 1936 two plants were producing chemical warfare agents: the Orgazid plant at Ammendorf near Leipzig, which began production of thiodiglycol in June 1936; and the Ergethan plant at Staßfurt, which started production of arsine oil in August 1936. The location of these and other factories is shown on the map in figure 6.1.

There were few chemical warfare munitions in Germany in 1936. Only one weapon filling station for mustard gas ammunition was functional—the 'smoke filling station Nebelfüllstelle', a code-name for the plant at Raubkammer (Munster). By 1937 a small plant for the manufacture of one tonne per day of N-mustard was also under construction at Ammendorf. Progress was not rapid, as a 13 January 1939 memo from the Inspector of the Nebeltruppen, Colonel Herman Ochsner, demonstrates: 'Chloroacetophenone producible, mustard, Blaukreuz, Grünkreuz in small quantities available, not yet producible by plants belonging to the Army'.[9] However, filling stations were available at Munster, Löcknitz (near Stettin) and Hohenbrunn (near Rosenheim).

There was a general shortage of raw material. Iron that had originally been allocated for storage tanks for mustard gas was confiscated by the government to be used for other purposes, and Colonel Ochsner noted that this made it impossible to build the storage tanks that he required. However, in the long-range programme presented by the Committee for Economic Development (Reichsstelle für Wirtschaftsausbau)[10] in August 1939 the chemical industry pledged to produce much greater quantities of chemical warfare agents than it had in the past. In 1939 the stockpile of mustard gas, for example, was 8000 tonnes, and it was expected that the chemical industry would be able to produce 118 000 tonnes of mustard gas by the end of 1942. These predictions were unrealistic; by 1945 the total amount of mustard gas produced in Germany was only 27 000 tonnes.

The shortage of basic raw material at the beginning of World War II worsened as the war progressed. This, together with the low priority given to the production of chemical warfare agents, may be why only a few plants for their production and a few ammunition filling stations became operational. In addition, there where shortages of chlorine, phosgene, sodium cyanide, ethylene, arsenic, and other chemicals. This may also have contributed to the limited utilization of the plants, an hypothesis which is supported by some documents from the period .[11]

By 1 September 1939 the production capacity for sulphur mustard (S-mustard) agents was 450 tonnes per month (tpm). As of 1 October 1939 supplies of S-mustard and chloroacetophenone were available as were small quantities of phosgene and very small quantities of Clark I and Clark II. The storage capacity for S-mustard at both Ammendorf and Raubkammer (Munster) was 3000 tonnes, but only the Raubkammer filling station was used. However, the expansion and completion of filling stations in Löcknitz and Hörpolding had

[9] Quoted in Brauch and Müller (note 1), pp. 160-61.

[10] The committee was under the leadership of Professor Carl Krauch, who had been on the executive board of I. G. Farben since 1925.

[11] The following discussion is based on Brauch and Müller (note 1).

Figure 6.1. Location of German plants which produced chemical warfare agents before and during World War II

begun. The production capacity for chloroacetophenone was 150 tpm, and the production capacity for phosgene was 350 tpm. By 1 November 1939 the production capacity for sulphur mustard had increased to 900 tpm.

A document from 2 December 1941 notes that an order to build a large-scale plant for the production of tabun with a capacity of 1000 tpm had been given at the beginning of the war, and production was supposed to commence in the spring of 1942. In addition, a test plant for sarin production was to be built in Dyhernfurth. The amount of raw material known to be available indicates that its capacity was approximately 100 tpm.

A 24 July 1942 document states that measures were to be taken so that it would be possible to produce 7000 tpm of chemical warfare agents by 1 March 1943. The process for production of ethylene oxide was to be refined so that 'the greater part of the chemical warfare agent production capacity, which cannot currently be used owing to other uses of ethylene oxide, should be made available so that a surplus of 1500 tpm can be produced'.[12]

A letter from the Army Ordnance Department to Minister of Armament and Ammunition Albert Speer on 6 August 1942, stressed the importance of the production of chemical warfare agents:

> In order to attain the superiority in the field of chemical warfare ordered by the Führer on 30 June 1942, it is necessary to classify projects related to 'mineral oil priority' [i.e., the highest priority] (see list [reproduced here as table 6.1]).
>
> The projects listed as nos 5–17 are already well under way; however, unless they are given higher priority it is uncertain if they can be completed according to schedule.
>
> If there are particular difficulties (as is expected for the projects listed as nos 1–14) at the appropriate time applications will be filed to take certain steps since it will not be possible for the Minister of Armament and Ammunition to give these projects DE priority.
>
> In addition, in order to attain the levels of production that have been scheduled 'mineral oil priority' should be given to repairing damage to materials and damage resulting from explosions, to making necessary changes and improvements in production and to repairing existing plants.
>
> Approval of such applications and agreement on the above paragraph is urgently requested.[13]

On 26 May 1943 Adolf Hitler ordered that the production of tabun had to be increased from 1000 tpm to 2000 tpm by the end of 1944, and the production of sarin from 100 tpm to 500 tpm. Owing to recognition of the difficulties inherent in attempting to meet the goals of the CW production programme, on 1 October 1943 an emergency programme was inaugurated that limited production to a few types of chemical warfare agents. This meant that by the end of 1943 the production plants for chemical warfare agents had to abandon many of their specialized pursuits. In addition, there were delays in obtaining the necessary materials and the purchase of new equipment was postponed. On 1 February 1944 Otto Ambros (a member of the directorate of I. G. Farben and an adviser on CW questions to the Ministry of Armament and Ammunition) complained that, despite the emergency programme, target dates were not being met.

In May 1943 the armed forces had requested that 1000 tpm of tabun be produced. It was envisaged that 350 tpm could be manufactured in May at the Dyhernfurth plant, and that starting in 1944 production levels of 1000 tpm could be attained. However, the actual production capacity was 700 tpm, and there was a stockpile of 6400 tonnes. By 20 March 1944 it was obvious that

[12] Quoted in Brauch and Müller (note 1), p. 175.
[13] Quoted in Brauch and Müller (note 1), p. 176.

Table 6.1. List referred to in the letter of 6 August 1942, from the Army Ordnance Department to Minister of Armament and Ammunition Albert Speer

No.	Factory	Project	Production and type of project
1	I. G. [Farben]	Dyhernfurth	1 000 tonnes per month (tpm) trilon 83 (tabun)
2	I. G. [Farben]	Dyhernfurth	100 tpm trilon 300 (aeroform)
3	I. G. [Farben]	Dyhernfurth	100 tpm trilon 46 (sarin)
4	I. G. [Farben]	Wolfen	Decontamination stoff 40
5	I. G. [Farben]	Gendorf	4 000 tpm D-mustard
6	Orgacid	Hahnenberg	400 tpm arsine oil
7	J. D. Riedel de Haen	Hahnenberg	500 tpm omegasalt (chloroacetophenone)
8	Ergethan	Staßfurt	Supplement to the 180 tpm arsine oil plant
9	I. G. [Farben]	Uerdingen	Supplement to the 200 tpm azine plant (adamsite)
10	Hagenuk	Barth	Plant for K-stoff smoke generators
11	I. G. [Farben]	Ludwigshafen	Predecessor of T9 (N-mustard)
12	I. G. [Farben]	Döberitz	130 tpm decontaminants for weapons
13	Auer	Oranienburg	Power plant
14	D. A. K.	Premnitz	90 tpm charcoal
15	Gasschutz-laboratorium	Spandau	Finishing of research laboratories for K-stoff, smoke powder, etc.
16	Lonal	Haselhorst	Finishing of the Clark plant (90 tpm)
17	Orgacid	Ammendorf	Finishing of the mustard plant
18	I. G. [Farben]	Leverkusen	Precursor for the new product trilon 46 (sarin)
19	I. G. [Farben]	Höchst	Precursor for a new chemical warfare agent (K-stoff)

Source: Quoted in Brauch, H. G. and Müller, R.-D., *Chemische Kriegsführung, Chemische Abrüstung* [Chemical warfare, chemical disarmament] (Arno Spitz: Berlin, 1985), p. 177.

doubling the output of tabun to 2000 tpm was impossible to achieve because of a shortage of phosphorus. It was therefore decided that construction would begin immediately on a test plant for the production of 100 tpm of tabun at Dyhernfurth and of a fabrication plant for 500 tpm of sarin at Falkenhagen. Production was envisaged to start at Dyhernfurth by the end of 1944 and at Falkenhagen by mid-1945.

Owing to the limits of the existing precursor plant, on 11 October 1944 Speer proposed limiting the production of Oxol mustard to 160 tpm and of direct mustard (Direkt-Lost, D-Lost or D-mustard) to 650 tpm.

By 2 November 1944 it was decided that the production of sarin should be accelerated since the available raw material made it easier to produce sarin than tabun. The production of tabun requires cyanide, of which there was a shortage. At the time the capacity for D-mustard production at Gendorf was 800 tpm. Production was scheduled to increase to 1800 tpm by January 1945, to 3000 tpm by February 1945 and to 4000 tpm by May 1945.

The production of chemical warfare agents was negatively affected by the fact that the various branches of the armed forces—the army, air force and the navy—did not remove their stocks of these agents from the storage depots at the production plants, which thus remained full. Additionally, there was disagreement among the branches of the armed forces about the importance of

Table 6.2. Requests from the General Staff for sulphur mustard

Year	Amount requested
1936	5 000 tonnes per month (tpm)
1937	7 500 tpm
1938	8 000 tpm
1943	4 000 tpm
1944	5 000 tpm

Source: Compiled by the author.

various chemical agents. For example, the Waffen-SS insisted that production of N-stoff take priority at Falkenhagen and that the ready product be stored there.

Production sites: an overview

The rate of production was based on the amount of CW agents requested by the army, air force and navy. The demands of the General Staff changed during the war and were related to the enormous consumption of materials for conventional warfare (for an example, see table 6.2). The planning for the construction of chemical warfare agent production facilities (mills) had to take this situation into account. The facilities that were in operation attempted to meet the production levels set by the three branches of the armed forces. Production was also dependent on the equipment and raw materials that were available to the plants. Because of shortages in manpower and materials, increases in the production of chemical warfare agents were not given the highest priority. This was particularly true for gunpowder and explosives.

The production plants for chemical warfare agents were owned and financed by the government and rented to private companies. The industrial plants that existed could be used for the production of chloroacetophenone and adamsite. The chemical precursors for these products were identical to those used in the manufacture of industrial products, and production of chemical warfare agents was possible with only slight alterations in the manufacturing process.

Initially, preparation for chemical warfare was of a preventive nature and the focus was on other types of weapons. Only enough chemical warfare agents were produced to mount a counter-attack in the event of a CW attack by the Allies. Plants were held in reserve to start production of chemical warfare agents only in the event of a chemical war. Tables 6.3 and 6.4 list the plants for the production of chemical warfare agents which were ordered to be built by the government (table 6.3) and those which were built but never functioned (table 6.4).

Production sites for chemical warfare agents

Chloroacetophenone, phosgene, nitrogen mustard (N-mustard), sulphur mustard, thickened mustard, tabun, sarin, soman, adamsite, Clark I, Clark II, arsine oil, excelsior and chlorotrifluoride were produced at various facilities. Appen-

Table 6.3. Plants which the German Government ordered to be built for the production of chemical warfare agents

Year ordered	Plant	Product	Initially planned capacity (tpm)	Year ready
1934	Ammendorf	Oxol mustard	300	1936
1935	Staßfurt	Arsine oil	180	1938
1937	Hahnenberg	Chloroacetophenone	600	1942
	Gendorf	Oxol mustard	800	1941
	Leese	Arsine oil	400	1941
	Ammendorf	N-mustard	50	1938
1938	Hüls[a]	Oxol mustard,	600	1941/42
		D-mustard (test plant)	200	1943
1939	Haselhorst	Clark I and II	90	1940
1939	Gendorf	D-mustard	4 000	1943 (partly)
	Wolfen	Phosgene (extended)	300	1940
	Dyhernfurth	Tabun	1 000	1942
1943	Dyhernfurth	Trilon 300 (aeroform, an alloy of As, Al and Mg),	100	Unfinished
		cyanogen chloride,	20	1944
		hydrogen cyanide	30	1944
1944	Falkenhagen	Sarin	500	Unfinished

[a] The plants at Hüls operated, but only on a test basis.

Source: Brauch, H. G. and Müller, R.-D., *Chemische Kriegsführung, Chemische Abrüstung* [Chemical warfare, chemical disarmament] (Arno Spitz: Berlin, 1985), pp. 262–63.

Table 6.4. Plants for the production of chemical warfare agents that were built but which never functioned

Start of construction	Date ready	Name and location	Product	Capacity (tpm)
1937	1942	Chemische Fabrik Hahnenberg (Riedel de Haen) at Hahnenberg	Chloroaceto-phenone	600
1937	1941	Lonal-Werke (Orgacid, prior to 1941) at Leese	Arsine oil	400
1937	1941	Orgacid at Gendorf	Oxol mustard	800
1938[a]	1941/42	Chemische Werke Hüls	Oxol mustard,	600
	1943	at Marl/Hüls	D-mustard	200 (test plant)
1942	1942	I. G. Farben at Ludwigshafen	Filling station for phosgene	120
1942	1942	I. G. Farben at Uerdingen	Filling station for phosgene	180
1943	1944	Anorgana at Dyhernfurth	Hydrogen cyanide,	20
			cyanogen chloride	20

[a] The plants at Hüls operated, but only on a test basis.

Source: Brauch, H. G. and Müller, R.-D., *Chemische Kriegsführung, Chemische Abrüstung* [Chemical warfare, chemical disarmament] (Arno Spitz: Berlin, 1985), pp. 262–63.

Table 6.5. Production of chemical warfare agents during the Third Reich

Chemical warfare agent	Quantity in tonnes (detailed in appendix 6B)	Quantity in tonnes according to CIOS source
S-mustard	27 597	27 250
Arsine oil	7 500	12 600
Tabun	12 000	12 000
Phosgene	5 900	5 000
N-mustard	1 928	2 000
Irritants	12 633	8 150
Total	**67 558**	**67 000**

Source: Combined Intelligence Objectives Subcommittee (CIOS), G-2 Division, Supreme Headquarters Allied Expeditionary Force (Rear), *Investigation of Chemical Warfare Installations in the Munsterlager Area, including Raubkammer*, APO 413, 1945.

dix 6B compiles a list of the plants which engaged in the production of chemical warfare agents including information about test plants and plant output.

In addition, other chemical warfare agents such as hydrogen cyanide, chlorocyanide, lewisite and some irritants were produced in small amounts for testing, training or special purposes. The production sites were probably laboratories like the army's gas laboratory at Spandau Citadel.[14]

Summary

The production quantities detailed in appendix 6B are summarized in table 6.5, which also includes data from a British source.[15] Some of the data corresponds but there are also major discrepancies which the author of this chapter is unable to explain.

Table 6.6 shows the quantities of chemical warfare agents that were requested by the General Staff as of 1 April 1944, and the amounts of these agents which existed at that time. In addition, table 6.6 compares production capacities with planned and actual stocks and presents a calculation of the monthly demand for these agents in the event of a chemical war. The production totals for chemical warfare agents are also given, based on information from the plants. These numbers allow comparison to be made to the amount of chemical warfare agents that it was estimated would be consumed monthly in a chemical war. The available stores were only sufficient for approximately six months, and existing production capacities would have met the demand for these weapons for only one-half month. Table 6.6 thus shows that Germany was not prepared for a long chemical war.

Code-names of chemical warfare agents and munitions

Various markings were used to identify different types of munitions. Although they contained the same chemical warfare agent, ammunition shells could be

[14] See chapter 5 in this volume.
[15] Combined Intelligence Objectives Subcommittee (note 1).

Table 6.6. The quantities of chemical warfare agents requested by the General Staff as of 1 April 1944, and the amounts of these agents which existed at that time

Figures are in metric tonnes.

Agent	General Staff request	Actual production: Filled in munitions	Actual production: At depots	Total	Monthly production capacity: Existing	Monthly production capacity: Planned	Estimated monthly consumption after start of chemical war	Factory estimates of total CW production
S-mustard	37 500	20 200	6 900	27 100	2 100	4 000	4 900	27 500
Arsine oil	4 500	1 700	3 600	5 300	400	400	535	12 600
Tabun	14 100	6 400	300	6 700	700	1 000	1 880	12 000
Phosgene	13 500	5 700	. .	5 700	700	1 400	1 500	5 000
N-mustard	2 050	1 050	600	1 650	50	100	340	2 000
Lachryma-tors[a]	9 800	6 050	5 450	11 500	1 080	1 120	1 180	8 150
Total	**81 450**	**41 100**	**16 850**	**57 950**	**5 030**	**8 020**	**10 335**	**67 250**

[a] Adamsite, diphenylchloroarsine, chloroacetophenone, etc.

Source: English summary from Combined Intelligence Objectives Subcommittee, British Intelligence Objectives Subcommittee and testimony of unspecified witnesses.

marked with a variety of colours and letters depending on the type of use for which they were intended. The coding for S-mustard was even differentiated based on the method of production.

Colour coding using cross codes

Munitions filled with chemical warfare agents were marked with coloured crosses using a system developed in World War I. This system aimed to provide a simple means of differentiating these shells from conventional shells. The following markings were used: a white cross for tear gas agents (lachrymators), a blue cross for vomiting agents, a yellow cross for blister agents (vesicants) and a green cross for 'choking' agents (lung injurants or asphyxiants, mostly lethal).

Generally, but not always, the colour of the cross indicated that a munition contained one or more agents from a particular class of chemical warfare agents. For example, a yellow cross was used for S-mustard; a blue cross for arsenic-containing agents, such as Clark I and Clark II; and a green cross for phosgene and diphosgene.

During the Third Reich the armed forces continued to categorize munitions according to this marking system to some extent, but munitions were no longer marked using the system of colour-coded crosses.[16]

[16] For example, during the Third Reich a white cross was used to indicate chloroacetophenone, a blue cross for arsenic-containing vomiting agents, a yellow cross for mustard agents and a green cross for lethal choking agents. In addition, a green cross was used for phosgene until the early part of World War II.

Table 6.7. A partial list of the colour-coding system for German CW munitions in World War II

Colour code	Chemical warfare agent	Type of explosive charge
Yellow ring	Mustard gas	Small burster
Double yellow ring	Thickened mustard gas	Small burster
Green ring	Mustard gas	Medium burster
Green ring 1	Nitrogen mustard	Large burster
Green ring 3	Tabun	Head burster
Green ring yellow	Mustard gas	Large burster
Blue ring 1	DM	Exterior burster
Blue ring 2	DA in arsine oil	Central burster
Blue ring 3	DM	Base ejection generator
White ring	CN/HE	Intimate [direct] mixture

Source: Combined Intelligence Objectives Subcommittee (CIOS), G-2 Division, Supreme Headquarters Allied Expeditionary Force (Rear), *Investigation of Chemical Warfare Installations in the Munsterlager Area, including Raubkammer*, APO 413, 1945.

An additional complication was added by the fact that during World War II Italian munitions filled with chemical warfare agents were marked with coloured crosses to differentiate them from German ones. In the Italian system a red cross, for example, meant that a munition was filled with chloropicrine.

Colour coding using ring codes

During World War II German munitions filled with chemical warfare agents were also marked using coloured rings, letters and numbers. The colour indicated the primary way in which the munition acted and did not necessarily identify its explosive charge or chemical fill.

The force of the explosive charges varied. A weak explosive charge distributes a chemical warfare agent in large drops. Mustard gas in a shell with a weak explosive charge was given a yellow marking to indicate the vesicant effect of the agent on the skin. A strong explosive charge creates an aerosol cloud. Mustard gas in a munition with a strong charge was marked with green to warn that the resulting aerosol cloud would rapidly damage the lungs.

A comprehensive listing of all of the markings used for the various types of munitions including bombs, mortar ammunition, and so forth is too extensive to include here. However, the examples listed below illustrate the marking system.

The coloured rings on the ogive (pointed end) and on the base of the shell were yellow, green, blue or white to indicate the physiological effect of the chemical warfare agent that would be released on detonation. Yellow (Gelb-Gb) indicated a vesicant; green (Grün-Gr) signalled a choking agent or one which had a systemic effect; blue (Blau-Bu) denoted a sternutator; and white (Weiss-W) indicated a lachrymator.

In addition to a coloured ring there were often numbers (1, 2 or 3) in the same colour. These were used to distinguish the various charges which had similar physiological effects but at different levels of effectiveness. Table 6.7 illustrates some of the coding that was employed.

Table 6.8. Mustard agents and mixtures used to fill munitions during the Third Reich

Code	Composition
O or OL	Oxol mustard or summer mustard (mustard prepared from thiodiglycol, Oxol)
D or DL	Mustard prepared from sulphur dichloride
OA	Winter mustard (Oxol mustard mixed with arsine oil)
OB	Winter mustard HP (a mixture of S-mustard and oxygen mustard)
OM, OKM	Winter mustard (mixed homologues of S-mustard and homologues of propyl mustard)
OR	Winter mustard (Oxol mustard mixed with anthracene oil, R)
ZO	Thickened mustard (prepared from Oxol mustard)
ZOA	Thickened mustard (prepared from OA)
ZOR	Thickened mustard (prepared from OR)

Source: Combined Intelligence Objectives Subcommittee (CIOS), G-2 Division, Supreme Headquarters Allied Expeditionary Force (Rear), *Investigation of Chemical Warfare Installations in the Munsterlager Area, including Raubkammer*, APO 413, 1945.

Six types of chemical shells were used, and the information that identified each type was stamped on the shell in the same colour as the ring on the shell. There was a code letter in the same colour as the ring on each shell, mortar bomb, rocket or aircraft bomb that was filled with chemical warfare agents; it indicated the exact nature of the chemical fill. On bombs, and occasionally on shells, the code letter was black. The following codes were used: A for chloroacetophenone; B for thiodiglycol mustard-arsine oil, 1:1 (winter mustard); C for thiodiglycol mustard-chlorobenzene, 4:1 (winter mustard); D for thickened mustard (made from 'B'); E for thickened mustard (made from a mixture of homologous mustards); F for phosgene; G for tabun; GA for tabun with 20 per cent chlorobenzene; N for diphosgene; K for nitrogen mustard; L for thiodiglycol mustard–anthracene oil, 2:1 (winter mustard); M for DM or DM with DA; N for DA in arsine oil, 40:60; O for thiodiglycol mustard (summer mustard); and P for hydrogen cyanide.

Coding for mustard mixtures

The markings for S-mustard mixtures which were used during the Third Reich have been described inadequately in the past. Table 6.8 attempts to fill this gap. It presents the codes that were used for the various types of mustard and the mixtures used to fill munitions. (A detailed compilation of the various markings for other chemical warfare agents already exists.[17])

Munitions filled with chemical warfare agents

Filling sites and depots

Munitions were filled with chemical warfare agents in several ways. Phosgene, a gas, was liquefied at the facility where it was produced and filled into the

[17] Kopecz, P., 'Verdachtsstandorte von Rüstungsaltlasten in Deutschland' [Suspected sites of old armament wastes in Germany], vol. 5, *Teilvorhaben Kampfstofflexikon* [Abridged warfare agents dictionary], Forschungsbericht 103 40 102, UBA-FB 93-029 (Umweltbundesamt: Berlin, 1993)

munitions. A similar procedure was used for tabun to avoid transporting the highly toxic product in railway tank-wagons. Mustard agents (S- and N-mustard) and arsine oil were distributed to armed forces depots (Munas)[18] where munitions were filled with them. Those depots had large storage tanks able to contain thousands of tonnes of chemical warfare agents. Solid warfare agents such as chloroacetophenone, adamsite, and Clark I were produced in special factories. Irritants were produced in an explosives plant in Güsen; smoke powder was prepared in Döverden; and smoke candles and smoke bombs were manufactured at the Pommersche Industriewerke in Barth.

Storage depots were located at Munster, Löcknitz, Hörpolding and Lübbecke, and there may have been storage depots at Hohenbrunn and Dessau. There was also a filling station for shells in the Orgacid factory at Halle Ammendorf, which ceased work after 1942. According to British reports, work on an additional filling station at Lübbecke was never completed. These reports also mention a filling station at Kapen, near Dessau. The most complete list of sites which are likely to contain old military wastes (including filling stations) is one compiled by the Federal Agency for Environmental Protection (Umweltbundesamt).[19]

The special nature of Munster

Munster holds a special position in German military history, and the history of the German armament programme for chemical warfare is inseparably connected to Munster in both World War I and World War II. Currently, the Federal Army uses Munster North (Munster-Nord) as a training ground.

When chemical weapons began to be used in World War I Munster was chosen as a suitable proving ground for chemical warfare agents and munitions, and construction of the 'Breloh-Camp' began there on 31 March 1916. Subsequent years saw additions to what became known as 'Gas Site Breloh' as facilities for the production of chemical warfare agents and munition filling stations were built. A firing range was also set up to conduct ballistic tests on ammunition. The Munster facility had an area of 6500 hectares, and the output from its plants met approximately one-quarter of the demand for CW munitions.

After the end of World War I large quantities of chemical warfare agents were assembled at Gas Site Breloh for destruction. On 24 October 1919 an accidental explosion destroyed 48 buildings (green cross and yellow cross factories), approximately 1000 tonnes of chemical warfare agents, one million shells, 230 000 land-mines and 40 railway tank-wagons loaded with chemical warfare agents. That explosion and the tests conducted during World War I contaminated Munster North with chemical warfare agents from filled shells and with arsenic compounds, and the contamination exists to this day.

Attempts were made to clear the debris after the 1919 explosion by blasting and burning, but this led to additional contamination of the soil with arsenic compounds.

[18] 'Muna' is the German word used during the Third Reich for munition filling stations.

[19] Umweltbundesamt, [Federal Agency for Environmental Protection], *Verdachtsstandorte von Rüstungsaltlasten in Deutschland* [Suspected sites of military wastes in Germany], Texte 8/93 (Umweltbundesamt: Berlin, 1993).

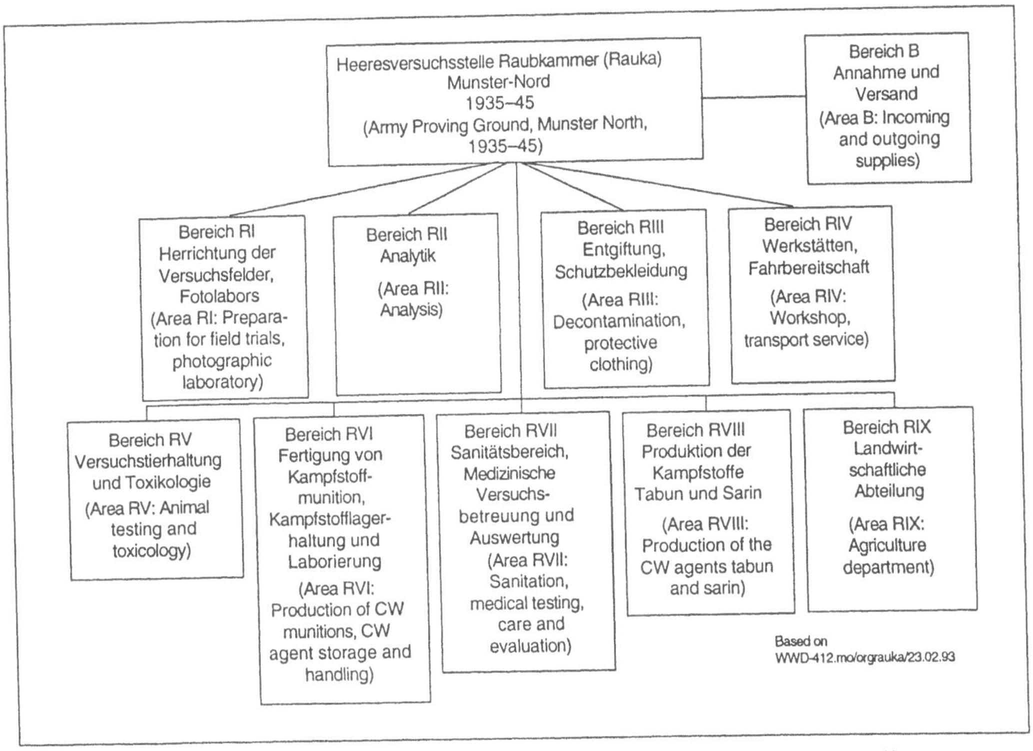

Figure 6.2. Organization of the Army Proving Ground (Heeresversuchsstelle Raubkammer) at Munster, Germany

During the Third Reich a new research and testing facility for chemical warfare agents, the Army Proving Ground (Heeresversuchsstelle Raubkammer), was established at Munster, and factories for the production of new agents—tabun, sarin and excelsior—were set up on a pilot scale. There was also a munition filling station, which was officially know as a 'smoke filling station' to mask its real purpose, and a proving ground for the air force. A large ammunition depot, the Muna Oerrel, was located nearby. In addition, there was a 10 000-hectare area for field trials, artillery field tests and bomb tests.

In the Third Reich scientific research on chemical warfare agents was carried out at the army's gas laboratory at Spandau Citadel, but technical and practical work was conducted at Munster. Owing to this, some filling installations for the various types of chemical warfare agents were located at Munster, including those for tabun and phosgene, although the latter were probably used primarily for testing. Figures 6.2 and 6.3 show the organization and location of the most important installations at the Army Proving Ground.

The effectiveness of munitions filled with chemical warfare agents

Data about the toxicity of chemical warfare agents that have been gathered from laboratory tests cannot easily be used to estimate the effect an agent will have in a shell. For example, if two such agents are compared, one of which is twice as toxic as the other, the one with greater toxicity will not necessarily have double the effect when it is used in a munition. Other factors such as the persistence of the chemical agent after detonation, speed of vaporization, likelihood of aerosolization, flammability, the manner in which the munition is constructed and the like are also important.

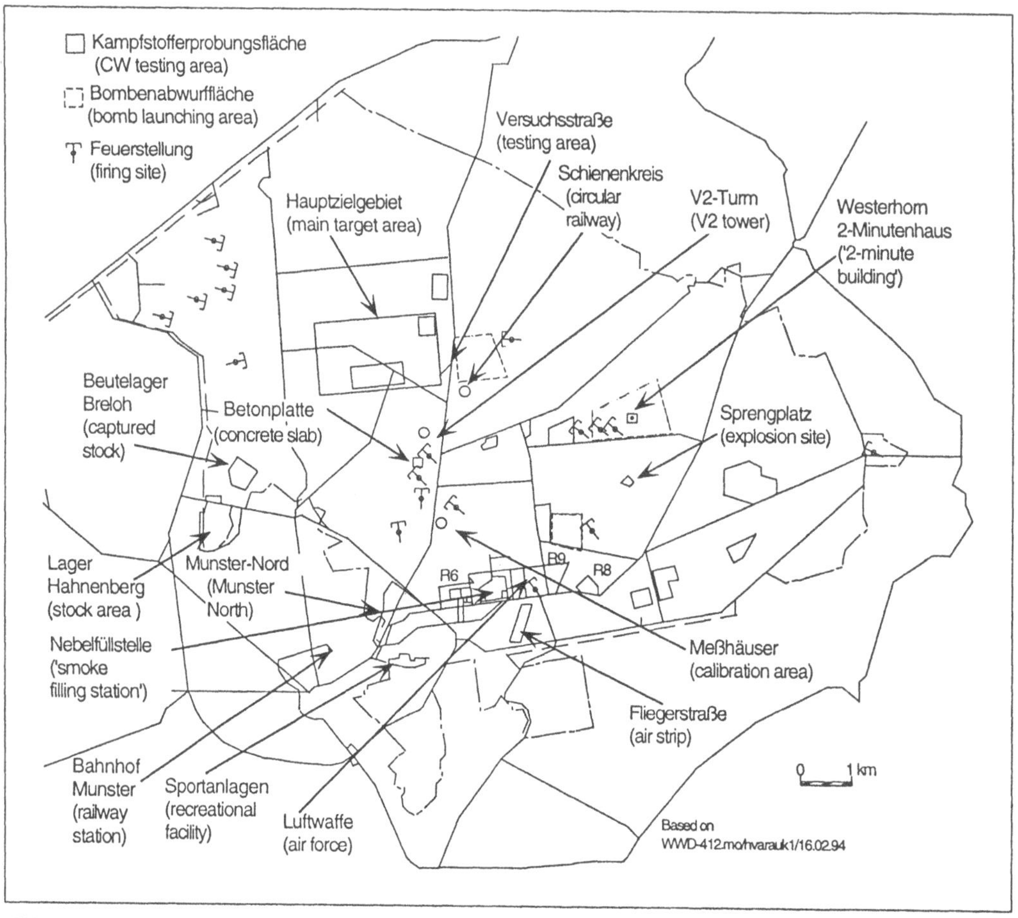

Figure 6.3. The Army Proving Ground (Heeresversuchsstelle Raubkammer) at Munster, Germany

The combination of these factors can be such that shells filled with chemical warfare agents may be surprisingly inefficient. Tests conducted during the Third Reich showed that, depending on the chemical warfare agent used, up to one-half of the agent might be decomposed or burned by the detonation. The undestroyed agent remained in the explosion crater, and only 10–40 per cent of the fill was disseminated as a cloud to contaminate the point of impact.

Tests were conducted using animals in order to obtain data about how munitions filled with chemical warfare agents would perform in combat. The agents were given numbers to indicate their effectiveness (lower numbers for the least effective agents), which facilitated comparison of munitions. Table 6.9, which shows the effect of chemical warfare agents in 15-cm shells, clearly demonstrates the superiority of sarin as a chemical warfare agent. Testing using the cylindrical chemical warfare (Kampfstoffbombe cylindrisch) KC 250 bomb produced similar results (see table 6.10).

The KC 250 bomb, the most frequently manufactured bomb for chemical warfare agents during the Third Reich, weighed 250 kilograms and contained approximately 80 litres of chemical warfare agent. Only two tests of the bomb were conducted using sarin, but they clearly demonstrated the superiority of sarin as a chemical warfare agent.

Table 6.9. The effectiveness of chemical warfare agents in 15-cm shells

Chemical warfare agent	Effective factor
Sarin	35
Tabun	12
N-mustard	7
S-mustard	7
Phosgene	6
Hydrogen cyanide	6
Cyanogen chloride	1

Source: Combined Intelligence Objectives Subcommittee (CIOS), G-2 Division, Supreme Headquarters Allied Expeditionary Force (Rear), *Investigation of Chemical Warfare Installations in the Munsterlager Area, including Raubkammer*, APO 413, 1945.

Table 6.10. The effectiveness of chemical warfare agents in the KC 250 bomb

Type	Chemical warfare agent	Effective area (square metres)	Effective factor
KC 250 Gb	Mustard	600	1
KC 250 II Gr	Phosgene	4 000	7
KC 250 III Gr	Tabun	10 000 (incapacitating) 3 000 (lethal)	17
KC 250 IV Gr	Sarin	20 000 (incapacitating) 10 000 (lethal)	33

Source: Combined Intelligence Objectives Subcommittee (CIOS), G-2 Division, Supreme Headquarters Allied Expeditionary Force (Rear), *Investigation of Chemical Warfare Installations in the Munsterlager Area, including Raubkammer*, APO 413, 1945.

The ranges given in table 6.10 for the 'effective area' are not large; the ranges for aerosol clouds of chemical warfare agents were much greater. The German armed forces conducted field tests at Munster North with adamsite smoke powder, a mixture of adamsite and explosive cotton, which resulted in clouds that drifted as far away as 60 km. After World War II field tests by the British army gave similar results.

Training and decontamination equipment

During the Third Reich chemical warfare agents were classified as top secret, and they were stored at only a few locations. The remnants of these agents or any contamination caused by them are therefore not widely distributed in Germany, and it is also not difficult to identify chemical warfare agents. However, the situation differs with respect to training and protection equipment and materials.

A great variety of these items was used (e.g., various testing vials, decontamination solutions for skin and weapons, 'smell pots', etc.), and they were commonly available for the protection of both civilians and members of the military. In addition, they were used for the education and training of troops.

These items continue to be found frequently in Germany, but knowledge about them, their intended use and the chemical composition of the materials in them is limited. When such items are found there is usually great uncertainly about how to deal with them. In order to address this situation the Federal Agency for Environmental Protection is in the process of producing and distributing a detailed catalogue describing these items.

III. The status of chemical warfare agents at the end of World War II

Stored chemical warfare agents

It is well known that in Germany at the end of World War II there were still several munition filling stations with storage tanks containing chemical warfare agents. Generally, the amount of agents that remained in these storage tanks is not known. However, the information about the Munster facility is relatively reliable. In 1945 British troops found approximately 4000 tonnes of mustard, which they burned. The mustard that was found at the Orgacid factory at Halle Ammendorf was later destroyed by the GDR. The munition filling station at Löcknitz was inspected by Soviets military authorities, but they did not release information about whether or not they had found chemical warfare agents and, if so, the quantity present.

A detailed analytical examination of these sites is needed in order to determine what remains of chemical warfare agents currently exist at the former production sites and munition filling stations that were destroyed by the Allies.

Stockpiles of munitions containing chemical warfare agents

In investigations conducted after World War II by the Allies information was obtained about the stocks of German munitions, both those which were filled and those which remained unfilled with chemical warfare agents.[20] Table 6.11 is a compilation of both the actual inventory and the amount requested by the General Staff as of 1 April 1944. Table 6.12 presents a comparison between the estimated monthly consumption of chemical warfare agents and the quantities captured by the Allies. There are significant differences in the quantities of shells reported to have existed, and the UK has conceded that mistakes in counting were made.

Cleaning up the munition depots

By the spring of 1945 it was apparent that Germany would lose the war and the problem of how to handle the munitions filled with chemical warfare agents became acute. All of the storage depots were expected soon to fall under the control of the advancing Allied forces. The best approach would probably have been to leave the munitions where they were and to provide information about them via diplomatic channels, thereby allowing control of them to be transferred without endangering anyone. Unfortunately, this was not possible.

[20] Based on information from an unpublished British report (see note 1).

Table 6.11. Actual and estimated inventory of munitions filled with chemical warfare agents as of 1 April 1944[a]

	General Staff request	Munitions filled with chemical warfare agents	Stock of unfilled munitions
Land-based ammunition			
Artillery shells	4 818 000	2 018 188	1 065 120
Rockets	300 000	155 657	169 370
Mortar cartridges	223 000	157 416	65 742
Land-mines	300 000	385 555	461 019
Bombs			
KC 50 II Bu	450 000	7 200	40 114
KC 250 W	. .	32 115	17 276
KC 250 II Gr	. .	57 185	63 920
KC 250 III Gr	108 000	76 032	19 119
KC 250 Gb	144 000	103 832	85 898
KC 250 II Gb	. .	38 977	81 468
Total KC 250	**252 000**	**308 141**	**267 681**
KC 500 W	18 000	100	16 086
KC 500 II Gr	54 000	. .	16 125
KC 500 Gb	36 000	. .	. .
KC 500 II Gb	54 000	. .	2 333
Total KC 500	**162 000**	**100**	**34 544**
KC 1000 II Gr	3 600	. .	. .
KC 1800 W	900	. .	347
KC 1800 II Gr	900	. .	786

[a] KC = Bomb filled with chemical warfare agent(s), cylindrical shape; 50, 250, etc. = 50 kg, 250 kg, etc. = Total weight; Bu = Blue ring; W = White ring; Gr = Green ring (II = phosgene, III = tabun); Gb = Yellow ring; and II Gb = Thickened mustard.

Source: English summary from Combined Intelligence Objectives Subcommittee, British Intelligence Objectives Subcommittee and testimony of unspecified witnesses.

Although it was evident that all of the weapons could not be removed to safe areas away from centres of population and that Allied bombing could create danger to German civilians, as late as 16 April 1945, Hitler refused to permit relocation of these munitions. Instead the Head of the Armed Forces High Command, Field Marshal Wilhelm Keitel, issued a directive that became known as the 'Keitel order'. It stipulated that: no modern chemical warfare agents were to be allowed to fall into the hands of the enemy; such agents were to be removed from the path of advancing enemy forces; any agents left behind were to be camouflaged; and facilities that worked with chemical warfare agents installations were to be destroyed when threatened by the enemy.

As early as 4 April 1945 Hitler had ordered that all chemical warfare agents and the munitions filled with such agents should be moved to more secure locations. As a result of this order still today the exact location of many of these agents and munitions remains unknown. There is thus little value in speculating on the validity of reports about the quantity or the transport of agents and munitions that took place at the end of World War II.

Table 6.12. Munitions filled with chemical warfare agents captured by the Allies[a]

	Stocks as of 1 Apr. 1944 (both filled and unfilled)	Munitions needed for 1 month of chemical warfare	Filled munitions captured in 1945
Land-based ammunition			
Artillery rounds	3 088 308	808 000	7 500 000
Rockets	325 027	50 000	88 900
Mortar rounds	223 158	. .	206 300
Munitions	846 567	50 000	179 150
Bombs			
KC 50 II Bu	47 313	50 000	25 000
KC 250 W	49 391	. .	32 500
KC 250 II Gr	121 105	. .	53 000
KC 250 III Gr	95 151	12 000	106 000
KC 250 Gb	189 730	16 000	150 000
KC 250 II Gb	120 445	. .	38 500
Total KC 250	**575 822**	**28 000**	**380 000**
KC 500 W	16 186	2 000	3 000
KC 500 II Gr	16 125	6 000	. .
KC 500 Gb	. .	4 000	. .
KC 500 II Gb	2 333	6 000	. .
Total KC 500	**34 644**	**18 000**	**3 000**
KC 1000 II Gr	. .	400	. .
KC 1800 W	347	100	. .
KC 1800 II Gr	786	100	. .

[a] KC = Bomb filled with chemical warfare agent(s), cylindrical shape; 50, 250, etc. = 50 kg, 250 kg, etc. = Total weight; Bu = Blue ring; W = White ring; Gr = Green ring (II = phosgene, III = tabun); Gb = Yellow ring; and II Gb = Thickened mustard.

Source: English summary from Combined Intelligence Objectives Subcommittee, British Intelligence Objectives Subcommittee and testimony of unspecified witnesses.

The fate of chemical warfare agents after World War II

There was an agreement among the Allies to destroy all German chemical warfare agents and the related facilities after the end of World War II. As mentioned above all of the munition filling stations and plants for the production of chemical warfare agents were relatively intact when the Allies took control of them. The Allies evaluated the facilities, removed the components which they could use and destroyed the chemical warfare agents or added them to their own stocks. The facilities were then blown up. This was done so effectively that today only a small amount of debris remains, and it is not possible to ascertain the former functions of the demolished buildings. Munitions filled with chemical warfare agents were handled in a similar manner. Owing to the activities conducted by the German forces in the spring of 1945, which are described above, it is impossible to know exactly how many munitions were destroyed, but a rough estimate can be made.

According to documents in the Federal Archives at Koblenz, Germany, the quantities of chemical munitions found, destroyed, captured or recycled in the four occupation zones by the end of 1947 were as follows: in the US zone 93 995 tonnes; in the British zone 122 508 tonnes; in the French zone 9250 tonnes; and in the Soviet zone 62 505 tonnes. Of the 288 258 tonnes of captured munitions, 269 000 tonnes were destroyed or otherwise used by the end of 1947. The destruction operations conducted in 1948 brought the total to approximately 300 000 tonnes. These figures indicate that most of these munitions were accounted for after World War II.[21]

These stocks were disposed of primarily by dumping them into the sea. Approximately 250 000 tonnes were dumped at sea, of which 42 000–65 000 tonnes were dumped into the Baltic Sea. A 1993 German report provides information about that dumping operation.[22]

Munitions filled with chemical warfare agents that could not be transported were blown up. Large quantities of these weapons were also transported to a variety of countries. For example, the following quantities were removed from the US zone and transported elsewhere: to Italy, 6090 tonnes; to Poland, 210 tonnes; to North America, 19 400 tonnes (most to the USA and some to Canada); to the former Czechoslovakia, 6103 tonnes; and to other countries, 2421 tonnes.[23] There were an additional 30 345 munitions filled with the chemical warfare agent N-mustard in Austria.[24]

According to an older report from the Federal Archives, such munitions were also removed from other zones (e.g., there are indications that after 22 May 1947, 3100 tonnes of munitions filled with chemical warfare agents were transported to France).[25]

It is obvious from the above that there cannot be complete certainty about the location of all of the chemical warfare agents produced by Germany. However, the available data provide a significant amount of information and indicate that most of the agents were removed from Germany or destroyed on-site.

IV. Chemical warfare agents as toxic armament wastes

It is well known that the industrial production of chemical warfare agents began during World War I. At first there was limited awareness of the way in which these agents work. In general, almost any substance was considered a potential chemical warfare agent if it was irritating, toxic and available in sufficient quantity. This approach led to the creation of chemical warfare agents that combined a large number of substances in relatively exotic mixtures that were then filled in munitions. As a result, it is difficult today to analyse and evaluate old munitions and production sites. Towards the end of World War I the large

[21] Brauch and Müller (note 1), pp. 277–78.

[22] *Chemische Kampfstoffmunition in der südlichen und westlichen Ostsee: Bestandsaufnahme, Bewertung und Empfehlungen*, Bericht der Bund/Länder-Arbeitsgruppe Chemische Kampfstoffe in der Ostsee [Chemical munitions in the southern and western Baltic Sea: compilation, assessment and recommendations, Report by the Federal/Länder Government Working Group 'Chemical Munitions in the Baltic Sea'] (Federal Maritime and Hydrographic Agency: Hamburg, May 1993).

[23] This information is based on a summary of a variety of sources: 'Die Munitionsbeseitigung nach dem 2. Weltkrieg in der US-Zone, insbesondere im Freistaat Bayern' [Munitions disposal after World War II in the US occupation zone, in particular in the free state of Bavaria]. The information about the removal of munitions from the US zone is based on a report from Truppenübungsplatz-Kommandantur Munster [Training area headquarters Munster], 1 Nov. 1991. See also Thamm, W., *Feuerwerker im Einsatz* [Fireworkers in action] (Biblio Verlag: Osnabrück, 1995), p. 39.

[24] See note 23.

[25] Quoted in Brauch and Müller (note 1), p. 278.

variety of different substances was reduced to a few particular chemical compounds.

When the disposal of military waste is studied currently it must be remembered that the first such problems stem from the World War I era. The procedures used then to produce, handle, test and ultimately dismantle munitions filled with chemical warfare agents and the facilities where they were produced would not meet the current stringent regulations that are designed to protect the environment. Consideration was not given to the environment during such activities at that time, and munitions containing chemical warfare agents were simply blown up. Some munitions contained materials which could not be decomposed, such as arsenic, and these materials continue to contaminate the soil and water yet today.

The same is true of the World War II period and of the subsequent demilitarization of chemical warfare agents and munitions. Lasting contamination was created by improper dismantling and destruction methods. It was still common practice to blow up munitions, to burn them in the open air without the use of filters, to bury them or to dump them into a body of water. The resultant contamination continues to create problems more than 50 years after the end of World War II.

Appendix 6A. A chronology of German research, development and preparation for chemical warfare agent production

1924 The Weapons and Materials Inspectorate begins collecting documents and asks for the co-operation of scientists.

1927 The Weapons and Materials Inspectorate Group is transferred to the Waffenamt (Wa), the Ordnance Department, and renamed Prüf S (Testing Activities). The group is not enlarged.

1928 Trials commence at Tomka, Russia (about 20 km west of Vol'sk) with approximately 30 personnel.

1932–33 Trials cease at Tomka.

1934 Major Dr Rüdiger takes over Prüf S. It becomes Wa Prüf 9 (Ordnance Department for Testing of Equipment) and begins to expand. Work focuses on gas protection and begins on detection and identification. A search for new CW agents is under way.

1935 Expansion continues. Field trials with improvised materials begin at Raubkammer (Munster). Construction begins at Raubkammer (Munster) and at the army's gas laboratory at Spandau Citadel, Berlin.

1936 CN and mustard gas are chosen to be the primary CW agents, and activities focus on developing and testing munitions for them. Factories, storage facilities and filling plants are planned for CN and mustard gas. Construction of the Spandau and Raubkammer (Munster) facilities nears completion. The search for new compounds continues. Tabun is discovered.

1937 Wa Prüf 9 now has 80 employees. R&D activities continue. The Spandau and Raubkammer (Munster) facilities work to develop their experimental technique.

1938 Lieutenant-Colonel Diplomaed Engineer Schmidt is appointed to head Wa Prüf 9. It now has 120 personnel, and the facility will not grow much larger. Work intensifies on the design of CN and mustard weapons so that manufacturing can begin. Activities next concentrate on DA, DC and DM. Tactical weapon trials begin at Raubkammer (Munster). Tabun experiments are conducted at Spandau and Raubkammer (Munster). Sarin is discovered. The Ammendorf plant begins to manufacture mustard and nitrogen mustard. The Bitterfeld plant starts to produce phosgene. The Seelze plant initiates CN production.

1939 The activities of 1938 continue. Experiments are conducted on arsenicals. Materials seized in Czechoslovakia and Poland are investigated. The Dyhernfurth plant commences production.

1940 The activities of 1939 continue. Other work is limited to that which can be concluded in 1941. Materials confiscated in France are investigated. The Haselhorst plant begins production.

1941 Work on viscous mustard intensifies. Air Force trials and charging of Air Force weapons receive special attention. Excelsior, an irritant, is studied but given low priority. The Gendorf plant starts production.

1942 Oberst (Colonel) Dr Walter Hirsch becomes head of Wa Prüf 9. A change to different raw materials, necessitated by supply problems, creates difficulties. Work on the D-mustard process is discontinued. Experiments are carried out in Norway on protection against gas and smoke in low temperatures as a result of combat experiences in the Russian winter. Smoke munitions are tested in North Africa for use in the African theatre of war. Experiments with substances such as hydrogen cyanide and cyanogen chloride gain importance owing to the possibility of their large-scale aerial use. Experimental work continues at Spandau and at the Army Proving Ground (Heeresversuchsstelle Raubkammer) at Munster. The emphasis is on filling trials with tabun and viscous mustard. Trials with close-combat munitions are also conducted. Production ceases except at the Gendorf and Dyhernfurth plants.

1943 An increase of personnel becomes necessary as Allied air superiority increases the likelihood of gas warfare against Germany. Heavy bombing attacks on Berlin lead to removal of a part of Wa Prüf 9 to Raubkammer (Munster). The need for protection against CW agents assumes greater importance as the possibility that the Allies will use gas warfare increases. The search for weapons that can be used in close combat intensifies. Experiments are conducted at Raubkammer (Munster) on tactical use of CW agents and the effects of weather. A series of field trials of tabun is held.

1944 Work focuses on defence. It becomes difficult to obtain raw materials. Experimentation with tabun and chemical agents of the sarin group intensifies. Trials are conducted with close-range weapons.

1945 There is a slight increase in the number of personnel; the increase is mainly in the area of defensive research. In early March Wa Prüf 9 and most of the Spandau personnel are moved to the Raubkammer (Munster) facility. The activities of 1944 continue. Production of sarin is scheduled to begin in mid-1945 at Dyhernfurth at a rate of 100 tonnes per month. The sarin plant at Falkenhagen is about one-quarter completed; it is designed to produce 500 tonnes per month.

Appendix 6B. German plants that produced chemical warfare agents and their total production output

Introduction[1]

Various chemical warfare agents were produced in German plants. The following compilation presents the agents that were produced and the plants which manufactured them.

Chloroacetophenone

Chloroacetophenone was produced by Riedel de Haen AG at Seelze (near Hanover). Construction of the facility began in 1934 and was completed in 1935. The facility had a capacity of approximately 140 tonnes per month (tpm). By 1942, when the facility was closed, it had produced a total of approximately 4000 tonnes.

Chloroacetophenone was also manufactured by I. G. Farben AG in its plant at Ludwigshafen. Construction began in 1934 and was completed in 1935. The plant had a capacity of approximately 70 tpm, and by 1944, when it was destroyed by air raids, it had produced 3133 tonnes.

Phosgene

Phosgene was manufactured by I. G. Farben AG at Wolfen Bitterfeld. The plant began to be built in 1939 and was ready by 1940. Its production capacity was 400 tpm, and it produced a total of 5900 tonnes.

Nitrogen (N-) mustard

Nitrogen mustard was manufactured by Orgacid GmbH, at Halle Ammendorf in a facility where construction began in 1937 and was completed by 1938. The facility could produce approximately 50 tpm, and it produced a total of 1928 tonnes. Its production activities are outlined in table 6B.1.

Sulphur (S-) mustard

Sulphur mustard was produced by Orgacid GmbH at Halle Ammendorf. Construction commenced in 1934 and was concluded in 1936. The capacity was 700 tpm, and the total production output was 24 097 tonnes (see table 6B.1).

Work on the Anorgana GmbH facility for the production of S-mustard at Gendorf began in 1939 and was completed by 1943. Its capacity was approximately 1500 tpm, and it produced a total of approximately 3500 tonnes.

Thickened mustards

Thickened mustards were highly viscous mixtures of mustard to which a thickener was added. They were designed to create long-lasting contamination on the battlefield and were produced only at the Army Proving Ground (Heeresversuchsstelle Raubkammer) at Munster. The amount produced is unknown, but it may have been several thousand

[1] Brauch, H. G. and Müller, R.-D., *Chemische Kriegsführung, Chemische Abrüstung* [Chemical warfare, chemical disarmament] (Arno Spitz: Berlin, 1985); Combined Intelligence Objectives Subcommittee (CIOS), G-2 Division, Supreme Headquarters Allied Expeditionary Force (Rear), *Investigation of Chemical Warfare Installations in the Munsterlager Area, including Raubkammer*, APO 413, 1945; Historical Division, Headquarters, European Command, *Chemical Warfare: A List of Reports of the Historical Division, Headquarters, European Command*, Frankfurt [no year] (manuscript); and an unpublished British report based on evaluations of CIOS and British Intelligence Objectives Subcommittee (BIOS) reports, investigations and interrogations and German material.

Table 6B.1. Production of sulphur mustard and nitrogen mustard at Halle Ammendorf

Figures are in metric tonnes.

Production period	Nitrogen mustard	Sulphur mustard
Prior to Apr. 1939	19	2 183
Apr. 1939–Mar. 1940	118	5 006
Apr. 1940–Mar. 1941	293	5 384
Apr. 1941–Mar. 1942	361	4 657
Apr. 1942–Mar. 1943	432	1 525
Apr. 1943–Mar. 1944	516	3 682
Apr. 1944–Mar. 1945	189	1 660
Total production	**1 928**	**24 097**

Source: English summary from Combined Intelligence Objectives Subcommittee, British Intelligence Objectives Subcommittee and testimony of unspecified witnesses.

tonnes. Two standard mixtures were used primarily: (*a*) a mixture of 91 per cent OA (a mixture of Oxol-Lost and arsine oil), 6 per cent tornesite and 3 per cent sulphur (S) wax; and (*b*) another mixture of 88 per cent OA, 9 per cent polystyrene and 3 per cent S-wax. In addition, a number of other mixtures were tested.

Tabun

Tabun was produced at the Anorgana GmbH facility at Dyhernfurth, where construction was begun in 1939 and completed in 1942. The production capacity was 1000 tpm, and the total production output was approximately 12 000 tonnes.

The Army Proving Ground at Munster had a pilot plant to test methods for the production of tabun. There is no information available about production capacity or total production.

Sarin

The Army Proving Ground at Munster also had a pilot plant for the production of sarin. Its total production was approximately 0.5 tonnes.

There are reports in the literature of a pilot plant for the production of sarin at the army's gas laboratory at Spandau Citadel in Berlin. However, nothing is known about its capacity or production output.

Soman

Soman was only produced under experimental conditions and on a laboratory scale.

Adamsite

Construction of the I. G. Farben AG plant for the manufacture of adamsite at Uerdingen began in 1939 and ended in 1940. The plant had a capacity of 200 tpm and a total production output of 3900 tonnes.

Clark I

Clark I was produced by Lonalwerke GmbH at Berlin Haselhorst. Work on the plant started in 1939, and the plant was ready by 1940. It had a production capacity of 90 tpm and a total production output of 1500 tonnes.

Clark II

Lonalwerke GmbH produced 100 tonnes of Clark II at Berlin Haselhorst for testing purposes.

Arsine oil

Arsine oil is an unrefined product which consists of arsenic trichloride, phenyldichloroarsine, diphenylchloroarsine (Clark I) and triphenylarsine. It was used to produce Clark I and as an admixture for S-mustard to lower the freezing point. It was then called winter mustard or winter-Lost.

It was produced by Ergethan GmbH at Löderburg Staßfurt in a plant that began to be built in 1934. Construction was completed in 1938, and the capacity of the plant was 270 tpm. The total production output from 1938 to 1 March 1944 was 7500 tonnes. After 1 March 1944 production concentrated on Clark I. Total production is estimated to have been approximately 12 000 tonnes.

Excelsior

Excelsior is an irritant similar to adamsite. It was produced in a test plant at the Army Proving Ground at Munster. The total production of excelsior was approximately 10 tonnes.

Chlorotrifluoride

In 1938 a plant was built at Falkenhagen for the production of chlorotrifluoride (N-stoff). The capacity of the plant was intended to be 50 tpm, but this was later reduced to 10 tpm. The total production of the plant was probably less than 50 tonnes. There were plans to use the plant later for the production of sarin.

Other chemical warfare agents

In addition, other chemical warfare agents such as hydrogen cyanide, chlorocyanide, lewisite and some irritants were produced in small amounts for testing, training or special purposes. The production sites were probably laboratories like the army's gas laboratory at Spandau Citadel.[2]

[2] See chapter 5 in this volume.

7. Chemical weapon production in the former Czechoslovakia

JIRÍ MATOUSEK*

I. Introduction

Chemical weapons were relatively effective in World War I, and because of this the Czechoslovak Army was eager to acquire such weapons. In 1919 the Referate for Gas Service was established in the Engineering Division of the Ministry of National Defence, and the first technical manual dealing with chemical defence was prepared in 1920.[1]

The starting point for the actual development of chemical defence and preparation for chemical warfare was a letter from General Eugène Mittelhauser, Chief of the General Staff, to the Minister of National Defence on 2 February 1922.[2] Mittelhauser urged that a Military Gas Service be developed to meet the potential threat posed by the German chemical industry and the research activities being conducted at the Research Centre in Munich. He stressed that in the future chemical weapons could be employed by air forces. In addition, the letter mentioned that Hungary was cooperating with Germany at the Munich Research Centre and that, during World War I, Hungary had produced gas shells at Magyarovár Sarvár.

The first official document to deal with both chemical defence and chemical weapons was the Plan of Organization of the Gas Service, which stated: 'Even if we have decided not to take the initiative as regards gas attack, we must be prepared to make a rapid response if we are attacked in this way'.[3] The plan noted that German preparations for chemical warfare violated Article 171 of the Treaty of Versailles,[4] and it questioned some of the data on the use of war gases which were presented in German manuals.[5] The plan established a chemical defence programme that provided excellent protection to both military troops and civilians and which also included a limited CW retaliatory capability.

[1] *Boj plynem a obrana proti nemu* [Gas combat and defence against it] T-str-1 (V-III-I) (MNO: Praha, 1920) (in Czech).

[2] On 20 Jan. 1919, the governments of Czechoslovakia and France concluded an agreement to send a French military mission to Czechoslovakia to assist in building up its armed forces. The mission was headed by General Maurice César Joseph Pellé, temporary Minister of National Defence of Czechoslovakia. General Mittelhauser, a member of the mission, later became Chief of the General Staff.

[3] General Staff, 'Plan of organization of the gas service', Ministry of National Defence, 9 July 1924.

[4] Article 171 states, in part: 'The use of asphyxiating, poisonous or other gases and all analogous liquids, materials or devices being prohibited, their manufacture and importation are strictly forbidden in Germany. The same applies to materials specially intended for the manufacture, storage and use of the said products or devices.' Brigham Young University, The World War I Document Archive, 'Peace Treaty of Versailles, Articles 159–213, Military, Naval and Air Clauses'. Version current on 4 Nov. 1996, URL <http://www.lib.byu.edu/~rdh/wwi/versa/versa4.html>.

[5] *Führung und Gefecht der Verbündeten Waffen: Ausbildungsvorschrift der Artillerie* [Command and action of the united forces: training regulations for the artillery].

* The author thanks the heads of the Military Historical Institute (VHÚ) and the Military Historical Archive (VHA), both located in Prague, and the personnel of the latter for assistance and access to documents related to the history of Czechoslovak defence.

Little else is known about the history of chemical weapons in Czechoslovakia in the period between World War I and World War II. The six-volume SIPRI study *The Problem of Chemical and Biological Warfare* contains one brief reference: 'The Czechoslovak weapons included mustard-gas land mines and irritant-agent candles, both arsenical and lachrymatory'.[6] This chapter provides additional information from historical documents, eyewitness accounts and personal experience.[7]

II. The development of chemical defence and retaliatory potential

Chemical defence

In Czechoslovakia preparation for chemical warfare emphasized chemical defence and protection. Initially, there were many different kinds of respirators since after World War I more than 100 000 respirators of varying design were brought to Czechoslovakia by soldiers of the Czechoslovak Legions (formed by volunteers who were deserters and prisoners of war of the former Austro-Hungarian army and from Czech residents in France, Italy, Russia or Serbia).

The first respirator of Czechoslovak design was introduced in 1923. The most advanced version was Model 35. By 1924 all soldiers were required to be given periodic training in a gas chamber, and protective suits were introduced in 1936. Czechoslovakia's rapidly developing industry was able to produce all of the necessary materials, components and equipment for chemical defence, and by 1938 the country was prepared for chemical attack by an enemy. Czechoslovakia exported some protective equipment to foreign countries—primarily to Yugoslavia, where the Model 35 respirator was produced under licence until 1960. When World War II began Czechoslovakia had the necessary equipment for chemical defence, including military respirators, several types of civil respirators, protective suits and ointment, detection kits, portable decontamination devices, special decontamination vehicles, ventilation and filtration equipment for shelters, and the like.

Chemical warfare agents and the means of their use

The Plan of Organization of the Gas Service also dealt with offensive measures, and chemical artillery shells and air bombs were the weapons chosen to be used. A decision to develop these weapons was taken in February 1938, and work was to have begun on them in 1939. However, when Germany occupied Czechoslovakia on 15 March 1939, the weapons were not yet ready.

Initially, modest use of chemical weapons in combination with other defensive systems was proposed to defend Czechoslovakia. It was felt that an adequate and effective defence could be achieved by surface contamination with mustard gas (sprayed from portable devices, vehicles and railway wagons) and by the use of chemical land mines.

[6] SIPRI, *The Problem of Chemical and Biological Warfare*, vol. 1, *The Rise of CB Weapons* (Almqvist & Wiksell: Stockholm, 1971), p. 293.

[7] The author served for more than 38 years in the Czechoslovak Army, working primarily with R&D for the Chemical Corps and the Medical Services.

This was in keeping with the national defence strategy: a system of many lines of well-armed fortifications in the border mountains and at other natural barriers close to the frontiers. If these fortifications had been combined with chemical contamination and other forms of defence (e.g., physical barriers, artillery, tank units and aircraft) German forces might not have penetrated them, particularly in the valleys between the mountains. Mustard gas would have been the chemical weapon of choice for such contamination.

The long-term plans to procure CW artillery shells and air bombs envisaged that phosgene and harassing agents would be used in them. A document issued by Division II/6 of the Ministry of National Defence on 15 December 1937 listed the quantities of chemical agents which the Czechoslovak Army had in storage. They included: 20 000 kg of mustard gas,[8] 66 kg of phosgene,[9] 2500 kg of diphenylchloroarsine,[10] 2732 kg of chloroacetophenone,[11] 1000 kg of adamsite[12] and 177 500 kg of fumic acid.[13]

Preparations for the use of chemical warfare agents were secret, and to mask their true identity the agents were referred to as 'stabilizers'. The primary chemical agents were also given code-names. For example, mustard was known as 'sulfonal', phosgene as 'chloral' and diphenylchloroarsine as 'sternit'. The equipment for chemical warfare was called 'attack chemical material' and included: (*a*) equipment which was already in use by military forces, such as the Model 36 portable smoke sprayer, the Model 35 aircraft smoke sprayer, the Model 36 portable spraying device, and three types of filling equipment for them; and (*b*) equipment which was at an advanced stage of development and soon to be introduced, such as a special railway wagon for spraying mustard gas, two types of smoke sprayers for tanks, a Model RA (rozstrikovací automobil) special spraying vehicle, smoke equipment for ships, a chemical filling and delivery vehicle, chemical land mines, 0.5 kg smoke candles, and 3 kg Model 36 smoke candles.[14]

Various types of portable and mobile spraying devices and equipment had dual or even triple uses. They were used to spray mustard gas, to generate smoke screens (utilizing the liquid chemical agent fumic acid) and to decontaminate (known at the time as 'asanation'). Prior to World War II there was only one true chemical weapon (i.e., equipment designed exclusively for that purpose): the Model 38 chemical land mine (known as the 'sulfonal pot' during its development). It was introduced on 21 June 1938, and had a volume of approximately 5.5 litres, was filled with 6.75 kg of mustard gas and stabilized with a 3 per cent turpentine oil solution. It could be buried and exploded (contaminating an area of approximately 200 square meters) or launched and detonated at a height of 6–8 meters (contaminating an area of approximately

[8] Military Historical Archive document VHA-1938-47/39-2, 27 July 1938, lists the actual amount of mustard gas stored as 70 tonnes (t), and the total capacity for bulk storage as 640 t. The programme for procurement listed the following figures: 2332 t of mustard gas, 15 t of diphenylchloroarsine, 30 t of adamsite, 15 t of chloroacetophenone, 980 t of fumic acid and 7800 t of lime chloride. Military Historical Archive document VHA-1938-47/39-4, 19 Aug. 1938.

[9] At the time phosgene was available only for testing respirator canisters.

[10] The arsenicals were imported from the USA.

[11] The chloroacetophenone was imported from Italy.

[12] See note 8.

[13] This mixture of chlorosulphonic acid and sulphur trioxide was the primary material used to produce smoke screens.

[14] Military Historical Archive document VHA-1937-47/31-4, 20 Feb. 1937.

300 square meters). In addition, there were chemical smoke candles that were used for training or combat (filled with irritant harassing agents).

III. Chemical defence institutions and preparation for the use of chemicals in war

Administrative structures

Both defensive and offensive CW activities were performed by the Gas Service, which was an independent part of the artillery. In 1937 a proposal was made to transform the Gas Service into the Chemical Corps.[15]

The Gas Service reported to the II/6 Division of the Ministry of National Defence of Czechoslovakia and consisted of four groups (A–D). Group A was responsible for organization and training; Group B for protection; Group C for attack; and Group D, which came into existence in 1937, for the production of decontamination, chemical warfare and smoke agents. Before 1937 the Gas Service consisted of Artillery Battalion 401 (later Regiment 401), located at Plumlov; a Chemical School at Olomouc; and 17 decontamination companies. In addition, there were CW specialists of varying military ranks.

Under the 1937 proposal, in wartime the Chemical Corps was to comprise: three chemical regiments with 9 chemical companies each (a total of 27 chemical companies); seven decontamination battalions with 3 decontamination companies each (a total of 21 decontamination companies); decontamination companies at the corps, division, brigade and mountain regiment levels (52 companies altogether); and decontamination patrols.

Decontamination companies were to be responsible for reconnaissance, decontamination and the use of smoke agents. Chemical companies were given these tasks and were also to spray chemical agents. Decontamination patrols were to carry out reconnaissance. The chemical companies were to be given four RA spraying vehicles, 320 chemical land mines, 12 tonnes of mustard gas, 10 tonnes of lime chloride, 1.1 tonnes of fumic acid and other equipment.

In February 1938 it was decided that by 15 February 1939, the first stage of mobilization, 6 chemical companies and 31 decontamination companies were to be functional. However, this had not yet taken place when Germany invaded Czechoslovakia.

R&D institutions

Chemical defence and the wartime use of chemical weapons were studied by the 4th [Chemical] Division (consisting of three departments) of the Prague Military Technical Institute, which was renamed the Military Technical and Aviation Institute in 1933.[16] In 1919 a depot was set up in Prague where chemical weapons were tested and stored and where troops were supplied with CW protection. It later moved to Olomouc and became the Military Chemical Institute,[17] where the Chemical School was also located until 1937. Most equipment

[15] Military Historical Archive document VHA-1937-87/43-2, 1937.
[16] Vojensky technicky ústav (VTÚ) and Vojensky technicky a letecky ústav (VTLÚ), respectively.
[17] Vojensky chemicky ústav (VCHÚ).

for defence against chemical warfare and for chemical warfare attack was developed at Lutín by the Czechoslovak company CHEMA.[18]

Production and storage facilities

Chemical defence equipment was generally produced by civil companies at a few factories (e.g., CHEMA at Lutín, TECHNA at Nové Mesto nad Váhom, FATRA at Napajedla and Eckhardt at Chotebor). Basic materials and components were supplied to these companies by the Czechoslovak chemical and engineering industry. However, dangerous materials which could be used in combat, such as munitions, explosives and chemical warfare agents, were concentrated at five military factories.[19]

Initially, explosives and munitions were produced at Military Factory VT-1 in Policka, but later they were manufactured by the civil company Explosia Semtín.[20] In the second half of the 1930s two new factories (Military Factory VT-2 and Military Factory VT-3) were built at Bohuslavice and near Nováky, respectively.

In 1931 a decision was taken to build Military Factory VT-4 (and later Military Factory VT-5) to produce chemical warfare agents, decontamination chemicals, smoke agents and other special chemicals and equipment.

Military Factory VT-4 was located in Zilina near the Slovak affiliate of the leading chemical manufacturer Spolek pro chemickou a hutní vyrobu Ústí nad Labem (known by the German name Aussiger Verein in the 1800s).[21] Construction of Military Factory VT-4 began in 1931. The factory produced centralite, and it was intended, initially, to produce mustard gas (from ethylene) and, later, to manufacture phosgene and fumic acid. The plant was projected to produce 1 tonne of both mustard and phosgene daily during war.

Experimental production of mustard at 50 per cent of production capacity started on 3 May 1937 and continued until 1938, when lack of storage forced a stop. Military Factory VT-4 was the only CW production facility in Czechoslovakia before World War II, since the larger Military Factory VT-5 was not ready to begin production until 1939.

Military Factory VT-5 was located at Zemianské Kostolany in the Lelovecká Valley not far from Nováky, where another Slovak affiliate (now Slovenské lucobné závody Nováky) of the Czech chemical manufacturer mentioned above was located. This factory was specially constructed and was also shielded from aircraft attack by its location in a narrow valley. In addition, it had many underground tunnels. In wartime it was intended to be the main military chemical production facility with a maximum daily output of 8 tonnes of mustard gas, 3 tonnes of phosgene, 1 tonne of diphenylchloroarsine, 1 tonne of adamsite and 0.5 tonne of chloroacetophenone. A production facility for toxic candles was to be established there in addition to one at the CHEMA facility at Lutín (where technical equipment for chemical warfare could be manufactured). CHEMA provided technical assistance for production planning at Military Factory VT-5.

[18] CHEMA [Firm for research, production and trade of equipment for defence against gases], Ltd, was located at Olomouc-Lutín.

[19] Vojenská továrna (VT) 1–5 [Military factories 1–5].

[20] Currently called Vychodoceské chemické závody Synthesia, Pardubice-Semtín.

[21] It is currently known as Povázské chemické Závody Zilina.

Military Factory VT-5 was a complex of 13 production facilities that never actually began manufacturing activities but which was intended for the electrolytic production of chlorine, liquefaction of chlorine, melting of caustic sodium and production of sodium, lime chloride, arsenic trichloride, diphenylamine and synthetic hydrogen chloride. End-products such as mustard, phosgene, chloroacetophenone, diphenylchloroarsine and adamsite were also planned to be manufactured. Production of lewisite was considered but decided against.

Metallic sodium and lime chloride were manufactured under a licence from the British company Imperial Chemical Industries, which provided the technical know-how. The necessary technology for the production of arsenical irritants was obtained from the US Ordnance Engineers, Cleveland, Ohio. Chlorine was to be produced at the rate of 2 tonnes per 24-hour period. Raw materials for the manufacture of arsenic trichloride were purchased, and 880 tonnes of particulate arsenic from the metallurgical production of lead and silver were transferred to the factory from the Central Bohemian mines in Príbram.

Construction proceeded quickly, and the buildings, underground tunnels and auxiliary structures were near completion in 1938. In February 1939, one month before Germany occupied all of Czechoslovakia, work ceased and nothing more was done beyond securing the existing facilities. Some tunnels were used for storage. Military Factory VT-5 was never used for the production of chemical warfare agents, but the land and facilities at the factory were used during World War II as a conventional weapon depot.

After World War II the facility was known as Military Repair Works 072 (Vojensky opravárensky závod 072) and was placed under the control of the Chemical Corps. It produced toxic chemicals for training, research and protection purposes and was designated as the single small-scale facility that a state is permitted under the Chemical Weapons Convention.[22]

In order to enable storage of agents in bulk (and filled in weapons in some instances) three storage facilities were built at Trneny Újezd, Ostrov u Blanska and Skala u Trencína. Reference is made to them in documents which deal with storage facilities that were considered important for national defence. The most important agent, mustard gas, was stored in bulk both in 50-litre glass bottles placed in metal frames that were filled with Fuller's earth, and in 200-litre nickel drums. Various pumping and filling equipment was connected to these storage vessels. Military Factory VT-5 was the most modern of the military factories and had the largest capacity for the storage of bulk mustard gas.

Training facilities

Training was conducted on a regular basis at two battalions of Artillery Regiment 401: (*a*) Battalion 1/401 in Plumlov near Olomouc, Moravia; and (*b*) Military Reservation Jince near Battalion 2/401 in Strasice, Bohemia. An additional battalion, Battalion 3/401, was to have been deployed near Military Factory VT-5 in Slovakia. The primary emphasis at these facilities was on decontamination, which included filling all types of spraying equipment and rinsing, decontaminating and subsequently destroying the products of decontamination. Such destruction was carried out in a relatively primitive manner, the details of

[22] Conference on Disarmament document CD/1048, 24 Jan. 1991.

which are presented in the following section. In contrast to the post-war period, prior to World War II environmental standards were poor at military chemical training sites.

Both conventional military training and exercises using actual chemical warfare agents were carried out in fields that were normally used for agriculture but which were not heavily cultivated and whose soil was poor. In general, training was done in fields where there were no crops so that the owner did not need to be compensated for economic loss.

IV. Destruction and disposal of chemical weapons

Destruction and disposal activities

Czechoslovakia destroyed and disposed of chemical weapons long before it began production activities. The first such operation was conducted in Olomouc in April–May 1921 by the predecessor of the Military Chemical Institute. Approximately 1350 steel gas bottles filled with phosgene, the property of the former Austro-Hungarian Army, were transferred to Olomouc for destruction.

Many of the bottles were leaking, and therefore the phosgene was disposed of in the simplest way possible—by release into the air. The bottles were then buried on state-owned land near what was known as Bunker No. 8, and covered with dry lime and soil. The colour of the unripe grain changed, and the farmer who was renting the field asked that the military authorities make restitution for damage to the crop. However, after several weeks the plants again turned green, and the farmer withdrew his claim.

In February 1922 the Olomouc facility was given the task of emptying 20 000 lead-lined 15-cm, Model 99/9 gas shells, which had also belonged to the former Austro-Hungarian Army. The shells were filled with T-Stoff (xylylbromide), which was removed by pumping it into glass bottles that were then stored in Bunker No. 22 at nearby Lazce.

A method was developed for the destruction and disposal of mustard gas in which it was pumped out under vacuum, rinsed with petroleum and decontaminated with lime chloride. Remaining solution or slurry were poured into 1-meter-deep pits which had a thick layer of lime chloride on the bottom. Mustard gas that had been used in training was also disposed of in this fashion. In addition, all storage facilities were equipped with so-called 'destruction pits' for similar decontamination in an emergency.

German removal of chemical weapons from Czechoslovakia

No traces of chemical warfare agents or contaminated areas have been found in Czechoslovakia that date from before World War II. The reason is that after Germany completely occupied Czechoslovakia all military equipment was confiscated and taken to Germany. This was so even in Slovakia despite the fact that, unlike the Protectorate of Bohemia and Moravia, Slovakia was formally independent, although a vassal state. It is unclear whether military equipment was removed from Slovakia before or during the military actions that took place during suppression of the national uprising in 1944. Similarly, except for some fragmentary and unimportant documents, there is no relevant information in existing historical archives which relates to the activities of the Military Technical and Aviation Institute in Prague.

Old chemical weapons discovered after World War II

As mentioned above, no old chemical weapons from World War I or before World War II have been found which were manufactured in Czechoslovakia. Some old German chemical weapons from World War II have been discovered, but these had been used for training purposes.

Soviet troops were stationed in Czechoslovakia in 1968–91. No stored chemical weapons were found after the withdrawal of Soviet troops except for some training land mines filled with chloropicrin which were found at the so-called Frunze barracks in Olomouc.[23]

V. Conclusions

Prior to World War II Czechoslovakia was one of a number of small states which not only had built up a high level of chemical defence, but also possessed a small retaliatory capability focused primarily on meeting the threat of the potential use of chemical weapons by Nazi Germany.

The Czechoslovak CW capability was oriented primarily towards defence and therefore was limited to land mines, various types of spraying devices and equipment for surface contamination. The main chemical agent was mustard gas, and several tonnes of it were produced and stockpiled before Czechoslovakia was occupied by Germany. Czechoslovakia also possessed harassing agents (such as adamsite, chloroacetophenone and diphenylchloroarsine) which were used in two types of smoke candles.

With the occupation of Czechoslovakia and the onset of World War II plans for a large chemical weapon production facility were abandoned. During World War II Germany removed all chemical weapons and related production and storage equipment from Czechoslovakia. No remnants of chemical weapons from World War I, World War II or the period between the world wars have been discovered in Czechoslovakia except for small quantities of training agents and equipment. These date from German occupation in World War II and from 1968–91, when Soviet troops were stationed in Czechoslovakia.

[23] Masek, I., Matousek, J. and Mican, J., 'Lessons learned from the polluted environment of the former Soviet garrisons on the territory of Czechoslovakia', Paper presented at the International Conference on Community–Army –Technology–Environment, Brno, Sep. 1993.

8. The history of chemical weapons in Poland

ZYGFRYD WITKIEWICZ and KAZIMIERZ SZARSKI*

I. Introduction

Chemical warfare agents were used in Poland in World War I by the German Army against Russian troops. During World War II large-scale German manufacture of chemical weapons was carried out in Poland. Prior to World War II Poland conducted research on protection against chemical weapons and this work continued after the war. A number of chemical warfare agents were studied, and some of these were manufactured. However, the manufacture of chemical warfare agents in Poland was never sufficiently large to have military significance.

This chapter presents the information that is currently available about various periods when chemical weapons were of significance in Poland. It is not all-inclusive owing to the fact that many of the relevant documents were lost or destroyed during World War II.

II. The use of chemical weapons in World War I

The first use of chemical weapons took place on 27 October 1914 near Neuve-Chapelle, France, when German forces fired 3000 rounds of 105-mm Ni-Schrapnells at British troops. German authorities disclosed after the war that the British soldiers were not affected by the CW agent.[1]

In World War I the first large use of CW on the Eastern Front occurred in Poland when Germany used the CW agent T-stoff in shells for 150-mm calibre heavy howitzers on 31 January 1915. T-stoff was a mixture of brominated aromatic hydrocarbons (mainly xylyl bromide).[2] The Commander-in-Chief of the 9th German Army took the decision to use T-stoff at the Rawka River near Bolimow, between the villages of Borzymowka and Wola Szydlowiecka. German artillery opened fire on the Russian positions before dawn, and approximately 150 howitzers fired a total of 18 000 shells filled with T-stoff at the Russian troops. A total of approximately 72 tonnes of T-stoff were employed.[3] The weather that day was cold with temperatures lower than minus 20°C, and as a consequence the concentration of T-stoff vapour was so low that the Russian forces did not notice that the CW agent had been used.[4]

[1] Trumpener, U., 'The road to Ypres', *Journal of Modern History*, no. 3 (1975), p. 464.

[2] SIPRI, *The Problem of Chemical and Biological Weapons*, vol. 1, *The Rise of CB Weapons* (Almqvist & Wiksell: Stockholm, 1971), p. 28.

[3] SIPRI (note 2), p. 132.

[4] *Die Aufzeichnungen des Generalmajor Max Hoffman* [The papers of General-Major Max Hoffman], vol. 2 (Verlag für Kulturpolitik: Berlin, 1929), p. 93.

* The authors are grateful to Colonel Dr Zbigniew Makles of the Military Institute of Chemistry and Radiometry for his assistance.

The second use of chemical weapons on Polish territory took place on the night between 30 and 31 May 1915 on the Rawka River at almost the same location as the first use. This time 12 000 cylinders filled with 264 tonnes of chlorine were deployed along a 12-km front.[5] After almost three weeks of waiting for a favourable wind German forces fired the gas cylinders at the Russian troops. It was assumed that all of the Russian troops had been killed since they were known to have poor protection against the use of gas.

However, the Russian forces held a strong position near the village of Keszyce, between the Rawka and Bzura rivers. The position was opposite the German left flank, and fewer gas cylinders had been placed there than in the other sectors. As a result the German infantry attacking Keszyce met heavy fire. On the other hand, further south at Bolimow Russian casualties were high, and infantry regiments no. 53 and 54 of the Siberian Division of the 2nd Russian Army were completely wiped out. The advancing Russian reserves also suffered high casualties, and approximately two-thirds of them sustained gas injuries. There were 8934 Russian casualties from the German gas attack, of which 1101 were fatalities.[6]

On the night of 6–7 June 1915 German forces carried out a third chemical attack against Russian forces at the same location as the first and second such attacks. This time chlorine gas cylinders were discharged in unstable wind conditions, and when the wind changed direction the chlorine cloud moved towards the German trenches causing a number of German casualties.

III. The interwar period

On 17 June 1925 in Geneva, Poland signed the Protocol for the Prohibition of the Use in War of Asphyxiating, Poisonous or Other Gases, and of Bacteriological Methods of Warfare.[7] Poland ratified the Geneva Protocol on 4 February 1929 and reserved the right to retaliate in the event that chemical weapons were used against it first by another state.[8] The preparation for defence against CW agents that occurred in Poland in the interwar period focused on protecting the civilian population. The Gas Defence Service of the Polish Army was established, initially with French assistance, and was based on a similar French unit.[9] Poland later developed its own independent programme and, for example, was granted a patent on an improved method of phosgene production.[10]

The few records which exist about the preparation for protection against chemical weapons are all of a technical or organizational nature.[11] Anti-chemical protection was organized by the Gas Chemical Section of Department III of the Artillery and Armament Service, and a Gas Training School for officers and non-commissioned officers was also established. Soldiers trained in gas masks and protective clothing, and there were exercises in the identifica-

[5] SIPRI (note 2), p. 31.
[6] Hanslian, R., *Der chemische Krieg* [Chemical warfare] (E. S. Mittler: Berlin, 1937), p. 100.
[7] Goldblat, J., SIPRI, *Agreements for Arms Control* (Taylor & Francis: London, 1982), pp. 135–36.
[8] Elemming, M., *Miedzynarodowe prawo konfliktow zbrojnych* [The international law of armed conflicts] (Artekon: Warsaw, 1991).
[9] Orski, K., *Tajne arsenaly* [The secret arsenals] (MON: Warsaw, 1974), pp. 90–107.
[10] Orski (note 9).
[11] Krauze, M. and Nowak, I., *Wspolczesne Wojska Chemiczne* [Contemporary chemical forces] (MON: Warsaw, 1983), p. 34.

tion of chemical weapons and the decontamination of troops and equipment. The civilian population was given information about the properties of chemical weapons and the procedures for protection against them in educational campaigns organized from 1922 to 1927 by the Society for Anti-Gas Defence and after 1927 by the League for Anti-Aircraft and Anti-Gas Defence.[12] The Ministry of Military Affairs and the League for Anti-Aircraft and Anti-Gas Defence prepared joint publications on chemical warfare and protection against chemical weapons. Some of these were published in 1930–34 in the League for Anti-Aircraft and Anti-Gas Defence monthly *Biuletyn gazowy* (gas bulletin) and in 1935–39 in the monthly *Przeglad obrony przeciwlotniczej i przeciwchemicznej* (review of anti-aircraft and anti-gas defence).

IV. Research activities

Scientific research was conducted at the Anti-Gas Institute, an institute subordinate to the Ministry of Military Affairs, in Warsaw in 1928–39. Research was done on protection against chemical weapons and on the properties of CW agents such as fluoro-organic compounds, sulphur mustard and lewisite.[13]

There were various divisions at the Anti-Gas Institute including those which studied chemical synthesis, protective clothing, absorptive substances, signal flares, toxicology and decontamination. The founder and manager of the Institute, Colonel Engineer Jan Wojnicz-Sianozecki, was also a professor of organic chemistry at Warsaw Technical University. Dr Engineer Zenon Martynowicz was another leading figure at the institute.[14]

Investigation of chemical warfare agents was conducted in 1935–39 at the Technical University, where 2-fluoroethyl alcohol, derivatives of fluoroacetic acid and the methyl ester of iodacetic acid were studied.[15] The important discovery of the highly toxic properties of fluoroacetates was made by Polish scientists working in the field of military chemistry. Dr Engineer Adam Sporzynski, who had studied fluoroacetates at the Technical University before World War II, continued his work in Great Britain together with British scientists during the war.[16] After World War II Professor Gryszkiewicz-Trochimowski, another scientist who had been involved in the CW research at the Technical University, conducted research on chemical agents in France.[17]

[12] Orski (note 9).

[13] During the investigation of lewisite a large number of geraniums were grown at the Anti-Gas Institute for camouflage purposes. (Their scent is similar to that of lewisite.)

[14] Orski (note 9).

[15] Those involved in the work included Engineer Eugeniusz Gryszkiewicz-Trochimowski, Engineer Stefan Chwalinski, Engineer Warnecki, Engineer Sobieranski, Professor Anna Chrzaszczewska, Dr Engineer Adam Sporzynski, Professor Dr Florian Swiderek, Engineer Marian Sagajlo and Professor Dr Marceli Struszynski. See Orski (note 9); and Lohs, K., *Synthetische Gifte* [Synthetic poisons], 4th edn (Militärverlag der DDR: Berlin, 1974), p. 192.

[16] Orski (note 9); and Saunders, B. C., *Some Aspects of the Chemistry and Toxic Action of Organic Compounds Containing Phosphorus and Fluorine* (CUP: Cambridge, 1957).

[17] Information supplied to the authors by Dr Frédéric Guir, Centre d'Études du Bouchet, Vert-Le-Petit, France.

V. Manufacturing activities

Sulphur mustard was manufactured in Poland before World War II in the town of Pionki near Kielce, but the exact quantity produced is not known.[18] It was produced primarily for training purposes and was used in chemical land-mines in a mixture of 5–15 per cent sulphur mustard. According to some sources artillery shells and land-mines filled with sulphur mustard were manufactured in the town of Skarzysko-Kamienna.[19] After World War II approximately 80 such chemical land-mines were found in Poland;[20] all were made of steel and labelled 'WP MP 1938'. They were 29 cm high and 21 cm in diameter, with a volume of more than 8 litres, taking into account the small container which held the explosive component. A few empty chemical land-mines still exist.

According to information provided by workers at the German chemical testing facility at Munster, in September 1939 the German Army captured approximately 3000 Gelbkreuzminen (chemical land-mines) in Poland, each of which contained approximately 5 kg of sulphur mustard, for a total of approximately 15 tonnes of sulphur mustard. The chemical agent was transported to Munster and in 1952 was buried in the vicinity of Jag.[21]

There is a large discrepancy between the German data on the amount of sulphur mustard contained in the chemical land-mines and the actual amount present. As mentioned above, there is still no explanation of the origin of these land-mines. However, some chemical warfare agents were produced on a laboratory, semi-technical or technical scale at the towns of Rembertow and Zegrze and at Fort Bema near Warsaw.[22]

VI. World War II

The use of chemical weapons

War correspondents for both the BBC and British newspapers reported that on 3 September 1939 German aircraft dropped a number of bombs filled with sulphur mustard on a Warsaw suburb.[23] In 1942 German military authorities acknowledged that the bombing occurred, but claimed that it was accidental.[24]

General Herman Ochsner, who was responsible for CW matters in the German Army, reported that the Polish Army used sulphur mustard against the German Army on 8 September 1939 in the Battle of Jaslo.[25] Polish sources do not corroborate these claims. A report by Lieutenant-Colonel Wladyslaw Zietkiewicz, Commander-in-Chief of the 1st Border Guard Regiment, was pre-

[18] Orski (note 9).

[19] Orski (note 9).

[20] Private communication to the authors from Colonel Dr Zbigniew Makles, Military Institute of Chemistry and Radiometry, Warsaw.

[21] Interview with Dr Bernd Appler at SIPRI–Volkswagen Conference, 'The Challenge of Old Chemical Scrap-Munition and Toxic Armament Wastes', Munster, Germany, 18–21 Oct. 1993.

[22] Orski (note 9). Professor Tadeusz Urbanski was involved in these activities.

[23] 'Poland's gallant fight against odds', *The Times* (London), 6 Sep. 1939; and 'German use of poison gas', *The Times* (London), 21 Nov. 1939.

[24] 'Germany and gas warfare: Berlin framing reply', *The Times* (London), 14 May 1942.

[25] SIPRI (note 2), p. 154; and Brauch, H. G. and Müller, R.-D., *Chemische Kriegsführung, Chemische Abrüstung, Teil 1: Dokumente aus deutschen und amerikanischen Archiven* [Chemical warfare, chemical disarmament, vol. 1, documents from the German and American archives] (Arno Spitz: Berlin, 1985), p. 241.

sented to the Polish Investigating Commission in Paris on 1 December 1939.[26] Zietkiewicz's regiment, which defended Jaslo on 8 September 1939, had been speedily mobilized and consisted of only two battalions with no artillery and no detachments capable of taking the action described by Ochsner.

The 1st Border Guard Regiment was approximately 30 km from Jaslo at Gorlice when, at 10 a.m. on 7 September, Zietkiewicz received orders to defend Jaslo. Prior to this the regiment had been retreating while fighting German forces. Some parts of the regiment reached Jaslo on the night of 7 September, while others did not arrive there until the morning of 8 September. The Battle of Jaslo began early on the morning of 8 September, and at midday the regiment received orders to withdraw from the town.

Although Polish sources do not reveal any evidence of the intentional use of chemical weapons by the Polish Army, there are reports that CW land-mines were detonated during the Battle of Jaslo. The Polish Army used CW land-mines for training purposes, and it is possible that some of those land-mines were accidentally discharged, thereby releasing sulphur mustard.[27]

The manufacture of chemical weapons by Germany

During World War II chemical weapons were manufactured in Poland by Germany at an I. G. Farben factory in Dyhernfurth (now Brzeg Dolny) on the Oder River.[28] Construction of the factory, which was known as Anorgana AG, began in 1940, and it was built primarily by prisoners of war. Prisoners also worked in the factory during the CW manufacturing period,[29] and some were forced to take part in experiments where they were placed in gas chambers and exposed to the nerve agent tabun.[30]

The main product of the Dyhernfurth factory was tabun, and 100 tonnes per month (tpm) were produced. A total of approximately 12 000 tonnes of tabun were manufactured by the plant.[31] In addition, two facilities were built that were intended to produce 600 tpm of sarin,[32] and some sarin may have been produced on a pilot basis.[33] The goal was to produce 100 tpm of sarin by 1945.[34]

Sources note that both hydrogen cyanide and cyanogen chloride were probably manufactured in Dyhernfurth at the rate of 20 tpm.[35] Phosgene was also

[26] Zietkiewicz, W., *Uwagi i spostrzezenia dotyczace przygotowan i dzialan wojennych* [Remarks and impressions concerning preparations and military operations], Paris, 1 Dec. 1939 (archived at the Polish Institute and Sikorski Museum, London, section BI 61/c).

[27] Brauch and Müller (note 25).

[28] See also chapter 6 in this volume.

[29] Czernik, H., 'Filie obozu koncentracyjnego Gross-Rosen w Brzegu Dolnym (Dyhernfurth I i II)' [Branches of Gross-Rosen concentration camp at Dyhernfurth], in *Studia nad faszyzmem i zbrodniami hitlerowskimi* [Studies of fascism and Hitlerite crimes], vol. 1, Acta Universitatis Vratislavensis no. 222 (PWN: Wroclaw, 1974), p. 227; and Groehler, O., *Der lautlose Tod* [The silent death] (Verlag der Nation: Berlin, 1978), p. 10.

[30] Czernik (note 29).

[31] SIPRI (note 2), p. 282; Krauze, M. and Nowak, I., *Bron chemiczna* [Chemical weapons] (MON: Warsaw, 1985), p. 42; Lohs (note 15), p. 209; and Franke, S., *Lehrbuch der Militärchemie* [Handbook of military chemistry], vol. 1 (Deutscher Militärverlag: Berlin, 1967), pp. 297–98.

[32] Lohs (note 15).

[33] Franke (note 31); and Stohr, R. (ed.), *Chemische Kampfstoffe und Schutz von chemischen Kampfstoffe* [Chemical warfare agents and protection against chemical warfare agents] (Militärverlag der DDR: Berlin, 1985), p. 41.

[34] Krauze and Nowak (note 31).

[35] SIPRI (note 2), p. 282; and Krauze and Nowak (note 31).

likely produced at Auschwitz (Oswiecim) at the rate of about 100 tpm, mainly for industrial purposes.[36] Other sources claim that sarin, hydrogen cyanide and cyanogen chloride were not manufactured at Dyhernfurth and that phosgene was not produced at Auschwitz because construction of the necessary facilities was not completed.[37]

The factory at Dyhernfurth was captured undamaged by the Red Army on 26 January 1945. German forces retook the factory on 5 February 1945 for one day, destroyed some equipment and withdrew.[38] It was recaptured by Russian forces who removed the remaining documents, equipment and CW munitions.

Polish authorities did not assume control of the factory until 1946. They found a quantity of phosphorus and approximately 60 tonnes of D4 (N, N-dimethyl phosphoramidic dichloride) base that had been produced while the factory was under German control.[39] D4 is a precursor used in the manufacture of tabun. A number of Soviet soldiers were poisoned after entering the factory,[40] and 42 Polish workers were affected by residual tabun contamination in the three years immediately after the war. Eleven of them were hospitalized, and 8 suffered serious poisoning.[41]

VII. The post-World War II period

After World War II some chemical agents were manufactured for protective uses. They were used in laboratories and training in such areas as investigating analytical methods, training soldiers and evaluating the effectiveness of equipment for protection against CW agents. Nearly all of the chemical agents manufactured were used and destroyed, mainly by chemical decontamination. Currently, some chemical agents with properties similar to CW agents are used for training purposes in Poland.

VIII. The remaining CW stockpile

Adamsite

There are 9.326 tonnes of adamsite from World War II in Poland. The adamsite was discovered immediately after the end of the war, and it is presumed to have been imported from Germany. However, owing to the fact that so much time has passed and because the adamsite has had to be repackaged this is not certain. On the other hand, adamsite was not produced in Poland before World War II, and statements from a number of witnesses certify it to be of German origin. In any case, neither the reason for importing the adamsite to Poland nor the identity of those who imported it is currently known.

[36] SIPRI (note 2), p 282; and Krauze and Nowak (note 31).
[37] Appler (note 21).
[38] See statements by factory workers and General Hans von Ahlfen in von Ahlfen, H., *Der Kampf um Schlesien* [The battle of Silesia] (Grafe und Unzer Verlag: Munich, 1962), p. 127.
[39] Makles (note 20).
[40] Orski (note 9), p. 76.
[41] Wroniewicz, Z., 'Jak zostalem lekarzem przemyslowym' [How I became an industrial medical doctor], eds B. Nowakowski *et al.*, *Lekarze Przemyslowi* [The industrial medical doctors] (Zaklad Narodowy im. Ossolinskich, PAN: Wroclaw, 1965).

In the late 1940s the adamsite was transported from a storage area for items from World War II in Lodz to the town of Modlin, and in the mid-1950s it was moved to the town of Ostrow Mazowiecka-Komorowo, where it is now stored.[42] The adamsite is stored in a safe facility in 219 steel barrels, each of which contains 50 litres of adamsite sealed in a polythene bag.

The adamsite ought to be destroyed, but destruction is expensive and difficult. The main problem is that even after the burning and chemical conversion of adamsite the remaining products contain toxic arsenic. Poland does not possess a factory capable of destroying the adamsite, and it has therefore proposed that Germany destroy the adamsite in its specialized incinerator at Munster. However, the issue remains unresolved.

The problem of destroying the adamsite, which is a toxic waste, may be resolved with the help of the Organisation for the Prohibition of Chemical Weapons. The possibility of international cooperation in destroying old toxic chemicals is addressed in Annex 2 of the Chemical Weapons Convention which states: 'If the Abandoning State cannot be identified or is not a State Party, the Territorial State Party, in order to ensure the destruction of these abandoned CWs, may request the Organization and other States Parties to provide assistance in the destruction of these abandoned chemical weapons'.[43] The following is also of relevance as regards the adamsite that remains in Poland: 'A State Party shall treat old chemical weapons that have been confirmed by the Technical Secretariat as meeting the definition in Article II, paragraph 5(a), as toxic waste. It shall inform the Technical Secretariat of the steps being taken to destroy or otherwise dispose of such old chemical weapons as toxic waste in accordance with its national legislation'.[44]

Other chemical wastes

Various dangerous chemicals have been discovered and destroyed in Poland. In 1957 a 60-tonne cistern containing chloro-arsenic compounds was found at Jastrowiec on Poland's Baltic coast. These compounds can cause blistering, and the cistern and its contents were destroyed by blowing them up.[45] In 1960, 56 cylinders filled with hydrogen cyanide were found in Belzec (the location of a Nazi wartime concentration camp) in the south-eastern part of Poland near Przemysl. The cylinders, which were of German origin, were first blown up and the hydrogen cyanide was then burned.[46]

In 1990 in Wizna in north-eastern Poland two steel barrels were found that were filled with a mixture containing approximately 15 per cent sulphur mustard. They are presumed to be from World War II and the property of the former Soviet Army. Their contents were destroyed using calcium hypochlorite.[47]

[42] Adamsite is a homogeneous fine, crystalline, granular substance that has a green-yellow colour.

[43] Annex 2, Part IV(B), section B, paragraph 6 of the CWC in Conference on Disarmament document CD/1173, 3 Sep. 1992. The text of the CWC and a list of its annexes is reproduced in *SIPRI Yearbook 1993: World Armaments and Disarmament* (Oxford University Press: Oxford, 1993), appendix 14A, pp. 735–56.

[44] Annex 2, Part IV(B), section B, paragraph 6 of the CWC (note 43).

[45] Makles (note 20).

[46] Makles (note 20).

[47] Makles (note 20).

In 1991 Polish and Soviet inspectors carried out an inspection of two Soviet facilities on Polish territory for the Conference on Disarmament (CD). No chemical weapons were found. The first inspection was at a central artillery depot of the Soviet Army Forces North Group (SAFNG) where various types of artillery and rocket ammunition were stored. The facility consisted of an administrative area with offices and living quarters and another area where ammunition stores and bunkers were located. The second inspection was held at a central chemical forces depot of the SAFNG where nuclear, biological and chemical defence equipment was stored. That facility included an administrative area with offices and living quarters and a storage area where chemical stores and a chemical laboratory were located. The inspection is described in a 1991 CD report.[48]

IX. Chemical weapons in the Baltic Sea

There is no special department in Poland which investigates the dumping of CW ammunition in the Baltic Sea. However, the Navy Hydrographic Service, the Maritime Office in Gdynia and the military institutions have collected some data. There are five regions where chemical ammunition was sunk in the Polish Economic Zone of the Baltic Sea: (*a*) region no. 1, 10–26 nautical miles east of Bornholm, an area of about 820 km²; (*b*) region no. 2, 67 nautical miles north-west of Memel, an area of about 1370 km²; (*c*) region no. 3, 26 nautical miles north-east of Kolobrzeg; (*d*) region no. 4, 24 nautical miles north-east of Ustka; and (*e*) region no. 5, 10 nautical miles north-east of the Hel peninsula.[49]

Since World War II there have been 18 cases of yperite contamination of beaches on the Polish coast and of fishing boats in the central and southern Baltic Sea. These occurred mainly in the 1960s and the 1970s. In five cases bombs and artillery shells filled with sulphur mustard floated onto beaches near the coastal towns of Darlowek, Jastarnia, Kolobrzeg, Mielno and the Hel peninsula. In these incidents 126 persons were scalded, and approximately 30 of them required hospitalization. The number of injured was large, and most of these were children playing on the beaches, some of whom attempted to dismantle the munitions. In the 13 cases involving fishermen, who accidentally caught bombs or CW munitions in their nets, 37 people were contaminated.[50]

There have been no recent casualties involving CW munitions, but there have been reports of items that may have been munitions becoming caught in fishing nets. These items have usually been thrown back into the sea, often far from where they were caught or near the coast. The majority of these cases have not been reported to either the Maritime Office or the Navy Hydrographic Service.

Prior to 1993 there were no plans to study chemical ammunition dumped in the Baltic Sea near Poland. However, the Maritime Institute in Gdansk now intends to conduct a pilot study in the Gulf of Danzig.

[48] Poland and the USSR, Joint report on a trial inspection on request, Conference on Disarmament document CD/CW/WP.354, 6 Aug. 1991.

[49] Poland, National report on war gases and ammunition dumped in the Polish Economic Zone of the Baltic Sea, Paper presented at Baltic Marine Environment Protection Commission, Helsinki Commission, Ad Hoc Working Group on Dumped Chemical Munition (HELCOM CHEMU), Vilnius, Lithuania, HELCOM CHEMU 2/2/4, 17 Sep. 1993.

[50] Private communication to the authors from Col. Dr Roman Jozwik, Department of Military Foreign Affairs, Warsaw.

The most recent cases of injuries from CW munitions dumped in the Baltic Sea occurred in the late 1970s, and it can be assumed that owing to decomposition the potential danger to coastal regions from wartime chemical weapons has diminished. Presumably, there is not a great threat to the environment within the Polish Economic Zone from disintegrating CW agents. However, individual items retrieved from the sea and abandoned on coastal beaches do present risks. Consequently, guidelines are now being prepared for fishermen about the handling and reporting of dangerous objects.

Additional information about Poland's problems with CW munitions that have been dumped in the Baltic Sea was provided at meetings of the Baltic Marine Environment Protection Commission of the Helsinki Commission in St Petersburg (19–21 April 1993) and in Vilnius (28–30 September 1993).[51]

X. Conclusions

Following World War I considerable attention was paid in Poland to preparation for protection against chemical weapons. Such preparations were probably more extensive before World War II than after, owing to the fact that chemical weapons were not used in the war.

However, the use of chemical weapons in Poland was contemplated. In the autumn of 1944 when Russian forces reached the banks of the Vistula River prominent Nazis, such as Robert Ley, Martin Bormann and Joseph Goebbels, urged Adolf Hitler to sanction the use of chemical weapons—primarily tabun—against the Russian forces. Hitler was inclined to allow their use and believed that the Western powers would not object to the use of CW to stop the advancing Russian troops. The German generals were apprehensive about the consequences of the use of chemical weapons and the idea was abandoned.[52]

In Poland knowledge is rather limited about the use of chemical weapons on Polish territory in World War I, their manufacture by Germany in World War II and Polish preparations for protection against CW use before World War II.

The authors of this chapter believe this chapter to be the first attempt to collect information about chemical weapons in Poland. It is, of course, only a brief outline, and additional, more comprehensive investigation of the issues connected with this topic is needed. Knowledge about the history of chemical weapons in Poland may contribute to better understanding of the Chemical Weapons Convention and of the need for its ratification and early entry into force. Poland is an active participant in these efforts,[53] and a regional conference on the CWC was organized in Warsaw in December 1993.[54]

[51] Baltic Marine Environment Protection Commission, Helsinki Commission, Ad Hoc Working Group on Dumped Chemical Munition (HELCOM CHEMU), Report of the 1st Meeting, St Petersburg, Russia, 19–21 Apr. 1993; and Baltic Marine Environment Protection Commission, Helsinki Commission, Ad Hoc Working Group on Dumped Chemical Munition (HELCOM CHEMU), Report of the 2nd Meeting, Vilnius, Lithuania, 28–30 Sep. 1993.

[52] Speer, A., *Erinnerungen* [Memories] (Verlag Ullstein: Frankfurt am Main, 1973), p. 421; and *Trial of War Criminals*, vol. 7 (His Majesty's Stationery Office: London, 1947), p. 1045.

[53] Poland ratified the CWC on 23 Aug. 1995.

[54] The Regional Seminar on National Authority and National Implementation Measures for the Chemical Weapons Convention was held in Warsaw, Poland, on 7–8 Dec. 1993.

9. Chemical munitions in the Commonwealth of Independent States and the surrounding seas

JUDITH PERERA

I. Introduction

In 1987 the USSR announced it had stopped producing chemical weapons, and in 1989 it declared that its CW stockpile was 40 000 tonnes. Russia, the member of the Commonwealth of Independent States which assumed control of the Soviet CW stockpile, now faces the enormous task of destroying these stocks in accordance with the Chemical Weapons Convention. Responsibility for this and for all future planning rests with the Presidential Committee on Matters Pertaining to the Chemical and Biological Weapons Conventions, which was set up in April 1992. Some Russian scientists have disputed the official figure of 40 000 tonnes, claiming that a more accurate figure for total production is approximately 70 000 tonnes.[1] In addition, they have stated that far greater amounts of CW have been dumped at sea and buried or hastily burned at or near production facilities. Under Article III of the CWC states are not required to declare CW stocks which were buried on land before 1977 or dumped at sea before 1985. These dates were arrived at in closed discussions and are widely believed to have been chosen primarily to accommodate Moscow.

This chapter discusses CW stockpiles, production and storage facilities, the dumping of weapons and wastes, environmental contamination including major accidents, and the problems of destruction and clean-up.

II. Chemical weapon stockpiles

In the mid-1980s Western estimates of Soviet chemical weapon stocks varied from 30 000 to 700 000 tonnes.[2] These figures were calculated on the basis of what was known about production facilities and Soviet military strategy. It is possible that these figures diverged so widely because some calculations included weaponry as well as agents, but that explanation is dismissed by Krause and Mallory in *Chemical Weapons in Soviet Military Doctrine*. They ascribe the difference in estimates to poor Western intelligence.[3]

In April 1987 then Soviet President Mikhail Gorbachev announced that the USSR had ceased CW production. At the end of 1987 the Soviet Foreign Ministry declared that Soviet stockpiles amounted to 'no more than 50 000 metric tonnes of chemical warfare agents'.[4] It was subsequently confirmed that this figure referred to the chemical agents alone and not to the weapons which con-

[1] Interview with Dr Vil Mirzayanov and Dr Lev Fedorov, *Novoye Vremya*, 27 Oct. 1992, pp. 4–9.

[2] Perry Robinson, J. P., 'Review of CW armament', *Chemical Weapons Convention Bulletin*, no. 4 (May 1989), p. 19.

[3] Krause, J. and Mallory, C. K., *Chemical Weapons in Soviet Military Doctrine* (Westview Press: Boulder, Colo., 1992), p. 155.

[4] TASS news agency, 27 Dec. 1987.

tained them.[5] The Soviet representative to the Conference on Disarmament in Geneva said, in April 1988, that the figure included both bulk storage and agents in filled munitions.[6] By the end of 1989 the figure was lowered to 40 000 tonnes after the September signing of the Memorandum of Understanding between the USA and the USSR,[7] which specified data exchange on CW. In 1990 the head of the chemical troops quoted the 40 000-tonnes figure.[8] Its accuracy is now generally accepted, and it was confirmed by Anatoly Kuntsevich, former Chairman of the Presidential Committee on Matters Pertaining to the Chemical and Biological Weapons Conventions, in May 1993.[9]

The 40 000 tonnes include 32 200 tonnes of nerve gases (sarin, soman and the V-agent nerve gases) and 7700 tonnes of lewisite, mustard gas and mustard–lewisite mixtures.[10] No mention was made of Russian stocks of riot control agents and psychochemicals. However, CS (2-chlorobenzalmalononitrile) grenades were displayed as standard chemical weapons at the Shikhany test site in 1987,[11] and CS, CN (chloroacetophenone, another riot control agent, believed to have been used at Tblisi in 1989) and various psychochemicals were listed as standard CW agents in the late 1970s.[12] These are expected to be treated under the CWC in future, and declarations on stocks may then be required. Chemicals listed as being held in reserve in addition to declared stocks include hydrogen cyanide, phosgene, diphosgene, adamsite and chloroacetophenone.[13]

According to Dr Vil Mirzayanov, a Russian chemist who worked for 25 years at the main chemical weapon research facility in Moscow—the State Union Scientific Research Institute of Organic Chemistry and Technology (GosNIIOKhT)—the figure usually cited within the chemical weapon community was 60 000–70 000 tonnes.[14] Mirzayanov was dismissed from his job in January 1992 several months after he wrote an article in *Kuranty* warning about poor safety at CW plants.[15] He was subsequently arrested and jailed for two weeks in October 1992 after publication of a more detailed article, which he co-authored with another scientist, Dr Lev Fedorov, in the weekly *Moscow News*.[16] All charges against Mirzayanov were dropped in March 1994.[17]

[5] Letter from Soviet Deputy Foreign Minister Victor Karpov to the London *Times*, *The Times*, 18 Feb. 1989, p. 11.

[6] Perry Robinson (note 2), p. 19.

[7] The Agreement between the United States of America and the Union of Soviet Socialist Republics on Destruction and Non-Production of Chemical Weapons and on Measures to Facilitate the Multilateral Convention on Banning Chemical Weapons is reproduced in *SIPRI Yearbook 1991: World Armaments and Disarmament* (Oxford University Press: Oxford, 1991), appendix 14A, pp. 536–39.

[8] Interview with Head of Chemical Troops, Col.-Gen. S. V. Petrov, *Izvestia*, 7 Aug. 1990, p. 6.

[9] Kuntsevich, A., 'Problems of Russia in the destruction of chemical weapons', Proceedings of the First Moscow Conference on Chemical and Biological Disarmament, Demilitarization and Conversion (MOSCON 93), Moscow, May 1993.

[10] Interview with Igor Vlasov, press secretary of the Committee on Matters Pertaining to the Chemical and Biological Weapons Conventions, Mayak Radio, Moscow, 4 Feb. 1994.

[11] Conference on Disarmament document CD/789, 16 Dec. 1989.

[12] Pikalov, V. K., 'Otravliaiushchie veshchestva' [Toxic agents], *Sovietskaya Voyenaya Entsiklopediia* [Soviet military encyclopaedia], vol. 6 (Voyenizdat: Moscow, 1978) (in Russian).

[13] Pikalov (note 12).

[14] Interview with Mirzayanov and Fedorov (note 1).

[15] *Kuranty*, no. 190 (10 Oct. 1991).

[16] *Moscow News*, no. 39 (20 Sep. 1992), p. 9.

[17] Stock, T. and De Geer, A., 'Chemical and biological weapons: development and destruction', *SIPRI Yearbook 1995: Armaments, Disarmament and International Security* (Oxford University Press: Oxford, 1995), chapter 10, p. 339.

Table 9.1. Chemical weapon production from before World War II to 1987

Figures are in metric tonnes.

	First generation					Second generation			
Period	H	L	AC	CG	DM	GB	GD	V	Total
Pre-war	6 000–8 000	. .	. .	. .	. .	. .	. .	. .	6 000–8 000
1939–45	77 400	20 600	11 100	8 300	6 100	. .	. .	. .	123 500
1958–87	. .	. .	. .	. .	. .	12 000	5 000	15 500	32 500
Total	**83 400–85 400**	**20 600**	**11 100**	**8 300**	**6 100**	**12 000**	**5 000**	**15 500**	**162 000–164 000**

H = Mustard gas, L = Lewisite, AC = Hydrogen cyanide, CG = Phosgene, DM = Adamsite, GB = Sarin, GD = Soman, V = V-agent nerve gases.

Source: Fedorov, L., *Problems of Chemical Safety (Russia)*, issue UCS-INFO.6e, 22 Oct. 1996.

It has been suggested that as much as 20 000–30 000 tonnes of chemical weapons were hastily dumped or destroyed in the 1980s in the run up to chemical disarmament.[18] Wartime stocks were disposed of in an earlier period of large-scale dumping in the 1950s.[19] According to some estimates, a total of between 500 000 and one million tonnes of chemical agents were manufactured in the USSR, most of which were buried or dumped in the sea when their service life ended.[20] There is some evidence to support these allegations in the form of a 1989 official document from the Central Committee of the Communist Party of the Soviet Union (CPSU), which is now in the possession of *Moscow News*. The document is stamped 'top secret' and recommends the immediate dumping 'in the places of past burials' of old chemical weapons totalling 112 523 tonnes.[21] This is thought to refer to munitions filled with chemical agents.

More recent data based on the research of Fedorov[22] suggest that at least 164 000 tonnes of chemical agents were produced up to 1987. These included 131 500 tonnes of first-generation chemical weapons, such as mustard and lewisite (see table 9.1). These were mostly disposed of after World War II in three waves. First, between 1946 and 1948 some 60 000–65 000 tonnes were dumped at sea along with captured German chemicals. Second, between 1956 and 1962 some 10 000–12 000 tonnes of first-generation weapons were disposed of, also mainly by sea dumping. Finally, in the late 1980s excess chemical weapons were disposed of in order to justify the declaration of a total stockpile of 40 000 tonnes.[23] At least 47 000–54 000 tonnes were thus either incinerated, buried or dumped at sea in this period.

[18] *Novoye Vremya* (note 1).
[19] Krause and Mallory (note 3), p. 19.
[20] Interview with Interfax News Agency, 19 Jan. 1993.
[21] *Moscow News*, 24 Jan. 1993.
[22] Fedorov, L., *Problems of Chemical Safety (Russia)*, issue UCS-INFO.6e, 22 Oct. 1996.
[23] Fedorov (note 22).

III. Chemical weapon production

The research, development, production and stockpiling of chemical weapons peaked in World War II. According to German military intelligence sources there were 80 production facilities in the USSR, of which approximately 29 were confirmed as such.[24] German intelligence also identified 42 depots where weapons were stored,[25] more than 20 research facilities and 9 test sites,[26] a few of which continued to serve as CW facilities in later years.

The production facilities produced a range of chemical weapons including blister gases (mustard, nitrogen mustard, lewisite, dichloroformoxime, ethyldichloroarsine and methyldichloroarsine), blood gases (hydrogen cyanide, cyanogen chloride), choking gases (phosgene, diphosgene and chloropicrin) and irritants (diphenylchloroarsine, diphenylcyanoarsine, adamsite, chloroacetophenone and bromoacetone).[27] Well over one-third of the total was mustard gas, but the technology used until 1943 produced an unstable compound which deteriorated and needed replacing every five years.[28]

Some of these factories still function as fertilizer and industrial chemical plants, and after World War II a few continued as CW plants (see table 9.2). One of the main production centres was at Chapayevsk in Samara, where production began in 1928. In the 1930s and 1940s a range of chemicals was produced, including mustard and lewisite. Other key pre-war factories were located at Berezniki, Dzerzhinsk, Kemerovo and Stalingrad (Volgograd).[29]

After World War II the USSR focused its main CW efforts on getting rid of old stocks of weapons, reorganizing the CW establishment and investigating the new nerve agents, including the V-agents which had been discovered in the UK in the early 1950s.[30] The first publication on these agents appeared in the open literature in 1957 in Sweden.[31] This did not escape the notice of Soviet researchers.[32]

Many facilities which had been switched to chemical weapon R&D and production during the war reverted to civilian work after the war. Soviet CW production centred mainly on the Volga basin and, to some extent, in the Urals area (see table 9.3). The research headquarters remained in Moscow at GosNIIOKhT on Shosse Entusiastov, with branches at Volsk (Saratov) and Volgograd.[33] Production of mustard gas and lewisite continued at Dzerzhinsk for several years after World War II, using a facility brought to the USSR from Germany.[34] The plant has recently been dismantled.[35]

[24] Perry Robinson (note 2), p. 21.
[25] Krause and Mallory (note 3), p. 106.
[26] Krause and Mallory (note 3), pp. 180–83.
[27] Perry Robinson (note 2), p. 20.
[28] Krause and Mallory (note 3), p. 47.
[29] Krause and Mallory (note 3), p. 59.
[30] SIPRI, *The Problem of Chemical and Biological Warfare*, vol. 1, *The Rise of CB Weapons* (Almqvist & Wiksell: Stockholm, 1971), pp. 74–75.
[31] Tammelin, L. E., *Acta Chemica Scandinavica*, vol. 2 (1957), pp. 1340–49.
[32] Interview with Vladimir Uglev, *Novoye Vremya*, 5 Feb. 1993.
[33] Interview with Vladimir Uglev, *Moscow News*, 31 Jan. 1993.
[34] *Novoye Vremya* (note 1).
[35] *Nezavisimaya Gazeta*, 30 Oct. 1992, pp. 1–2.

Table 9.2. Main chemical weapon facilities during World War II

Site (region)	Production	Storage	Research	Test site
Alexin (Tula)	H, HN3, AC	. .	. .	. .
Asha (Chelyabinsk)	H, L, AC, CK, CG, DP, DM	. .	. .	. .
Berezniki (Perm)[a]	H, L, AC, DP, A, DA, DM	. .	X	. .
Usolye (Perm)	H, L, AC, DP, A, DA, DM	2000 tonnes	. .	. .
Chapayevsk (Samara)[b]	H, L, CG, DP, PS	X	. .	. .
Syzran (Samara)	H, L, CG, DP, PS	X	. .	. .
Dzerzhinsk (Gorky)[a]	H, HN3, AC, CK, CG DP, A	X	. .	. .
Gorokhovetsk[c] (Gorky)	H, HN3, AC, CK, CG DP, A	X	. .	X
Grozny (Chechen–Ingush Republic)	CK, PS, DA, DM	. .	. .	. .
Irkutsk (Irkutsk)	H, CG	X	X	. .
Lake Baikal (Irkutsk)	H, CG	. .	. .	X
Ivanovo (Ivanovo)	CG	. .	. .	. .
Kambarka (Udmurt Republic)[a]	. .	L	. .	. .
Kazan (Tatarstan)	H, CG	. .	. .	. .
Kemerovo (Kemerovo)[b]	H, L, AC, CG, DP, PS, DM, CN	. .	. .	. .
Kineshma (Ivanovo)	DM	. .	. .	. .
Kirovsk (Kola)	H	. .	. .	. .
Leningrad[d]	H, AC, CK	. .	X	. .
Luga	H, AC, CK	. .	. .	X
Magnitogorsk (Urals)	HN3, ?CX, AC, CK	. .	. .	. .
Moscow[e]	H, L, CG, DP, PS, DM	. .	X	X
Novomoskovsk (Tula)	H, L, AC, CG, DP, PS, DA, DC, DM, CN	. .	. .	. .
Shchelkovo (Moscow)	H, AC	. .	. .	. .
Tomsk (Tomsk)[f]	H, L, AC, CG, DP, DA, DM, CN	. .	. .	. .
Volgograd[a] [Stalingrad]	H, L, CG, DA, DC, DM	. .	. .	. .
Sarepta (Volgograd)	H, L, CG, DA, DC, DM	X	. .	. .
Volsk (Saratov)[a]	H, L, ED, MD	. .	X	. .
Shikhany[a] (Saratov)	H, L, ED, MD	X	. .	X
Voskresensk (Tula)	H, L, AC, CK, CG, DP	. .	. .	. .

X = Applicable for this site. Vesicants: A = Arsine, CX = Dichloroformoxime, ED = Ethyldichloroarsine, H = Mustard gas, HN3 = Nitrogen mustard, L = Lewisite, MD = Methyldichloroarsine; Blood gases: AC = Hydrogen cyanide, CK = Cyanogen chloride; Choking gases: CG = Phosgene, DP = Diphosgene, PS = Chloropicrin; Irritants: CN = Chloroacetophenone, DA = Diphenylchloroarsine, DC = Diphenylcyanoarsine, DM = Adamsite.

Table 9.2 *contd*

[a] Current CW facility.
[b] Now civilian chemical plant producing fertilizer, pesticide and dyes.
[c] Dzerzhinsk had 3 CW facilities, of which this was one.
[d] Leningrad had 2 CW factories, one of which relocated in World War II; research was conducted at the Technological Institute of Leningrad University.
[e] Moscow had 5 chemical weapon factories. The main factory was at Shosse Entusiastov, where research and testing were also done. At least 8 other research institutes, including the main chemical training centre of the Red Army, conducted CW activities.
[f] Factory was set up with US help in 1942–43.

Sources: Perry Robinson, J. P., 'Review of CW armament', *Chemical Weapons Convention Bulletin*, no. 4 (May 1989); Krause, J. and Mallory, C. K., *Chemical Weapons in Soviet Military Doctrine* (Westview Press: Boulder, Colo., 1992); and private communication with Dr Lev Fedorov.

In addition to CW agents, the USSR also produced riot control agents and herbicides for military use. Riot control agents were produced at Berezniki and Slavgorod (Altai–Siberia)[36] and at Volsk, where psychochemicals were also produced.[37] Herbicides for military use, similar to the Agent Orange used by the USA in Viet Nam, were made for a period in the 1960s at a chemical plant in Ufa (Bashkortostan).[38]

At the close of World War II the Soviet CW establishment was surprised to discover that Germany had been working on far more potent chemicals—the nerve gases sarin, soman and tabun. Soviet forces occupied what had been two German nerve gas factories on the Oder River and continued to operate one of them for a short while[39] before setting up their own pilot facility at Dzerzhinsk near Chernoreche, 30 km west of Gorky (Nizhny Novgorod), in a facility which had produced mustard and several other chemical weapons throughout the war. The plant at Volgograd was switched to the production of sarin and soman. The production of soman, the main product from the 1960s onwards, continued until 1987.[40] The Volgograd facility also produced a small pilot batch of a new highly toxic nerve agent known as 'Novichok'.[41]

A CW plant—production facility no. 3—was opened at the new factory city of Novocheboksarsk in Chuvashia in December 1972 to make V-gas. Production stopped in 1987, putting some 2500 people out of work. Most of them were assigned to other parts of the factory where dyes, lacquers and paints are produced. A few remained at the facility making antioxidants, chelates and plasticizers.[42] Uglev stated that the plant has also produced a binary V-gas.[43]

[36] *Moscow News*, 15 Nov. 1992.
[37] *Sobesednik*, no. 3 (Jan. 1993); and *Nezavisimaya Gazeta* (note 35).
[38] Interview with Lev Fedorov, *Moscow News*, 15 Nov. 1992; and *Nezavisimaya Gazeta* (note 35).
[39] Interview with Lev Fedorov (note 38), p. 12.
[40] *Nezavisimaya Gazeta* (note 35).
[41] *Novoye Vremya* (note 1); *Novoye Vremya* (note 32); and *Moscow News* (note 33). See also Stock, T., 'Chemical and biological weapons: developments and proliferation', *SIPRI Yearbook 1993: World Armaments and Disarmament* (Oxford University Press: Oxford, 1993), chapter 7, pp. 266–67.
[42] *Krasnaya Zvezda*, 24 Sep. 1992.
[43] *Novoye Vremya* (note 32).

Table 9.3. Postwar chemical weapon facilities

Figures are in metric tonnes.

Site (region)	Factory	Storage	Research	Testing
Baku (Azerbaijan)	. .	?	. .	. .
Berezniki (Perm)	H, PS	. .	. .	. .
Pochep (Bryansk)	. .	OPs (5 000–7 000) ?Novichok	. .	. .
Buyanka (Far East)	. .	?	. .	. .
Chapayevsk (Kuybyshev)[a]	. .	H, L	Training centre	. .
Dzerzhinsk (Gorky)	A, H, L, ?T	. .	. .	. .
Gorny (Saratov)[b]	. .	L (225), H (690, L/H (210)	. .	. .
Kambarka (Udmurtia)[b]	. .	H (175), L (6 400)	. .	. .
Maradykovsky (Kirov)	. .	OPs (5 000–7 000)	. .	. .
Kizner (Udmurtia)	. .	OPs (5 000–7 000)	. .	. .
Moscow[c]	H, OPs (small-scale)	. .	Main CW research centre (Novichok)	Lab-scale (Novichok)
Novocheboksarsk[b] (Chuvashia)	V-gas, CS, V-(bin)	. .	. .	. .
Leonidovka (Penza)	. .	OPs (5 000–7 000)	. .	. .
Shchuchye (Kurgan)	. .	OPs (5 000–7 000)	. .	. .
Shikhany-2 [Volsk-18] (Saratov)	. .	DM, various	. .	Main test site for Agent Orange and Novichok
Slavgorod (Altai)	Riot gas	. .	. .	. .
Ufa (Bashkortostan)	Agent Orange	. .	. .	. .
Ustyurt (Uzbekistan)	. .	. .	. .	Test of Novichok
Volgograd	Sarin, soman, Novichok	. .	Branch of Moscow centre	. .
Volsk-17 [Shikhany-1] (Saratov)	?BZ, riot gas (CS, CN, others), OPs	. .	Branches of Moscow centre	. .

OPs = Organophosphorus compounds/nerve gases: Novichok (new highly toxic OP), sarin, soman, tabun, V-(bin) = Binary chemical, V-gas; Vesicants: A = Arsine, H = Mustard gas, H/L = Mustard and lewisite mixture, L = Lewisite; Irritants/riot control agents: CN = Chloroacetophenone, CS = 2-chlorobenzalmalononitrile, others; Other agents: Agent Orange = 2,4,5-T (2,4,5-trichlorophenoxyacetic acid and 2,4-D (2,4-dichlorophenoxyacetic acid) mixture (herbicide), BZ = 3-quinuclidinyl benzilate (psychochemical), PS = Chloropicrin (choking gas).

Table 9.3 *contd*

[a] CW destruction facility built. Closed because of local protest and now training centre.

[b] Site proposed for new destruction facility.

[c] State Union Scientific Research Institute of Organic Chemistry and Technology at Shosse Entusiastov.

Sources: Collated from Russian press reports in *Izvestia, Komsomolskaya Pravda, Krasnaya Zvezda, Kuranty, Moscow News, Nezavisimaya Gazeta, Novosti, Novoye Vremya, Rossiyskiye Vesti, Sobesednik* and *Soviet Weekly* and from the Interfax News Agency, ITAR-TASS, Mayak Radio, Russia Radio, Soviet television and TASS.

IV. Storage depots

Russia has declared that it possesses seven storage depots for chemical weapons. At Kambarka in Udmurtia there are 6400 tonnes of lewisite, and at Gorny in Saratov there are 1125 tonnes of mustard and lewisite.[44] In August 1990 US experts visited the Kambarka site where the liquid lewisite is stored in 80 huge containers made of 11-mm thick steel.[45] Nerve agents are stored at Kizner (Udmurtia), Shchuchye near Kurgan, Maradykovsky (Kirov), Leonidovka (Penza) and Pochep (Bryansk).[46] The nerve agents are all stored in munitions, and some of the first-generation chemicals are stored in bulk (see table 9.4). The nerve agents are stored in missile artillery munitions at two sites and in aviation munitions at three sites, while the mustard and lewisite are mainly at two sites belonging to the chemical forces (see table 9.5).[47]

Chemical weapon storage depots are also said to exist at Baku (Azerbaijan),[48] and at Buyanka, 100 km from the Chinese border.[49] It is possible that mustard gas and lewisite, which had been previously stored at Chapayevsk, were moved to the Buyanka depot when it was converted to a fertilizer plant in the 1960s.[50] Small stocks of Novichok were also reportedly stored in Bryansk.[51] It is possible that some VX may be stored at a depot in the Transbaikal area. (Fedorov refers to the dangers of transporting VX from Transbaikal to Bryansk.[52])

The main CW test site is at Shikhany (Saratov), and small stocks of chemical weapons are probably maintained there, including 3400 tonnes of irritants. Some 1200 tonnes of mustard were incinerated there in the 1960s.[53] Tests using Novichok were reportedly conducted as recently as early 1992 on the Ustyurt plateau near Nukus in Uzbekistan.[54]

[44] Sobesednik (note 37).

[45] Soviet television, 27 Aug. 1990; and *Wall Street Journal Europe*, 25 Feb. 1993.

[46] Interview with Igor Vlaslov (note 10).

[47] Interview with Alexander Gorbovsky, Department Chief of Committee on Matters Pertaining to the Chemical and Biological Weapons Conventions, *Nezavisimaya Gazeta*, 27 Jan. 1994.

[48] *Novoye Vremya* (note 1).

[49] *Economist Foreign Report*, 12 July 1984, pp. 5–6.

[50] *Novoye Vremya* (note 1).

[51] *Novoye Vremya* (note 32).

[52] *Moscow News*, 3 Jan. 1993.

[53] Fedorov (note 22).

[54] Interview with Vil Mirzayanov, *Baltimore Sun*, 16 Sep. 1992; and *Novoye Vremya* (note 1).

Table 9.4. Distribution of chemical weapon stockpile by storage mode

Figures are percentages.

Chemical agent type	In munitions and devices	In bulk containers
VX	100	. .
Sarin (GB)	100	. .
Soman (GD)	100	. .
Mustard	. .	100
Lewisite	2	98
Mustard–lewisite mixtures	40	60
Phosgene	100	. .

Source: *Special Federal Programme on the Destruction of Chemical Weapons Stockpiles in the Russian Federation*, Government Decree no. 305, 21 Mar. 1996.

V. Environmental contamination on land

The Russian chemical weapon programme left a dangerous legacy of environmental pollution as a result of accidents, poor safety procedures at production plants and the disposal and dumping of wastes and old weapons. The full extent of contamination at the main production facilities is not known, but some details have begun to emerge.

At Chapayevsk, for instance, the technology used was primitive. Throughout World War II shells were filled by pouring chemicals into them from a kettle, and closed technology was not introduced until the 1950s.[55] Most local people have lost a relative who died working there—there are thousands in the local graveyards. 'During the war young men unfit for the frontline and girls arrived by the trainload—and just as quickly became ill and died', says Fedorov.[56]

After the war, when the plant was converted to produce fertilizers and other chemicals, the mustard gas and lewisite which were not transferred to the Far East were taken out and buried close to the nearby village of Pokrovka. Approximately 1200 tonnes were simply poured into trenches, covered with bleach or lime and buried. Villagers tell of the ground heaving and releasing heat in the 1980s.[57] In 1992 Fedorov asked to take samples from the site, but he was refused permission by chemical troops head Colonel-General Stanislav Petrov.[58] No reason for the refusal was given. According to Fedorov, the local people have gained little from the conversion of the plant, which now spews out dioxins as a result of its production of pesticides.[59] He says that only 8 per cent of the children in the area are born completely normal. Ecological tests conducted in 1993–94 showed arsenic concentrations in the soil at the plant to be 8500 times the maximum permissible concentration. In areas of the town surrounding the plant concentrations are 2 to 10 times the maximum permis-

[55] *Novoye Vremya* (note 1).
[56] *Izvestia*, 3 Dec. 1992, p. 3.
[57] *Izvestia* (note 56); *Sobesednik* (note 37); and *Novoye Vremya* (note 1).
[58] *Novoye Vremya* (note 1).
[59] *Sobesednik* (note 37).

Table 9.5. Russian storage depots

Figures are in metric tonnes.

Site	Authority	Total volume (%)	V-gas	GB	GD	H	L	H/L	CG
Kambarka (Udmurtia)	CF	15.9	. .	. .	. .	. .	6400 tonnes	. .	. .
Gorny (Saratov)	CF	2.9	. .	. .	. .	Total of 1125 tonnes for H, L and H/L			. .
Kizner (Udmurtia)	GRAU	14.2	X	X	X	. .	X	. .	. .
Maradykov-sky, (Kirov)	VVS	17.4	X	X	X	. .	. .	. .	. .
Pochep (Bryansk)	VVS	18.8	X	X	X	. .	. .	. .	. .
Leonidovka (Penza)	VVS	17.2	X	X	X	. .	. .	. .	. .
Shchuchye (Kurgan)	GRAU	13.6	X	X	X	. .	. .	. .	X

X = Applicable for this site. V-gas = V-agent nerve gases, GB = Sarin, GD = Soman, H = Mustard gas, L = Lewisite, H/L = Mustard–lewisite mixture, CG = Phosgene, CF = Chemical forces, GRAU = Main missile and artillery administration, VVS = Air force.

Sources: Fedorov, L. *Problems of Chemical Safety (Russia)*, issue UCS-INFO.6e, 22 Oct. 1996; and *Special Federal Programme on the Destruction of Chemical Weapons Stockpiles in the Russian Federation*, Government Decree no. 305, 21 Mar. 1996.

sible concentration. Toxic products of the hydrolysis of lewisite are still found in the soil at the plant.[60]

The wartime conditions were similar at Dzerzhinsk, which also did not adopt closed technology until the 1950s. Large amounts of mustard are believed to be buried on site, and wastes were dumped into the Oka River and pumped underground untreated.[61] Leaking weapons were taken to the nearby 'air station' at Kotelnikovo where they were burnt in the open air, leaving the surrounding forest 'rust coloured', according to Fedorov.[62] He has said that the soil at and around the plant 'must be contaminated with arsenic compounds'.[63]

The chemical combine at Volgograd is also responsible for considerable pollution. Measures to decontaminate wastes from shop 34, which produced sarin and soman, were inadequate at times, according to Fedorov and Mirzayanov. The concentration of these compounds in the effluent was sometimes hundreds of times the permitted maximum.[64] The plant stored its waste in the open in a large toxic lake. In the spring of 1964 the Volga River turned white from Volgograd to Astrakhan with dead fish floating in it after flooding caused the toxic lake to overflow into the Volga. The disaster was not explained at the time.

60 Fedorov (note 22); and *Segodnya*, 10 Aug. 1994.
61 *Nezavisimaya Gazeta* (note 35).
62 *Novoye Vremya* (note 1).
63 *Nezavisimaya Gazeta* (note 35).
64 *Novoye Vremya* (note 1); and *Nezavisimaya Gazeta* (note 35).

Ten years later there was a fire at the VX plant in Novocheboksarsk, higher up the Volga, and shop 83 was almost destroyed. Official reports noted that several firemen were poisoned.[65] The full effect of the accident and the number of casualties were never made public, but according to a report by the Moscow Institute of Biophysics, tens of tonnes of toxic material were dispersed over a 30-km radius. There is a high incidence of mental retardation among children born in the contaminated zone, and a commission has been set up to investigate the consequences of the accident.[66] Officials admit that there have also been 'several unpleasant incidents connected with disturbances to the manufacturing cycle' but say these had no serious consequences.[67]

No full-scale environmental impact assessment has been conducted, but there is clearly a perceived problem. During production a 2-km safety zone was established around the site. In mid-1994, seven years after production stopped, this was extended to 4 kilometres.[68]

At parliamentary hearings on chemical weapons in May 1996, representatives of the Chuvash Republic reported that in recent years 82 people had died who formerly worked at the plant. Their average age was 49. The main causes of death were cancer and cardiovascular disease. Novocheboksarsk has been declared a zone of ecological disaster.[69]

As a result of herbicide production Ufa is so heavily contaminated with dioxins that water has to be brought in by truck every year during the spring floods, and local people climb the nearby hills in search of relatively clean water from springs.[70] The situation became critical in March 1990 when wastes from the herbicide factory—mainly dioxins and phenols—leaked into a reservoir which supplies water to the town, leaving its one million inhabitants without drinking water.[71] Massive protests followed with requests that the town be declared an ecological disaster area.[72] In August 1990 an explosion and fire at the plant released dioxins and phenols into the Ufa River. Phenol concentrations were reported to be 37.5 times permissible levels.[73] A decision taken at that time to end herbicide production at the plant was never put into effect.[74] Water samples analysed in 1992 by the M.M. Shemyakin Biochemistry Institute of the Russian Academy of Sciences reportedly found dioxin levels to be 15.5 times the permitted maximum.[75]

Mirzayanov and Fedorov have expressed particular concern about contamination at GosNIIOKhT's Moscow centre and the danger the institute may pose to Moscow residents. Mirzayanov says that there are no filters on the ventilation systems. Decontamination procedures are inadequate, and wastes are stored in barrels in the open air.[76] Formal systems for monitoring emissions are totally ineffective, according to Mirzayanov. 'All plants have such a system. We know

[65] *Krasnaya Zvezda* (note 42).
[66] *Moscow News*, 12 Mar. 1993.
[67] *Krasnaya Zvezda* (note 42).
[68] Fedorov (note 22).
[69] *Problems of Chemical Safety (Russia)*, issue UCS-INFO.1e, 10 June 1996.
[70] *Moscow News*, 17 May 1992.
[71] *Novosti*, 14 May 1990.
[72] *Moscow Radio*, 9 Apr. 1990.
[73] *Moscow News*, 1 Sep. 1990.
[74] *Moscow News* (note 70).
[75] *Moscow News* (note 70).
[76] *Moscow News* (note 16).

they discharge thousands of tonnes of waste into the atmosphere, but laboratory analysis shows the reverse. This will go on until they are caught red-handed by monitoring agencies with a real interest in protecting the environment'.[77] He says the decontamination department destroyed toxic substances in reactors which have no filters: 'Moscow's air was being constantly polluted'.

In February 1980 fire-fighters rushed to put out a blaze at GosNIIOKhT which had been noticed by passers-by. Guards tried to stop them, but they went in anyway. The fire had started in an incubator where tests were being run involving 400 grams of nerve gas. 'The water went all over the fire-fighters and they were enveloped in smoke', says Mirzayanov. 'No-one knows what became of them or whether they are alive today. I am pleased if they are—just one gram of that substance is enough to kill 100 people. The smoke also went all over Moscow and nobody bothered to find out where it had settled'.[78]

In the past GosNIIOKhT worked on mustard gas and ran a production unit during the war. In 1941 'in face of the German attack mustard gas and filled shells were hastily buried on the site'.[79] Large amounts were simply poured into holes in the ground. 'The entire territory is contaminated and substances are certainly filtering through the soil and getting into the water that everyone uses', says Mirzayanov.[80] Chromatographic analysis of water in Block 7, taken from an artesian well on the site, revealed a range of sulphur and chlorine-containing compounds similar to mustard gas. 'The institute's leadership is well aware of this which is why any excavation on the territory is carried out like a combat operation and with the use of gas masks', according to Mirzayanov and Fedorov.[81]

At the heart of the Volga facilities is the main CW test site at Shikhany, 100 km north of Saratov. Safety procedures are no better there. Wastes from GosNIIOKhT in Moscow are taken to Shikhany and buried in open pits,[82] and old chemical weapons, including adamsite, have reportedly been dumped in the Kiselevsk ravine on the site.[83]

VI. Dumping at sea

The USSR took part in the postwar international operation to dispose of Germany's chemical weapons. According to official figures, captured German CW munitions totalled 311 200 tonnes, of which the USSR was responsible for disposing of 70 500 tonnes.[84] Some of these munitions were destroyed at storage sites or at special areas, and some of the chemical agents were used in the civil chemical industry. However, approximately 60 per cent were dumped at two sites in the Baltic Sea. These munitions contained approximately 12 000 tonnes

[77] *Kuranty*, 23 Jan. 1993.
[78] *Kuranty* (note 77).
[79] *Novoye Vremya* (note 1).
[80] *Kuranty* (note 77).
[81] *Moscow News* (note 16).
[82] *Moscow News* (note 16).
[83] *Moscow News* (note 52).
[84] Russian Federation, National report of the Russian Federation: complex analysis of the hazard related to the captured German chemical weapon dumped in the Baltic Sea, Paper presented at Baltic Marine Environment Protection Commission, Helsinki Commission, Ad Hoc Working Group on Dumped Chemical Munition (HELCOM CHEMU), Vilnius, Lithuania, HELCOM CHEMU 2/2/1Rev.1, 27 Sep. 1993.

of chemical agents—60 per cent mustard gas, 20 per cent arsenic-containing substances and 20 per cent irritants.[85]

There are reports that the total amount of chemical weapons dumped at sea was greater than the official figures of 300 000 tonnes dumped in the Baltic Sea and 200 000 tonnes disposed of in the Skagerrak. In addition, reports suggest that after 1947 some old Soviet chemical weapons were dumped at other unauthorized sites in the Baltic and in the seas between Norway and Iceland.[86] The USSR also dumped some of its old mustard gas and lewisite weapons in the Pacific Ocean and the Arctic Ocean in the late 1940s.[87] Fedorov says there are a total of 12 such dumping sites, including three in the Baltic.[88]

Munitions filled with mustard gas and lewisite were dumped by the USSR in the White Sea (Beloye More) in the Arctic in the 1950s in ships leaving from Severodvinsk.[89] Leakage from these weapons is thought to be responsible for the mass death of starfish and other marine animals which occurred in 1990. In May and June 1990 six million dead starfish were washed ashore in the Dvinskaya inlet of the White Sea together with 7500 crabs, 30 seals and 10 belugas.[90] An official investigation failed to find the cause of their death, and the only conclusion was that a toxic substance containing sulphides was responsible.[91] However, scientists in Archangelsk reported that the initial analysis had identified mustard gas as the cause.[92]

In addition, the contents of approximately 50 trains, each with 60 railway cars of CW aerial bombs, were dumped in the Barents Sea off Murmansk (by ships which departed from the port of Linakhamari) in the early 1960s.[93] The bombs were reportedly loaded at a station near Penza,[94] which suggests that they probably came from the Chapayevsk facility or its depot at Syzran. The ports of Tallinn and Paldiski (Estonia) were used for Baltic dumping, and the ports of Posyet and Nakhodka were used for Pacific operations.[95]

Finally, there is the CPSU document mentioned above which suggests that a further 112 523 tonnes of old chemical weapons produced in 1952–64 may have been disposed of at sea in 1989–90 at the sites of previous dumping.[96] Fedorov estimates that well over 500 000 tonnes of chemical weapons have been buried or dumped and need to be recovered and properly destroyed. He says that clean-up operations are needed in Chuvashia, Gorky, Ivanovo, Moscow, Perm, Samara (Kuybyshev) and Volgograd as well as at the sites of dumping at sea.[97]

[85] Russian Federation (note 84).
[86] *Svenska Dagbladet*, 23 Mar. 1992; see also chapters 19 and 20 in this volume.
[87] *Novoye Vremya* (note 1).
[88] Interview with Interfax News Agency (note 20).
[89] *Novoye Vremya* (note 1).
[90] TASS news agency, 18 July 1990.
[91] Sosnov, A., 'Toxic wastes kills crabs, starfish', *Science in the USSR*, May/June 1991, p. 38.
[92] Private communication with the author, Nov. 1992.
[93] *Izvestia*, 3 Dec. 1992.
[94] *Komsomolskaya Pravda*, 8 June 1991.
[95] Fedorov (note 22).
[96] *Moscow News* (note 21).
[97] Interview with Interfax News Agency (note 20).

VII. Chemical weapon destruction

Russia faces serious problems in winning public acceptance of its programme for the destruction of its 40 000 tonnes of chemical weapons. Only a mobile complex for the destruction of faulty chemical munitions (KUASI) functions, and thus far it has destroyed over 200 tonnes of damaged weapons.[98] The first and only 'purpose-built' demolition facility is located 12 km from Chapayevsk. It was prevented from beginning operation in 1989 because of public protests, including rallies, picketing and a petition which collected 60 000 signatures organized by the local 'Initiative' ecology movement.[99]

The Chapayevsk facility cost 50 million roubles and took three years to build. Chapayevsk was one of nine sites considered for the facility.[100] It was designed to process 350 tonnes a year of nerve gas, which was to be transported there by rail. The technology involved was to include both hydrolysis and incineration, with solid wastes being vitrified and buried and liquid wastes cleaned and discharged.[101] The plans were cleared with the local authorities, but the public was never informed.[102]

Three government commissions visited the plant at the request of the local authorities and found it safe, provided no serious accident occurred. However, the local council voted against its operation, and in September 1989 the government agreed to convert it into a training facility for CW destruction.[103] According to Mirzayanov, GosNIIOKhT's design for the facility included no filters on the ventilation system.[104]

The government appears to have learned some lessons from the Chapayevsk experience. All subsequent proposals and plans for CW destruction have been accompanied by promises to improve the infrastructure and social conditions of the areas selected for new facilities and by statements that nothing will be done without local public approval. President Boris Yeltsin issued a decree on 12 June 1992 concerning immediate measures to prepare for CW destruction. It promised benefits for workers involved in CW disposal: social facilities and medical centres to be built within 15 km of any CW disposal facility and compulsory state insurance. According to Kuntsevich, at least 15 per cent of the overall cost of destruction will be spent on social support.[105]

The Special Federal Programme on the Destruction of Chemical Weapons Stockpiles in the Russian Federation was finally adopted in 1995 by government decree and approved on 21 March 1996 after several years of discussion and debate.[106] The programme is expected to cost 16 600 billion roubles over

[98] USSR, Complex for the destruction of faulty chemical munitions (KUASI), Conference on Disarmament document CD/CW/WP.369, 8 Oct. 1991.

[99] *Izvestia*, 14 June 1989; and *Izvestia*, 30 Aug. 1989.

[100] Solovyev, Col. V., 'History of the establishment of the facility for destruction of chemical weapons in Chapayevsk', Paper presented at the International Conference on Banning and Destruction of Chemical Weapons, Moscow, 9–11 Apr. 1991.

[101] *Kuranty*, 15 Feb. 1994.

[102] Solovyev (note 100).

[103] *Soviet Weekly*, 16 Sep. 1989.

[104] *Moscow News*, 20 Sep. 1992.

[105] *Izvestia*, 23 Sep. 1992.

[106] *Special Federal Programme on the Destruction of Chemical Weapons Stockpiles in the Russian Federation*, Government Decree no. 305, 21 Mar. 1996.

Table 9.6. Proposed deadlines for destruction of the Russian chemical weapon stockpile

Figures are in thousand metric tonnes.

Year	Quantity to be destroyed
1998	0.11
1999	0.31
2000	2.40
2001	6.09
2002	9.19
2003	10.20
2004	8.03
2005	3.67

Source: *Special Federal Programme on the Destruction of Chemical Weapons Stockpiles in the Russian Federation*, Government Decree no. 305, 21 Mar. 1996.

15 years and will be financed mainly from the budget. Some 1700 billion roubles should be recouped through the commercial use of the products of destruction (such as pure arsenic).

The programme of destruction extends from 1995 to 2009. Phase 1 will involve the destruction of blister agents stored in bulk (at Gorny and Kambarka), and Phase 2 will be the destruction of filled munitions. The weapons will be destroyed in the regions where they are stored. Actual destruction is to begin in 1998 (see table 9.6).

Testing of a destruction facility at Gorny will begin in 1998, and destruction will take place from 1999 to 2002. Testing of a destruction facility at Kambarka will start in 2000, with destruction taking place from 2001 to 2005. The programme is less specific about the destruction of the filled munitions which is to take place at the same time and to be completed by 2005. The following four years will be spent in decommissioning the destruction facilities and cleaning up the sites.

However, the programme is already facing delays because of financial difficulties and opposition from some local authorities. Some foreign aid has been forthcoming from Germany for the Gorny and Kambarka facilities and from the Netherlands and Sweden for Kambarka, while the USA has agree to help with the facility at Shchuchye.[107]

Despite parliamentary approval at federal levels, there was considerable resistance in Cheboksary, the Chuvash capital near Novocheboksarsk, which was chosen as the site for one of several destruction facilities. It was proposed initially that 3.5 million CW projectiles, which are now stored in Kizner (Udmurtia) and Shchuchye (Kurgan), be transferred to Cheboksary.[108] The site was chosen for nerve gas disposal for several reasons. It has a pool of experts from the period when VX was made there. Some of the equipment which was formerly used to make VX could be used for nerve gas disposal; at present it is standing idle and costing 135 million roubles a year to maintain.[109] Moreover,

[107] Private communication with Sergey Kisilev, Deputy Representative of Russia at the Organisation for the Prohibition of Chemical Weapons, Oct. 1996.

[108] *Izvestia* (note 105).

[109] *Krasnaya Zvezda* (note 42).

the technology would be familiar as the factory already has a small destruction facility which was used for destroying old and faulty weapons.[110] There are also approximately 200 employees with experience of this technology.[111] However, this plan has now been dropped, and a special facility will be built at Kizner.

Opponents argued that transporting old chemical weapons by rail to the plant would be dangerous, especially as the railway is in need of repair. Moreover, the expert personnel are no longer employed there—most of them have retired or are severely ill.[112] In December 1992 the Chuvash Supreme Soviet ruled against allowing the destruction facility,[113] and the following month Tatarstan refused to permit any chemical weapons to be transported through its territory,[114] effectively blocking the routes from Udmurtia and Kurgan.

Aware of the growing problems, President Yeltsin issued a statement on 20 April 1993 in which he appealed to Chuvashia, Saratov and Udmurtia to join in elaboration of the CW destruction programme. He repeated that a considerable part of the funds set aside for the programme would be spent on health, housing, culture and infrastructure. He said the programme would be based on four principles: unconditional provision of protection for the population and environment; unconditional social protection; the use of advanced technologies to minimize risk; and minimal transport of toxic agents across Russian territory. He further added that destruction would take place only after the State Committee on Ecology had approved both the programme as a whole and each proposed facility.

The government has now included measures in the destruction programme to ensure environmental protection for the regions involved and additional health monitoring. In line with Yeltsin's assurances, the regions are also to get additional infrastructure including utilities and roads and development of social facilities.

Russia is keen to minimize cost by phasing the destruction of its chemical weapons and by converting existing facilities to civil use wherever possible. Conversion to civil production has begun at Novocheboksarsk and Volgograd. At Volgograd alone this is estimated to have saved Russia 42.3 billion roubles.[115]

VIII. Conclusions

Russia must destroy 40 000 tonnes of chemical weapons from its CW programme and dismantle a network of facilities. In addition, vast areas of its territory and seas have been left severely contaminated by the CW programme. There is a need for detailed investigations to properly assess the full extent of this problem before any plans for clean-up operations can be drawn up. The Volga basin is the worst affected area, and it has been selected as the site of new facilities for destruction of the 40 000 tonnes of chemical weapons.

[110] *Rossiyskiye Vesti*, 22 Sep. 1992.
[111] *Krasnaya Zvezda* (note 42).
[112] *Moscow News*, 12 Mar. 1993.
[113] *Moscow News* (note 112).
[114] Interview with Interfax News Agency (note 20).
[115] Ivanov, V. P. and Semenov, V. V., 'Prospects for chemical weapons industry conversion in Russia', Proceedings of the First Moscow Conference on Chemical and Biological Disarmament, Demilitarization and Conversion (MOSCON 93), May 1993.

Decades of secrecy have led to a crisis of confidence, and local opposition to these plans can be expected. In 1992 and 1993 the Russian Government undertook a major investigation of nuclear waste dumping around its coasts. Similar action is apparently needed to locate the sites of CW dumping at sea and burial on land.

The cost of the destruction and clean-up programme is beyond the capability of Russia at the present time of economic restructuring and financial crisis. International support is vital to progress. The USA committed $55 million to help Russia as part of the so-called Nunn–Lugar programme.[116] In addition, Russia can expect to use for the CW programme some of the 10 million Deutschmarks which Germany set aside for 1993 to support Russia's plans to destroy its weapons of mass destruction. Continuing international support and flexibility, however, will be necessary if Russia is to fulfil its obligations under the CWC.

[116] The Soviet Nuclear Threat Reduction Act, passed by the US Congress in Nov. 1991, provided up to $400 million for the 'transportation, storage, safeguarding, and destruction of nuclear and other weapons in the Soviet Union . . . and to assist in the prevention of weapons proliferation'. In 1992 Congress added an additional $400 million when it passed the Former Soviet Union Demilitarization Act. The Nunn–Lugar programme is discussed in Kile, S. and Arnett, E., 'Nuclear arms control', *SIPRI Yearbook 1996: Armaments, Disarmament and International Security* (Oxford University Press: Oxford, 1996), chapter 14, pp. 640–42.

Part V. Chemical weapon disposal and destruction

10. The destruction of old and obsolete chemical weapons: past experience

RONALD G. SUTHERLAND

I. Introduction

The discovery, dismantling and safe destruction of chemical weapons has been a significant problem since the time of their first use in World War I. Until 1969 both agents and munitions were destroyed by methods now considered to be environmentally unsound. These included open-pit burning, land burial and ocean dumping, all of which are now no longer in use owing to a variety of environmental and safety legislation. In June 1969 the US National Academy of Sciences (NAS) in its review of the disposal of CW agents and munitions recommended that ocean dumping be abandoned, that the nerve agent GB (sarin) be destroyed by chemical neutralization and that incineration be used for the blister agents H and HD. In 1972 the US Congress prohibited ocean disposal, and in 1984 the NAS recommended that incineration be the disposal methodology used for the destruction of the US stockpile which consisted of the nerve agents VX and GB, the mustard gas agents H, HD and HT and the hallucinogen BZ, together with residual quantities of lewisite. Hence the NAS Committee was in agreement with the US Army view that thermal destruction was the most appropriate technology. The NAS Committee had considered other alternatives such as chemical neutralization and the nuclear explosion of agents and munitions in underground cavities in detail[1] and also explored some experimental technologies, but none of them met with approval although further research was recommended especially on in-shell combustion.

II. Chemical destruction technology

The US Army has carried out extensive investigations on neutralization and incineration of CW agents and munitions. Other methodologies investigated were: in-shell combustion, steam pyrolysis, drain-in-furnace and underground combustion, and caustic hydrolysis.[2]

Chemical neutralization technology[3]

The US Army examined the neutralization of the nerve agent GB as a potential route for its industrial scale disposal. Sarin is water soluble, and its hydrolysis should be straightforward. The reaction is dependent on concentration, pH and temperature. The initial products are hydrogen fluoride and isopropyl methyl-

[1] Committee on Demilitarizing Chemical Munitions and Agents, Board on Army Science and Technology, Commission on Engineering and Technical Systems, National Research Council (NRC), *The Disposal of Chemical Munitions and Agents* (National Academy Press: Washington, DC, 1984).

[2] See NRC (note 1), pp. 71–72.

[3] Program Executive Officer, Program Manager for Chemical Demilitarization, US Army, *Chemical Stockpile Disposal Program: Final Programmatic Environmental Impact Statement*, vol. 3, appendix D, 21016.5401 (US Army: Aberdeen Proving Ground, Edgewood, Md., Jan. 1988).

phosphonic acid. The reaction is pH dependent, and GB is most stable at pH 7. Alkaline hydrolysis produces isopropyl methylphosphonic acid, which is further converted to sodium methylphosphonate. Under certain circumstances the reaction was marginally reversible; under industrial hydrolysis conditions, with excess sodium hydroxide, traces of GB could still be detected for up to two weeks. Not only did this increase costs, but led to increased salt wastes of approximately 2.3 kg for every 0.45 kg of GB destroyed instead of the projected 0.675 kg.[4] The organic salts contained no traces of CW agent but did contain sodium fluoride, phosphonates and heavy metals, and so could not be placed in landfills without expensive precautions. The spray-drying process also produced minute traces of GB, and over $7 million was spent by the US Army to avoid this problem. Approximately 3.7 million kg of GB were processed at the Rocky Mountain Arsenal (RMA) and at the Chemical Agent Munitions Disposal System (CAMDS) at Tooele, Utah. Project Eagle (Phase II) involved the neutralization of 21 114 M34 cluster bombs (1873.2 x 10^3 kg) by: (*a*) declustering bomblets, (*b*) deactivating the fuse, (*c*) punching and draining the bomblets, (*d*) chemically decontaminating the bomblets, (*e*) incinerating the explosives and thermally decontaminating metal parts, and (*f*) storing GB for neutralization. The Expanded Project Eagle (Phase II) involved emptying and neutralizing GB from tank storage (171.5 x 10^3 kg) and from 2422 one-ton containers (1362.8 x 10^3 kg) by neutralization and thermal decontamination. Honest John Warheads (M190s) and M139 bomblets were destroyed by stripping them of explosives and removing and punching the bomblets to allow the GB to drain (34.7 x 10^3 kg).

The actual neutralization process involved pumping GB to a holding tank, then to a mixing tee where it met an 18 per cent solution of sodium hydroxide. The brine flowed to a tank containing agitators to assist heat removal via heat exchangers. When neutralization was determined to be complete by sampling and testing, the brine was transferred to a spray dryer to evaporate the water and reduce the brine to salts and water vapour which was scrubbed before atmospheric discharge. The salts were packed in drums and any waste water transferred to the lagoon. About 10 million kg of sprayed dried salts became subject to the Resource Conservation and Recovery Act (RCRA) and were placed in an approved landfill in 1986.

The neutralization–demilitarizing processes carried out at the CAMDS during 1979–1981 involved the disposal of M55 rockets and 155-mm and 105-mm projectiles; 13 951 M55 rockets were drained of GB. The rockets were then disassembled, the explosive/propellant incinerated and the metal fragments thermally decontaminated. Between 1981 and 1982, 12 673 155-mm and 105-mm projectiles were drained of GB and the empty projectiles decontaminated. The total amount of GB to be neutralized was 27.4 x 10^3 kg. The Agent Destruction System (ADS) at Tooele was modelled on the RMA configuration but modified to reduce the problems of foaming and line plugging previously encountered; for example, the caustic solution and agent were mixed in the tank and excess heat was removed by a reactor recirculation and cooling system (also necessary for the VX process). Another change was the use of drum dryers instead of spray dryers. The former were more cost effective and involved smaller vol-

[4] See NRC (note 1), p. 69.

umes of air. Two major problems remained: (*a*) traces of GB were present in the brines; and (*b*) the process still took significantly longer than expected. The certification standard for the brine was no more than 2 ng of GB per ml. This was not met in neutralization reactions, which lasted 10–20 days and occasionally 40 days, even if excess caustic solution was added. Measures taken to prevent excess heat production sometimes lowered temperatures sufficiently to cause precipitation and hence clogging in the pipelines. The US Army thus abandoned industrial-scale neutralization owing to the: (*a*) technical complexity compared to incineration, (*b*) hydrolysis reversal problems leading to traces of GB (*c*) quantity of waste produced, and (*e*) additional capital costs involved. The total amount of GB neutralized in these projects totalled 3800 x 10^3 kg. VX was not neutralized on an industrial scale because of the problems encountered with GB. VX is much more stable than GB,[5] is only poorly soluble in water, and its primary hydrolysis product also exhibits acute mammalian toxicity.

Chemical incineration technology[6]

If the toxicity of a substance is related to its chemical structure, then thermal decomposition of an organic compound will lead to the loss of toxic characteristics. Incineration of hydrocarbons leads to the formation of carbon dioxide and water; in addition chemical warfare agents will produce hydrogen fluoride (GB), phosphorus pentoxide (GA, GB and VX), nitrogen dioxide (GA, VX), hydrogen chloride (H, HD, HT) and sulphur dioxide (H, HD, HT). All of these gases can be removed by scrubbing. In principle, incineration is an environmentally safe method of toxic waste disposal assuming that the temperature used is sufficient to degrade the organic compound to simple inorganic compounds. Incineration has been considered as a destruction methodology also because it was a necessary adjunct to the chemical neutralization process in that there was a need to thermally decontaminate metal parts. Incineration can also be used to deactivate the explosive and propellant components of chemical weapons. By 1988 the US Army had incinerated 2.8 million kg of CW agents (60 000 munitions) at the RMA and the CAMDS.

Project Eagle (Phase I) was carried out at the RMA in 1972–74 and involved the incineration of 2.79 million kg of mustard gas (H, HT) stored in one-ton containers in two campaigns; a third exercise involved the thermal destruction of 21 000 CW agent identification sets (CAIS) containing 17 055 kg of various chemical warfare agents. The process involved: (*a*) preheating the one-ton containers, (*b*) draining the mustard gas agent, (*c*) incinerating the CW agent, and (*d*) thermal decontamination of the containers.

The primary incinerator was a modified hydrazine furnace and relied upon residence time (i.e., the time the material spends in the furnace to ensure combustion) to ensure complete combustion. Two ton container furnaces were used to incinerate any residue remaining in the containers. The pollution abatement system (PAS) consisted of two caustic quench and scrubber systems connected to an electrostatic precipitator and stack. Waste water and scrubber brines were

[5] Trapp, R., SIPRI, *The Detoxification and Natural Degradation of Chemical Warfare Agents*, SIPRI Chemical & Biological Warfare Studies no. 3 (Taylor & Francis: London, 1985).

[6] See *Chemical Stockpile Disposal Program: Final Programmatic Environmental Impact Statement* (note 3).

spray dried to form waste salts. The facility was monitored for mustard gas during incineration and the perimeter monitored for nitrogen dioxide, sulphur dioxide, hydrogen chloride and suspended particulates. A total of 6.4 million kg of waste salts were produced. As some samples showed excess levels of arsenic and lead, all of these residues had to be treated as hazardous waste and placed in an appropriate landfill.

The CAIS destruction posed unique problems in that each set could contain from 1 to 5 agents. The packages were first incinerated in a deactivation furnace followed by passage through a decontamination furnace. The deactivation system was modified by the addition of an afterburner to deal with glass residues, and the PAS system had to be modified to deal with arsenic oxide. All work areas had to be monitored for 9 different CW agents. The sprayed dried salts were held in a secure landfill; the electrostatic precipitator residues contained arsenic, lead and zinc and also had to be placed in a hazardous waste landfill.

The CAMDS facility at Tooele is a test and evaluation centre that examines processes and equipment for the destruction of chemical weapons and, as part of this, has destroyed significant quantities of agents. It incinerated 36 740 kg of GB and 21 770 kg of VX between 1979 and 1990. The facility uses three furnaces: a deactivation furnace, a metal-parts furnace (MPF) and a liquid incinerator. The deactivation furnace is used to incinerate residual agent and for the destruction of explosives and propellants. The MPF is used to decontaminate metal parts, one-ton containers and projectiles which have residual agent. The liquid incinerator burns CW agents and organic matter in waste liquids.

The deactivation furnace is a rotary kiln furnace with a heated discharge conveyor, scrap conveyor, cyclone separator, afterburner and PAS system. The charge[7] of munition bodies and the like is fed from an explosive containment cubicle, and the PAS system removes acid gases and particulate matter. The furnace exhaust is monitored for oxygen, carbon monoxide, carbon dioxide, sulphur dioxide, nitrogen dioxide and traces of CW agent. Ambient air is checked at eight perimeter monitoring stations. GB can be removed at 99.9 per cent efficiency with a level of 99.9998 per cent with the afterburners. The M55 rocket tests involved 18 300 munitions, and the GB removal efficiency was rated at 99.999994 per cent. A polychlorinated biphenyl (PCB) incineration test conducted in 1986 as a test of the system's ability to meet the solid PCB criteria of the Toxic Substances Control Act (TSCA) showed that the Environmental Protection Agency (EPA) procedure was not sensitive enough to demonstrate the required 99.9999 per cent removal efficiency; compared with commercial incinerators the deactivation furnace emissions were an order of magnitude less. Tests of the brine from this furnace also showed them to be non-RCRA hazardous, but ash samples exceeded the RCRA standards for cadmium content and cyanide activity.

The metal parts furnace is used to thermally decontaminate metals but can be used to destroy bulk agent and has destroyed both GB and VX. Tests were carried out using on-site incineration, but it was decided that agent-injection incineration was superior. GB could be processed at 4 kg/min with removal

[7] The 'charge' is the technical term for the amount of material from the furnace.

efficiencies of 99.99999 per cent. VX injection showed that it could be incinerated at flow rates of 3 kg/min with efficiencies of 99.9999998 per cent.

The CAMDS liquid incinerator is designed to burn drained CW agents and organic matter in waste water. In one year of operation it incinerated 17 205 kg of GB; the unit has one-third the capacity of the incinerators to be used at the CW stockpile disposal plants. The CAMDS shares a PAS with the MPF, and they cannot be used simultaneously. The PAS scrubber brines and the resulting salts have been found to be non-RCRA-hazardous. No organic priority pollutants have been detected.

Johnston Atoll Chemical Agent Disposal System

The CAMDS, which became operational in 1979, was the US Army's prototype chemical demilitarization (chemdemil) facility. The plant, described in the preceding section, served to evaluate various destruction processes to be incorporated in a full-scale industrial operation.[8] The first full-scale plant, the Johnston Atoll Chemical Agent Disposal System (JACADS),[9] has been built on Johnston Atoll in the Pacific south-west of Hawaii and is at the stage of equipment testing and systemization. The JACADS process uses thermal destruction to destroy the CW agents contained in munitions. The munition propellants and explosives are also thermally destroyed, and the metal parts are thermally decontaminated. All associated dunnage (i.e., packing material, particularly wooden pallets) is also burnt.

The disposal process involves: (*a*) disassembly of explosives and propellants, (*b*) destruction of explosives and propellants, (*c*) removal of toxic agent from the cavity of the munition, (*d*) disposal of the toxic agent, (*e*) disposal of munition bodies (shells, etc.), and (*f*) disposal of process wastes. These criteria must be used to develop a process that ensures the safety and security of the personnel, and there must be total containment of the toxic agent before its destruction. The disposal criteria are only part of a triad which involves safe storage, transportation and disposal of CW agents and munitions. There are three munitions categories—rockets and missiles, projectiles and cartridges, and bulk items—stored at Johnston Atoll. The first category of munitions can be processed without the removal of the explosive, while removal of the explosive is the first step in the disassembly of munitions in the second category. Bulk items such as bombs, spray tanks and one-ton containers do not contain explosives in storage.

The first step in the disposal process is the removal of the munitions and so on from storage to a munition-processing centre. Munition processing involves a rocket shear machine to shear rockets and explosives, a projectile/munition disassembly machine to remove explosives by a reverse assembly process, a multi-purpose machine that drains agent from projectiles, and a bulk drain station to punch and drain bombs, containers and spray tanks.

[8] For a detailed description see *Chemical Stockpile Disposal Program: Final Programmatic Environmental Impact Statement* (note 3), appendix C, 'Demilitarization plant design'.

[9] *Johnston Atoll Chemical Agent Disposal System: Final Second Supplemental Environmental Impact Statement*, vols 1 and 2 (US Army: Aberdeen Proving Ground, Edgewood, Md., June 1990); Greenpeace, *Greenpeace Review of JACADS, Dec. 1988* (Greenpeace International: Washington, DC, Aug. 1989); and Demilitarization and disposal of U.S. chemical warfare agents and munitions, Conference on Disarmament document CD/CW/WP.265, 11 Dec. 1989.

Four furnaces—the liquid incinerator (LIC), the deactivation furnace system (DFS), the MPF and the dunnage incinerator (DUN)—are located so that there is a gravity feed of munition components into the furnaces. All four furnaces are prewarmed by firing with propane, and fuel oil is used during operations.

The LIC system is a two-stage refractory-lined incinerator designed to destroy GA, GB, VX, mustard gas (H, HD, HT) and lewisite. The drained agent is pumped from the intermediate tank, atomized by a nozzle and mixed with combustion air. Propane is used to maintain combustion at or above 1370°C with the flue gases being transported to an afterburner maintained at a minimum temperature of 1090°C before ducting to the PAS. Spent decontamination solution is likewise incinerated to remove organic substances. The salts are removed from the system and sent to a brine reduction area. At JACADS, destruction requirements are a destruction efficiency of 99.99999 per cent for GB and VX and 99.99995 per cent for H.

The DFS is used to dispose of explosives and propellants in the chemical munitions. Fuses, booster pellets and supplementary charges are fed whole to the furnace while bursters and propellants are preprocessed by mechanical shear. Pieces enter via a blast-gate valve into the stainless steel rotary kiln which operates at a gas temperature of 540–980°C with a residence time of at least 12 minutes; it can process 68 kg/hr of explosives. After leaving the kiln, the pieces travel on a heated discharge conveyor for at least 15 minutes. The flue gases exit through an afterburner maintained at 1200°C to ensure complete combustion of agent vapours, hydrocarbons and any other organic substances from the combustion of fibreglass.

Munition metal parts are thermally decontaminated in the MPF system which maintains all metal parts at a temperature of 538°C for at least 14 minutes. Metal parts for projectiles and bulk items have a residence time of approximately 60 minutes. Exhaust gases exit into the PAS through an afterburner maintained at 1090°C.

The DUN deals with wooden pallets, laboratory solid waste, protective clothing, charcoal, ventilator filters and other combustible packing materials. The refractory lined furnace operates at 870°C. The two-chambered incinerator has a throughput rate of 454 kg/hr and gases exit through an afterburner to a PAS.

Each furnace system has its own PAS which is designed to scrub the combustion products and any resulting heavy metal impurities. The afterburner flue gases are quenched with water or a sodium hydroxide solution in a quench tower and the temperature of the gases is reduced to about 80°C. The gas then passes to the baghouse where submicron particulates are removed before moving to a counterflow caustic packed-bed scrubber to scrub for the remaining acid gases. The function of the final demister system is to remove P_2O_5 and any particulates not captured previously. The dunnage system requires only a quench tower and a baghouse to remove particulates; wastes are processed by type with the use of caustic solution when acid gases have to be quenched (plastics). The dunnage environmental standards are: (*a*) 99.99 per cent efficiency for principal organic hazardous constituents, (*b*) 99 per cent efficiency for removal of HCl emissions, and (*c*) particulate matter discharge not greater than 180 mg/m^3 (i.e., at least 99 per cent of the particulates).

The spent brines (liquid wastes) from the PAS systems are discarded when the specific gravity reaches 1.08–1.15 and dried to salt wastes in the brine-reduction area, which consists of an evaporator and two steam-heated rotary double-drum dryers. The brines are collected and certified as CW-agent free before the drying process commences. The waste salt is then prepared for storage in a hazardous landfill.

The instrumentation in the process-control area runs the completely automated chemdemil process; it has to be able to: (*a*) monitor process conditions, (*b*) provide data to show compliance with regulatory requirements, and (*c*) control and shut down processes as required. The control operating system (CON) can hold data (8000 points) on all plant operating information for 10 minutes and can store up to 60 minutes of data permanently. All work areas are monitored by an automated continuous agent monitoring system (ACAMS) that can measure CW agents at concentrations less than the allowable limits established by the Army Surgeon General; the allowable work area agent concentrations (time weighted average) are GB (0.0001 mg/m), VX (0.00001 mg/m) and H (0.003 mg/m).

Table 10.1 summarizes the experience of the US Army in the destruction of chemical weapons on an industrial scale between 1974 and 1991; this involves all major agents in the US stockpile and approximates 6.8 million kg of CW agent.

JACADS is the prototype chemdemil facility for all other US stockpiles of chemical weapons. Modifications to its plant design will relate only to the size and composition of the stocks at the various mainland (OCONUS) locations. The JACADS programme was established in 1981, and agent disposal commenced in 1990 to destroy the US CW stocks in the Pacific and those from Germany. The initial operational verification test (OVT) was designed to show safe operation in the destruction of CW representative of all weapon types. By February 1991 the system had destroyed 7565 rockets, drained 34 881 kg of agent and incinerated 32 793 kg of agent. The capital costs were $254 million, 'systemization costs' of $100 million and operating costs of $88 million, with the destruction of the remaining stocks expected to cost $450 million. The total cost of destruction of the US CW stockpile is now in the range of $8 billion.

The JACADS plant has been plagued with minor problems which have resulted in shut downs and an extension of the operational test period to March 1992, but none of the problems involved the incinerators themselves. Greenpeace continues to be very critical of the chemdemil process, and contends that the environmental assessments which have been published are inadequate for approval of the incineration operations. They suggest that highly toxic emissions are possible and that they could be subject to bioaccumulation. They also claim that the transportation of dried brines will be potentially hazardous and that environmental impact on the Johnston Atoll bird sanctuary remain unaddressed. In general, they believe that incineration of any waste material is an environmentally unsound technique. Acceptance of their views would lead to continued above-ground storage of chemical weapons while a search was made for an alternate destruction methodology which was demonstrated to be environmentally safe.

Table 10.1. Summary of the experience of the US Army in industrial-scale chemical agent/munition disposal, 1974–91

Operation	Description	Date	Agent	Site[a]	Process[b]	Quantity (1000 kg)	(1000 lbs)
Project Eagle, Phase I	Ton containers	July 1972– Mar. 1974	H	R	I	2 008.5	4 428.0
Project Eagle, Phase I	Ton containers	July 1972– Mar. 1974	HD	R	I	777.5	1 714.0
Project Eagle, Phase II	M34 cluster bombs	Oct. 1973– Nov. 1976	GB	R	N/I	1 873.2	4 129.6
Project Eagle, Phase II (expanded)	Underground storage tanks	Sep. 1974– Nov. 1974	GB	R	N	171.5	378.0
Project Eagle, Phase II (expanded)	Ton containers	May 1975– Nov. 1975	GB	R	N/I	1 635.0	3 604.5
Project Eagle Phase II (expanded)	Honest John warhead (M139)	Apr. 1975– Nov. 1976	GB	R	N/I	34.7	76.5
Chemical agent identification sets disposal	Chemical agent identification sets	May 1982– Dec. 1982	[c]	R	I	16.6	36.7
M55 rocket disposal		Sep. 1979– Apr. 1981	GB	C	N/I	58.1	128.0
Agent injection incineration tests		Apr. 1981– Jan. 1984	GB	C	I	5.1	11.2
Agent injection incineration tests	Ton containers	June 1981– Aug. 1984	VX	C	I	3.6	7.9
155-mm projectile disposal		July 1981– July 1982	GB	C	N	27.4	60.5
105-mm projectile disposal		Mar. 1982– July 1982	GB	C	N	–	–
In situ agent incineration		Oct. 1982– Dec. 1983	GB	C	I	8.0	17.6
M55 rocket incineration		Nov. 1985– Nov. 1986	GB	C	I	1.0	2.3
Liquid incinerator test		Aug. 1985– Aug. 1986	GB	C	I	17.2	37.9
Agent BZ disposal		May 1988– Sep. 1989	BZ[d]	P	I	42.6	94.0
Liquid incinerator test		Sep. 1989– Oct. 1989	VX	C	I	18.1	40.0
JACDS	M55 rockets	June 1990– Mar. 1991	GB	J	I	35.1	77.2
Total						**6 733.2**	**14 843.9**

[a] R = Rocky Mountain Arsenal, C = Chemical Agent Munitions Disposal System, P = Pine Bluff Arsenal and J = JACADS.

[b] N = Agent neutralization only; I = Incineration of agent and explosive (and/or metal parts thermal decontamination); N/I = Agent neutralization and explosive incineration (and/or metal parts thermal decontamination).

[c] Agents include: phosgene, chloropicrin, mustard gas, lewisite, cyanogen chloride, nitrogen mustard and GB.

[d] The incapacitating agent BZ is not lethal.

Source: This table appears in *Disposal of Chemical Agents and Munitions Stored at Anniston Army Depot, Anniston, Alabama: Final Environmental Impact Statement* as table 2.6 (US Army: Aberdeen Proving Ground, Edgewood , Md., May 1991).

The destruction of 3-quinuclidinyl benzilate

The complexities of the destruction of a nonlethal agent are illustrated by the time and cost involved in the chemdemil of 3-quinuclidinyl benzilate (BZ).[10] The initial planning took place in 1976 with the programme required to destroy 1570 BZ munitions, 4760 kg of pure BZ and about 637 tons of contaminated residues from manufacturing, clean-up operations and initial testing associated with process development. A four-phase programme was developed in 1977. The laboratory development and verification of the process lasted to 1981 when incineration was selected over chemical neutralization. The second phase involved the design of the equipment and facilities which took until 1982. Phase three was the construction of the plant and the systemization of the destruction plant process, which was completed in mid-1988. Destruction then took place with all munitions destroyed by 1989; all other waste was detoxified by incineration by early 1990. Plant closure was completed in June 1990.

It took 14 years to complete this programme at a total cost of $163 million broken down as follows: process design ($20.4 million), construction ($8.8 million), equipment ($31.0 million), systemization ($26.6 million), operations (May 1989–January 1990, $51.6 million) and closure ($5.2 million). In addition, the total department support charges were $19.3 million.

CW destruction in Canada[11]

Surplus CW agents, primarily mustard gas, were shipped to the Defence Research Establishment Suffield (DRES) at the end of World War II for eventual disposal. In 1975–76 over 700 tons of sulphur mustard gas were destroyed in a hydrolysis facility especially designed for that purpose. The products of the alkaline hydrolysis were then incinerated and the residues placed in a landfill. Empty or contaminated containers were then detoxified by burning. The processes were safe and met the environmental and safety standards of the era but could not be used in Canada today. In subsequent years bulk agent stocks and

[10] See USA, Report on the destruction of 3-quinuclidinyl benzilate, BZ, Conference on Disarmament document CD/1074 (also issued as Conference on Disarmament document CD/CW/WP.336, Mar. 1991).

[11] Destruction and disposal of Canadian stocks of World War II mustard agents, Conference of the Committee on Disarmament document CCD/434, 16 July 1974; *Project Swiftsure: Environmental Protection Plan for the Disposal of Hazardous Waste at the Defence Research Establishment Suffield*, 3 vols (Department of National Defence, Defence Research Establishment: Suffield, May 1990); and McAndless, J. M., 'Project Swiftsure: destruction of chemical agent waste at Defence Research Establishment Suffield', Paper presented at US National Research Council Committee on Alternate Chemical Demilitarization Technologies, Washington, DC, June 1992.

various wastes remaining after previous disposal activities were stored at protected sites awaiting eventual disposal. By 1989 the toxic waste included: 12 tons of mustard gas (liquid or thickened); 1.5 tons of lewisite; 0.3 tons of nerve agents (G and V classes); 300 tons of contaminated scrap, containers and ordnance; 4 tons of contaminated waste (decontaminating solution, etc.); and 40 tons of nonhazardous scrap.

In 1989 the Minister of Defence acted on the recommendations contained in the Barton Report[12] and directed that all CW agent waste be destroyed by 1992. As required by Canadian environmental legislation, the potential health, safety and environmental impacts were identified and examined. A comprehensive Environmental Protection Plan was developed for review by the public and concerned government departments (provincial and federal), and a very successful campaign was mounted to ensure that the public was informed and its enquiries answered. The project began in 1989 and was carried out in three phases. Phase I involved the neutralization of nerve agents by DRES; phase II involved the awarding of a contract to supply and operate neutralization and incineration technologies to destroy mustard gas, lewisite, nerve agent hydrolysates and the various categories of scrap metal. The Environmental Protection Plan (EPP) was developed for government and public review and input, and the incinerator was installed with test burns to ensure that it met appropriate emission standards. Phase III, in 1991, involved the neutralization and incineration of CW wastes and the final dismantling of the incinerator equipment as promised to the local citizens.

Swiftsure disposal technology relied on: (*a*) direct incineration of mustard gas as the only practical choice, especially for that which had aged or been thickened with polymeric material; and (*b*) neutralization of all other agents prior to incineration or alternative disposal. Nerve agent containers were processed in a drill and drain apparatus and neutralized by 20 per cent potassium hydroxide in methanol. The neutralization wastes were then incinerated. The methodology for mustard gas incineration involved draining the preheated containers into polyethylene-lined boxes in which the mustard gas was frozen before transport to the incinerator where it was loaded through the solid-waste feeding system. The problems with thickened or aged mustard gas required puncturing the containers and this was done by shaped perforated charges under cold ambient conditions (–5°C or less). The punctured items were sealed in polyethylene-lined boxes for direct loading into the incinerator.

The destruction of lewisite involved chemical neutralization with alkaline peroxide solution that contained sodium or calcium chloride, which leads to the formation of sodium or calcium arsenate and the evolution of acetylene and dissolved chloride salts. These solid residues were stabilized in concrete and disposed of in a landfill.

The incinerator could handle liquid and solid-waste feeds at a capacity of 1.5 tons per hour. The combustion chamber operated at 900°C with the secondary combustion chamber at 1200°C. After the test burns, the chemical waste was incinerated over a three-month period using natural gas.

[12] Barton, W. H., *Research, Development and Training in Chemical and Biological Defence within the Department of National Defence and the Canadian Forces* (Department of National Defence: Ottawa, Canada, 1988).

Leaking munitions

The age of the US stockpile has led to problems with leakage of some munitions (mostly M55 rockets).[13] Procedures have been developed to detect and contain leaking munitions. The cost of containment is approximately $4000 per munition. In 1979 the US Army instituted a drill and transfer system (DATS) to dispose of leaking munitions. DATS is a transportable system that removes CW agent by drilling through the munition wall. The agent is transferred to a container for safe storage. The munition casing is then decontaminated and the casing fractured by detonation in a fragment-containing chamber. Scrap metal is then stored for thermal decontamination and the spent decontamination liquid subsequently dried. DATS has a capacity of 3 to 6 rounds per day, and over 900 leaking munitions have been processed.

KUASI is the system designed by the USSR to handle faulty munitions containing sarin, soman or VX without explosives. Approximately 4000 munitions were destroyed between 1980 and 1990 using this system. The process involves opening, draining and decontaminating the opened munition in the first chamber. The drained munition is decontaminated by washing with ethanolamine followed by 1 per cent alkali solution and finally thermally decontaminated in a furnace. The drained agent is then processed by heating with a reagent at the required temperature for 30–45 minutes. The reaction mixture is then transferred to a tank truck and diluted with water or alkali solution (1:1) before transfer to a liquid-waste incinerator.

Other CW destruction

At the 1991 session of the *Ad Hoc* Committee on Chemical Weapons, it was decided to hold a meeting devoted to the various aspects of CW destruction.[14]

At that meeting, the former USSR reported that it preferred a two-stage process.[15] Sarin, soman and mustard are reacted with ethanolamine to produce a detoxified product which is then incinerated; VX is reacted with a 1:1 mixture of ethylene glycol and orthophosphoric acid before incineration. The processes are discontinuous, and the reaction takes about one hour with a further 1.5 hours required for analysis to disclose traces of agent in the material to be incinerated. The munition casings are also incinerated. The USSR indicated that 300 000 kg of toxic substances had been destroyed but that an industrial-scale facility had not yet been built. New processes have been developed for dealing with arsenicals (lewisite) by detoxification with a melt of elemental sulphur to give a polymer which can be buried. The USSR also examined chlorination to produce arsenic trichloride.

A British review paper briefly discussed hydrolysis of nerve agents by 10–20 per cent sodium or potassium hydroxide with the addition of ethanol as agent solubility decreases (GD, GF).[16] The UK noted problems related to the volume

[13] *Chemical Stockpile Disposal Program Final Programmatic Environmental Impact Statement* (note 3), appendix E; and USSR, Complex for the destruction of faulty chemical munitions (KUASI), Conference on Disarmament document CD/CW/WP.369, 8 Oct. 1991.

[14] For a review of that meeting see Canonne, P. (France), Report of experts' meeting on the destruction of chemical weapons, Conference on Disarmament document CD/CW/WP.377, 9 Dec. 1991.

[15] Conference on Disarmament document CD/CW/WP.367, 7 Oct. 1991.

[16] Conference on Disarmament document CD/CW/WP.373, 21 Oct. 1991.

of hydrolysates produced. The advantage of the ethanolamine process used by the former USSR is that the products are organic and more amenable to incineration. Mustard gas can be hydrolysed either by aqueous/alcoholic caustic solutions or by 10–20 per cent chlorine bleach solutions. The UK prefers to incinerate the resulting hydrolysate from mustard gas since destruction efficiencies of 99.99999 per cent removal of organics can be attained.

A review of German experiences described the capture of arsenicals by the oxidation of arsenic (III) to arsenate (V) by permanganate and its subsequent precipitation as ferric arsenate with its disposal in landfill.[17]

An Italian paper described problems associated with the destruction of mustard gas-phenyldichloroarsine (Y-PhDA).[18] Italy developed a destruction process in which the agent was emulsified by a surfactant and then oxidized by hydrogen peroxide. The product of the reaction was then mixed with sand and cement to form a concrete which stabilized the arsenical residues. The costs of developing the destruction plant were $2.1 million with operating costs of $370 000 or about $20 to destroy each kg of agent. Two other plants are planned; one for adamsite and the other for obsolete CW rounds.

Table 10.2 summarizes the collective experience on the destruction of chemical weapons.

Alternative technologies

There has been some interest in alternative technological methods on the part of Greenpeace,[19] which is opposed to incineration in all its aspects, and by government agencies as the cost of incineration escalates. Greenpeace argues that incineration is inherently unsafe and should not be used for the detoxification of CW. It should be noted that the US Army has spent large sums of money on an examination of the destruction of unitary chemical weapons and embarked on its incineration programme over a decade ago. The Chemical Weapons Convention requires that all chemical weapons be destroyed within a 10 year period after entry into force. The US–Soviet Memorandum of Agreement of 1 June 1990 also includes the destruction of CW as one of its provisions.[20] The problems of the destruction of CW must therefore be solved in the near future and cannot await the results of an extended research programme into potential destruction technologies.

In principle, the many alternate technologies suggested are capable of detoxifying chemical weapons, but most are still years away from industrial level of operation. The following methodologies have been suggested or investigated to some extent and have been reviewed by Greenpeace: biological cleavage of the carbon–phosphorous bond, supercritical water oxidation, steam gasification,

[17] Conference on Disarmament document CD/CW/WP.374, 31 Oct. 1991.

[18] Conference on Disarmament document CD/CW/WP.375, 20 Nov. 1991.

[19] Greenpeace, *Alternative Technologies for the Detoxification of Chemical Weapons: An Information Document* (Greenpeace International: Washington, DC, 1991); *Presentations to the Committee on Alternative Chemical Demilitarization Technologies* (US Government Printing Office: Washington, DC, June 1992).

[20] The 'Agreement between the United States of America and the Union of Soviet Socialist Republics on Destruction and Non-Production of Chemical Weapons and on Measures to Facilitate the Multilateral Convention on Banning Chemical Weapons' is reproduced in *SIPRI Yearbook 1991: World Armaments and Disarmament* (Oxford University Press: Oxford, 1991), appendix 14A, pp. 536–39.

Table 10.2. Experience in the destruction of chemical weapons[a]

Site	Agent	Amount	Method	Date
RMA, USA	H	3071 tons	I	1969–74
RMA, Tooele	GB	4188 tons	N	1973–76
Tooele	GB	38 tons	I	1981–86
JACADS	GB	40 tons	I	1991
Tooele	VX	8 tons	I	1984
JACADS	VX	54 tons	I	1992
JACADS	HD	56.5 tons	I	1992
DRES, Canada	H	700 tons	N/I	1974–76
DRES	H	12 tons	I	1990–91
DRES	VX, GA, GB	0.3 tons	N/I	1990–91
Munster, Germany	H, etc.	70 tons/year	I	1980—>
Shikhany, Russia	GB, GD, H	300 tons	N/I	1980–90
Shikhany, Russia	VX	30 tons	N/I	1980–90
Porton Down, UK	H	small	I	1970
Nancekuke, UK	GB	20 tons	N	1967–68
UK (several)	H	6000 tons	I	1958–60
Iraq (UN)	H	500 tons	I	1992–93
Iraq (UN)	GB, GB/GF	70 tons	N	1992–93

[a] CW agents destroyed in 18 campaigns in 1958–93 totalled 16 000 tons. There were: 11 incinerations, 3 neutralizations and 4 combined incinerations/neutralizations.

Source: Compiled by the author from literature sources.

wet air oxidation, molten salt processes, reductive degradation, dechlorination, photochemical processes, electrochemical processes, neutralization technology, chemical reprocessing and thermal processes.

In addition to the above potential processes, France described a chemical ammunition destruction process in which a dispersive charge is ignited in a closed aqueous pool thereby destroying the explosive and toxic charges simultaneously.[21] A new process for the disposal of hazardous waste—the Eco Logic Process (ELI)—has been demonstrated by a Canadian company; it uses hydrogen at 850°C or higher to reduce contaminants; a process that should be applicable to chemical warfare agents.

More recently, the Committee on Alternative Chemical Demilitarization Technologies held an extensive symposium where a large number of presentations were made on potential methodologies for the destruction of chemical warfare agents. The following technologies were discussed and indicate those that could be developed if the need arises: biological technologies; chemical technologies; medium-temperature technologies, including supercritical water oxidation; high-temperature technologies, including plasma technologies; and foreign programmes (overviews of British, Canadian and Iraqi experiences). In June 1992 the US House of Representatives adopted an amendment to its 1993 Defense Authorization Bill which removed a date for completion of the chem-demil programme, established a Chemical Demilitarization Advisory Commission and required the Department of Defense to: (*a*) develop an alternate CW

[21] See Biseau, M., Preliminary technical study for a new destruction process, reported in Conference on Disarmament document CD/CW/WP.377, 9 Dec. 1991. See also chapter 11 in this volume.

disposal plan, (*b*) present a revised CW disposal concept plan, and (*c*) establish an international consultation and exchange plan for chemdemil.

Work in this area becomes more urgent since some US states such as Kentucky appear to be on the verge of banning incinerator technologies for all purposes. This would mean that the chemdemil programme proposed by the US Army could not take place at the Lexington–Blue Grass ammunition depot. A recently released Office of Technology Assessment (OTA) suggested that the most useful alternative technologies to incineration would be chemical neutralization, supercritical water oxidation, steam gasification and plasma arc pyrolysis.

Selected bibliography

Letter dated 3 August 1992 from the Representative of the United States of America addressed to the President of the Conference on Disarmament transmitting the agreement between the Department of Defense of the United States of America and the President's Committee on Conventional Problems of Chemical and Biological Weapons of the Russian Federation concerning the Safe, Secure and Ecologically Sound Destruction of Chemical Weapons, Conference on Disarmament document CD/1161 (CD/CW/WP.426), 5 Aug. 1992.

Committee on Alternative Chemical Demilitarization Technologies, *Workshop on Alternate Technologies Study* (National Research Council: Washington, DC, June 1992).

Barrass, B.C., 'Destruction of the Iraqi chemical arsenal', Fourth International Symposium on Protection against Chemical Warfare Agents, Stockholm, Sweden, 8–12 June 1992.

Gee, J., 'The destruction, removal or rendering harmless of Iraq's chemical warfare capability', *Disarmament*, vol. 15, no. 2 (1992), pp. 77–93.

Leonov, G. S. and Sheluchenko, V. V., 'Principal technological and environmental aspects of the destruction of chemical weapons', *Disarmament*, vol. 15, no. 2 (1992), pp. 94–100.

A report on the U.S. chemical weapons (CW) destruction experience at Rocky Mountain Arsenal, Colorado, Conference on Disarmament document CD/CW/WP.381, 14 Feb. 1992.

Johnston Atoll Chemical Agent Disposal System (JACADS), Conference on Disarmament document CD/CW/WP.382, 14 Feb. 1992.

U.S. chemical weapons (CW) destruction safety and environmental requirements, Conference on Disarmament document CD/CW/WP.383. 14 Feb. 1992.

A report on the destruction of 3-quinuclidinyl benzylate (BZ), Conference on Disarmament document CD/1074, 20 Mar. 1991.

Greenpeace, *Alternative Technologies for the Detoxification of Chemical Weapons: An Information Document* (Greenpeace International: Washington, DC, 1991).

Aroesty, J., *Comments on Implementation: Contingency Options for Chemical Weapons Demilitarization* (Rand Corporation: Santa Monica, Calif., July 1991).

Experts Meeting on Technical Aspects of CW Destruction, Conference on Disarmament, Geneva, Oct. 1991. This meeting was chaired by Dr J. Ooms, organized by Dr P. Cannone and reported on in Conference on Disarmament document CD/CW/WP.377, 9 Dec. 1991.

Main technological aspects of the destruction of chemical weapons, Conference on Disarmament document CD/CW/WP.367, 7 Oct. 1991; and Environmental aspects of the destruction of chemical weapons, Conference on Disarmament document CD/CW/WP.368, 7 Oct. 1991 (papers presented by the USSR to the Experts Meeting on Technical Aspects of CW Destruction, Geneva, Oct. 1991).

U.S. Chemical Weapons Destruction Program: A Presentation to the CW Committee at the CD by Brigadier General W. L. Busbee (US Army: Aberdeen Proving Ground, Edgewater, Md., Mar. 1991).

Disposal of Chemical Agents and Munitions Stored at Anniston Army Depot, Anniston, Alabama: Final Environmental Impact Statement (US Army: Aberdeen Proving Ground, Edgewater, Md., May 1991), and related documents of July 1989, Dec. 1989 and Feb. 1990.

Johnston Atoll Chemical Agent Disposal System: Final Second Supplemental Environmental Impact Statement, Final Second Supplemental Environmental Impact Statement for the Storage and Ultimate Disposal of the European Chemical Munition Stockpile and Comments Received from the Public and Agencies, vols 1 and 2 (US Army: Aberdeen Proving Ground, Edgewater, Md., June 1990).

Project Swiftsure: Environmental Protection Plan for the Disposal of Hazardous Waste at the Defence Research Establishment Suffield, 3 vols (Department of National Defence, Defence Research Establishment: Suffield, May 1990).

Brigadier General D. L. Benchoff, 'Removal of chemical munitions from Germany', *Press Release*, Pirmasin, Germany, Mar. 1990.

Lohs, Kh., Lundin, S. J. and Stock, T., *The Destruction of Chemical Weapons and Chemical Warfare Agent: Chapters reprinted from the SIPRI Yearbooks 1978 and 1979, with an introduction compiled by Thomas Stock* (SIPRI: Solna, Sweden, 1990).

Greenpeace, *Greenpeace Review of Johnston Atoll Chemical Agent Disposal System (JACADS), Dec. 1988* (Greenpeace International: Washington, DC, Aug. 1989).
Chemical Stockpile Disposal Program: Final Programmatic Environmental Impact Statement, 3 vols (US Army: Aberdeen Proving Ground, Edgewood, Md., Jan. 1988).
Information presented to the visiting Soviet delegation at the Tooele Army Depot, Nov. 1987, Conference on Disarmament document CD/830, 19 Apr. 1988.
Information on the presentation at the Shikhany military facility of standard chemical munitions and of technology for the destruction of chemical weapons at a mobile facility, Conference on Disarmament document CD/789, 16 Dec. 1987.
Chemical stockpile disposal program, Conference on Disarmament document CD/711, 9 July 1986.
Committee on Demilitarizing Chemical Munitions and Agents, Board on Army Science and Technology, Commission on Engineering and Technical Systems, National Research Council (NRC), *The Disposal of Chemical Munitions and Agents* (National Academic Press: Washington, DC, 1984).
Destruction and disposal of Canadian stocks of World War II mustard agents, Conference on Disarmament document CCD/434, 16 July 1974.

11. The technical challenge of dismantling and destroying old and abandoned chemical weapons

FRÉDÉRIC GUIR

I. Introduction

The destruction of the existing chemical weapon stockpiles was one of the most difficult issues in the final stages of the negotiations on the Convention on the Prohibition of the Development, Production, Stockpiling and use of Chemical Weapons and on their Destruction (the CWC).[1] In hope of totally removing the CW threat an obligation was incorporated into the Convention (Article IV and Part IV(A) and (B) of the Verification Annex) to destroy all kinds of CW of whatever vintage, including so-called 'old and abandoned' chemical weapons.

Obviously, it was thought that the destruction of old and abandoned CW would be the last step in freeing the world of the CW threat. While Russia and the USA have the largest declared stockpiles many other parties to the CWC will also find that the mandatory destruction of these weapons presents a significant challenge. For many States Parties the destruction of old chemical munitions and the related toxic wastes will be the greatest challenge of the CWC.

The compulsory destruction of old and abandoned chemical weapons will be the most frequent type of CW destruction under the Convention. In France the Centre d'Etudes du Bouchet (CEB), the French Research Centre on Nuclear, Biological and Chemical Defence, was contracted to study the problem and to find ways to carry out the necessary destruction. This research centre was designated by the Ministry of Defence to coordinate all chemical disarmament efforts. This chapter outlines its approach to the problem of dismantling and destroying old and abandoned chemical weapons.[2]

II. The scope of the task

In order to explain the difficulties of this task and their character it is necessary to determine the true nature of the problem. A definition of old and abandoned chemical weapons can serve as a starting point.

[1] The text of the CWC is reproduced in *SIPRI Yearbook 1993: World Armaments and Disarmament* (Oxford University Press: Oxford, 1993), appendix 14A, pp. 735–56.

[2] This chapter is based primarily on the handbooks written for training courses for OPCW inspectors at the French Training Centre for the Prohibition of Chemical Weapons (CEFFIAC), which held its first courses in the autumn of 1993.

Definitions

Chemical weapons[3]

In the context of old and abandoned chemical weapons, a chemical weapon can be defined as a hard metallic container filled with a chemical warfare agent, often fitted with an explosive component consisting of a burster (containing a secondary explosive) and a fuse (containing primary and secondary explosives). The discussion here excludes conventional munitions and bulk chemical munitions without explosive mechanisms since the former contain no toxic chemicals and the latter do not represent an explosive hazard. Although there is a need to destroy them, old chemical shells with no explosive component are merely small chemical containers and outside the scope of this chapter, as are conventional explosive munitions.

Old and abandoned chemical weapons

Old and abandoned chemical weapons[4] can be defined in various ways. Article II of the CWC defines *old chemical weapons* as those produced before 1925, or between 1925 and 1946, but which have deteriorated to such an extent that they can no longer be used as chemical weapons. Under the CWC chemical weapons must be declared if they were owned or possessed by a State Party, if they were owned or possessed by another state and located in any place under the jurisdiction or control of another state, or if they have been transferred or received after 1 January 1946. Old weapons which meet the above definition would not fall into that category.

The Convention defines *abandoned chemical weapons* as those which were abandoned by a state after 1 January 1925 on the territory of another state without the consent of the latter. The implication is that the weapons were left unattended.

Chemical weapons which fit the definition of old or abandoned have thus been left unattended for 50 years or more. These munitions have not been maintained or repaired. They are generally so heavily corroded that it is impossible to separate the explosive component from the chemical container using normal disassembly procedures without destroying one or both parts.

An old or abandoned chemical weapon consists of two inseparable parts—one chemical and one explosive. This combination makes the destruction of these weapons particularly difficult and presents problems that would not exist if the parts could be dealt with separately.

Safety aspects

Owing to the nature of old and abandoned chemical weapons particular safety procedures must be observed. The hazards created by each part of the weapons must be considered individually so that proper safety precautions can be taken during dismantling and destruction.

[3] This definition of a chemical weapon is not identical to that in Article II of the CWC (see note 1).
[4] See CWC (note 1), Article II, paragraphs 5 and 6.

Safe handling of explosives[5]

The basic principle of safe handling of explosives is 'non-confinement'. Explosives can only be handled safely in an open area where no people are present. If an explosion occurs in a confined area, combustion can cause a devastating detonation. Accidental detonation can be disastrous, and the energy of a shock wave can only be dissipated safely in a large open area, designed for detonation, from which everyone has been evacuated. For this reason all explosive or solid propellant plants have ample safety zones which are determined based on the quantity of explosives handled in the facility.

Buildings where activities involving explosives are carried out are designed to have at least one thin wall that will be blown out in the event of an accident, releasing the shock wave of a detonation. Such a building is not airtight.

Safe handling of chemical warfare agents[6]

In contrast to the safety procedures for work with explosives, the basic principle for safe handling of chemical warfare agents is total confinement. There are two types of barriers that can be used to confine chemicals: (*a*) active systems such as filters, scrubbers, decontamination solutions, and the like; and (*b*) passive barriers, including air locks, sealed containers, and so on.

Safe handling requires that a worker not come in contact with toxic chemicals, and in order to accomplish this at least one airtight barrier is needed between the worker and the chemicals. Generally, two airtight barriers are used to ensure a high level of safety. For example, the chemical is contained in a sealed ampoule while protective equipment such as a gas mask is worn by the worker. The mask acts as an active barrier to prevent toxic vapours from penetrating the facial protection and entering the respiratory tract of the wearer.

Incompatible safety requirements

The above discussion demonstrates the conflicting requirements for the safe dismantling of old and abandoned chemical weapons. Safe handling of the explosive component of these weapons calls for working in an open area, while an area with at least one airtight barrier is needed to dismantle the chemical component. This is a problem which must be addressed in order to meet the requirements of the various provisions of the CWC.

The complex nature of dismantling chemical weapons helps to explain past disinterest in dealing with old and abandoned CW; it was simply easier to bury the weapons than to attempt to dismantle them. This is also the reason why old and abandoned chemical weapons are found in such a bad state. Dismantling these munitions is highly dangerous, and there has been little interest in the task of destroying the scrap metal and the toxic waste. Nevertheless the job must be done and with adequate attention to safety.

New safety rules are needed for the dismantling and destroying of old and abandoned chemical weapons. Neither the procedures for safe handling of the explosive component of chemical weapons nor those for safely removing the

[5] Boyer, F., 'Introduction to explosive science and safety', Chemical munition course, CEFFIAC, Sep. 1993.

[6] Froment, D., 'Manufacture of chemical agents', Chemical agent course, CEFFIAC, Sep. 1993.

chemical component address all of the risks encountered with old and abandoned CW, and a new safety code cannot be created by merely adding the two together.

Old and abandoned chemical weapons are a legacy from the past. Before attempting to solve the problem of their destruction, it is necessary to investigate the nature of the problem presented by chemical warfare munitions from both World War I and World War II and the status of French preparation to meet chemical warfare prior to its invasion by Germany in World War II.

History of the problem[7]

The retrieval of old or abandoned chemical weapons is an acute problem for countries like France which were the site of major battles in both world wars. In those wars chemical munitions—fired and unexploded, or unfired but abandoned—were buried deep in French soil, accidentally or otherwise. Although less than 10 per cent of the munitions retrieved each year on former World War I battlefields are old or abandoned chemical weapons, a solution must be found to the problem.

Prior to World War I Germany had not planned to use toxic gas on the battlefield. The German offensive attack was expected to succeed rapidly and the need to consult the chemical industry for help in developing new weapons was not contemplated. However, after the French victory at the Marne, when the battlefield stabilized on a line of trenches that provided shelter from conventional munitions, a search began for a way to render the trenches unsustainable.

The German forces used chlorine for the first time in April 1915 near Ypres, disabling 15 000 soldiers of whom 5000 died. This event was a violation of the 1899 and 1907 Hague Conventions and Declarations[8] and the starting point of the 'gas war' in which new toxic compounds were weaponized and used. The chemists of the time continued to seek new weapon agents, and increasingly lethal discoveries were made. By May 1915 phosgene mixed with chlorine was in use, and in September 1915 the French forces used Clairsite[9] in the assault on Champagne.

The protective equipment used by the military improved rapidly, rendering the use of gas ineffective. The decision was therefore made to shift to a more effective lethal gas. In July 1916 the French forces used hydrogen cyanide stabilized with a mixture of chloroform, arsenic trichloride and tin tetrachloride. Phosgene shells were first used by France in March 1916 at Verdun, and mustard gas was used by French forces in July 1918.

Table 11.1 lists the primary chemicals used by both the French and the German armies and the quantities produced during the war. The agents were mixtures of chemical warfare agents and smoke generators, such as arsenic trichloride, tin tetrachloride, titanium tetrachloride or inert solvents. The quantities of shells fired and of CW shells delivered to the Allied Forces by the French Army are presented in tables 11.2 and 11.3.

[7] Leblanc, A., 'History of chemical warfare', Chemical warfare history course, CEFFIAC, Sep. 1993.

[8] For the texts of the Hague documents and an assessment, see Goldblat, J., *Arms Control: A Guide to Negotiations and Agreements* (Sage Publications: London, 1994).

[9] Clairsite was the French code-name for perchloromethyl mercaptan.

Table 11.1. Quantities of chemical warfare agents produced by France and Germany in World War I

Figures are in metric tonnes.

Chemicals	France	Germany
Chlorine	23 900	38 600
Phosgene	15 800	10 700
Vincennite	4160	
Diphosgene	. .	12 700
Yperite	1968	9000
Clark I	. .	4 200
Chloropicrine	493	7130
Bromoacetone	481	. .
Acrolein	183	. .
Arsenic trichloride	2710	. .
Tin tetrachloride	4116	. .
Titanium tetrachloride	191	. .

Source: Unpublished data of the French Ministry of the Interior.

The French military handbook from the period between World War I and World War II stated: 'Determined to respect the international commitments to which France has subscribed, the French government will try at the beginning of a conflict and in consultation with its allies, to obtain from the hostile government a commitment not to use chemical warfare. If this commitment is not met, it will take appropriate action.[10]

Following World War I France focused on defence, and in 1922 research laboratories devoted to chemical weapon defence were established at the Artillery Pyrotechnic Facility at Le Bouchet. By 1939 it comprised several departments covering CW synthesis, dispersion, protection, physiology, therapeutics, and toxicology and semi-industrial production facilities. Between the two world wars the Le Bouchet synthesis laboratory prepared more than 1300 compounds including nitrogen mustards, which were also studied by Germany and the USA.

At the outbreak of World War II in 1939 France possessed only two mustard gas production plants. They were located at Sorgues and Angoulême; the Sorgues plant also had an adamsite (DM) production unit. New production units were designed for lewisite and trichlorotriethylamine (HN-3) as alternatives to mustard gas, but they were never built. Phosgene would have been produced by the civil chemical industry.

War led to confusion in France, as elsewhere, and it is difficult to track the French chemical weapon stockpile during this period. Germany took it over after France fell. The French CW stockpile was not reconstituted after 1946, and no additional chemical warfare agents were produced for a CW stockpile.

[10] *Manuel d'instruction militaire Français de l'entre-deux-guerre* [French military handbook for the interwar period] (Ministère del la Guerre: Paris, 1925–39).

Table 11.2. Total quantity of chemical weapon-filled shells fired in World War I

Country	Quantity
France	16 million
Germany	33 million
Great Britain	4 million
USA	1 million

Source: Unpublished data of the French Ministry of the Interior.

Table 11.3. Quantity of chemical weapon-filled shells delivered to the Allied forces by France in World War I

Country	Quantity
USA	948 000
Belgium	193 000
Italy	92 000
Romania	50 000
Portugal	46 000
Greece	12 000
Russia	12 000

Source: Unpublished data of the French Ministry of the Interior.

Types of chemical warfare munitions

The types of chemical warfare munitions on French soil vary tremendously but can be categorized by three characteristics: (*a*) the chemical agent in the munition (chemical structure, current status, constituent chemical elements, reactivity, composition, etc.); (*b*) the design of the munition (type, calibre, mechanical structure, filling level, wall thickness, etc.); and (*c*) the status of the munition (explosive status, corrosion status, transportability, etc.).

It is obvious that there is no universal destruction process that can be used for such a variety of munitions. The most appropriate method of destruction must be chosen based on the particular combination of the three parameters listed above. At best, a destruction process may be found which is applicable to a range of munitions if properly designed. The following section outlines a proposal expected to accommodate a large variety of munitions regardless of their status and to solve the problem in one step: simultaneously dismantling the munition and destroying its components.

III. A proposed universal solution: the firing pool[11]

The firing pool technique utilizes the fact that a munition contains an explosive (which is intended to disperse the chemical agent on the battlefield) inside a

[11] The firing pool technique discussed here was designed by Société Nationale des Poudres et Explosifs (SNPE) Ingénierie and based on an unpublished laboratory study at the CEB. See also Report of experts' meeting on the destruction of chemical weapons, Conference on Disarmament document CD/CW/WP.377, 9 Dec. 1991, p. 4.

burster tube. A firing pool is a large pool capable of withstanding repeated explosions. It is filled with an aqueous decontamination solution, and the chemical weapon to be destroyed is placed in the centre of the pool and submerged. One chemical weapon can be exploded every 15 minutes.

In the firing pool the metal of the munition is destroyed by the use of controlled explosive charges while the explosion delivers the chemical agent into a decontamination solution. Extrapolated from laboratory-scale experiments, a firing pool with a 12-metre (m) diameter, a depth of 6 metres and filled with approximately 500 m^3 of decontaminant should be able to destroy approximately 100 tonnes of pure mustard gas.

The great advantage of this process is its ability to accommodate both the explosive and the chemical components of CW munitions. However, is it able to meet all of the safety requirements outlined above?

Safe handling of explosives

The firing pool technique was designed initially for explosive experiments with conventional weapons and meets the requirements for the safe handling of explosives. In order to ensure the detonation of the inner burster charge, two additional explosive charges are fitted on opposite sides of the munition. This means that the pool must be sufficiently large so that it can contain the blast from the inner burster and the two additional charges.

Old and abandoned chemical weapons were designed with an explosive charge that was just sufficient to make the shell burst, resulting in large pieces of irregularly shaped scrap.

Safe chemical decontamination

If dispersion were the ideal, then dispersion of a chemical warfare agent in a decontamination solution would be a satisfactory means of destruction. However, it is very difficult to assess the viscosity of some agents after 70 years in a shell. Mustard gas, for example, is prone to polymerize, and this polymer sticks to even a fragmented shell. As noted above, the fragments will be large and few in number. This implies that their decontamination by the action of the decontamination solution alone will be totally unsatisfactory in terms of current environmental standards. In the case of mustard gas, when the scrap is removed from the decontamination bath it will still be contaminated by a significant amount of chemical agent, and air filtration equipment will need to be used to prevent the spread of contamination.

In addition, the exhaust gas from the blast shell necessitates a huge filtration unit in order to ensure the maintenance of a permanent negative pressure over the firing pool. Moreover, the gas will be contaminated by the chemical agent that is dispersed, and the exhaust gas will not be thoroughly decontaminated by the solution owing to bubble formation during the underwater explosions.

However, the firing pool technique is particularly suited to the detoxification of chemical warfare agents because the excess decontamination solution causes instantaneous decomposition of the agent liberated in the bath. This should be remembered when large quantities of untransportable bulk agent must be destroyed. A simple decontamination pool or vessel may be adequate for complete detoxification of a bulk agent if the amount of decontamination solution is always maintained at a level above a value that has first been determined by laboratory experiments.

Summary

The success of a destruction process is best measured by how it deals with detoxification of the resulting waste. It is difficult to carry out decontamination of solid waste (scrap from the munition) and gaseous waste (exhaust gas) in a firing pool. On the other hand, a firing pool is excellent for the decontamination of liquid waste (CW agent detoxification), provided the agent is not viscous or thickened. A detoxifying bath can operate at maximum efficiency in cases where a munition can be opened in an appropriate amount of decontamination solution. The chemicals released from the munition casing will be destroyed quickly and contamination will not occur since the walls of the opened munition will be bathed by decontamination solution. In addition, this method offers an advantage as regards compliance with the Chemical Weapons Convention since the chemical warfare agent will not be recovered.

Unfortunately, the firing pool technique does not completely meet chemical containment requirements since, at the stage when scrap is removed from the pool, some of the toxic compound sticks to the metallic parts and then escapes from the process without being decomposed unless a specific cleaning stage is added prior to removal of the scrap from the pool. Safe chemical decontamination must ensure environmental safety, and an acceptable destruction process must prioritize the safe decontamination of chemicals in secure dismantling and destruction facilities.

IV. The destruction process

In order to find the best approach to the destruction of old and abandoned CW it can be useful to look at the various steps from retrieval of a chemical weapon to its disposal. These steps include: (*a*) detection, recovery and assessment of explosive status, (*b*) cleaning, (*c*) transportation, (*d*) chemical identification, (*e*) storage, (*f*) dismantling, (*g*) destruction, and (*h*) waste disposal.

The first step is explosive ordnance disposal (EOD): determination of whether or not a munition is a chemical munition and its categorization as an explosive object (i.e., assessment of the degree of explosive risk presented). Extensive knowledge of CW munitions and long experience are needed to make such assessments, and these munitions must be addressed on a case-by-case basis since they vary greatly. Only an EOD specialist can define the explosive status of a given munition. After a munition has been categorized it will likely be moved to another site for destruction. Identification of the chemical fill of the munition will then need to be made.

The technical challenge of the destruction of old chemical weapons lies in finding a destruction process that is compatible with the type of old or abandoned CW munition to be destroyed. It is paramount that munitions be sorted according to their chemical fill to ensure safe handling. The destruction process involves the following steps: (*a*) cleaning, (*b*) chemical identification, (*c*) storage, (*d*) dismantling, (*e*) destruction, and (*f*) waste disposal.

The CEB is currently working on problems such as designing a non-intrusive identification device and an automated cutting device, discovering detoxifying solutions that act on a broad spectrum of chemicals and finding analytical methods that are designed to detect traces of chemicals.

Non-intrusive identification

In order to perform the identification of a munition that is not opened a neutron activation method is currently used for chemical identification. This seems the most promising method to sort munitions according to chemical fill and yields several groups of chemically identical munitions. Each group can then be treated further following specified procedures. Neutron activation analysis has been designed to be used in equipment that can easily be transported to various sites as needed.

Automated cutting devices

The search for a munition-opening device that will function in corrosive liquids like detoxifying solutions led to consideration of water-jet cutting technology instead of the use of a standard saw. A water jet can easily process several sizes of munitions. It can be used, first, to saw munitions into workable pieces and, then, to clean them at a lower jet pressure in order to ensure complete decontamination of the scrap generated by the process. This could solve the problems raised by polymerized derivatives of mustard gas, which do not flow out of a munition that is pierced by the use of the conventional drilling technique. A fully automated water-jet cutting device that is fitted in a small decontamination pool will soon be tested at the CEB to evaluate its efficiency.

Detoxifying solutions

Many chemical mixtures were used as warfare agents in World War I, and the chemical reactivity of a proposed detoxifying solution must be tested on all of them to ensure its applicability. Arsenical products, for example, must be treated separately to avoid contamination of the detoxifying solution. This illustrates the problem of the disposal of wastes that contain inorganic arsenic.

The cost of destruction of a given quantity of munitions is determined by the size of the containment facility needed for the opening and dismantling units and by the complexity of the disposal of the products of decomposition. Chemical munition destruction facilities conduct detoxification by using either a thermal reaction in an incinerator or a chemical reaction in a detoxifying solution. Direct incineration necessitates monitoring the entire process for both toxic agents and the usual industrial pollutants. This requires a large, costly pollution abatement unit, which usually relies on a wet scrubbing process that utilizes a detoxifying solution which yields additional liquid and solid waste.

In a chemical decomposition destruction facility the detoxifying solution contains all of the decomposition products that result from the destruction reaction.

If the decomposition reaction is irreversible and the decomposition products are not on Schedule 1 of the CWC, the exhausted detoxifying solution can be regarded as normal industrial chemical waste and can be transferred to an industrial waste disposal unit outside the destruction site. This type of destruction facility requires little pollution abatement because all of the chemicals are contained in the detoxifying solution. In addition, if the munitions are opened in the decontamination solution there is no need for a large containment unit or extensive filtration equipment.

However, the problem of safe handling of explosives remains. The cutting device will be used on munitions which may be in such poor condition that some of the explosive can pollute the decontamination solution. This solution cannot be treated by a normal waste disposal facility but only by direct incineration. It thus appears that a process must be employed in which munitions are first dismantled, then incinerated. This is an expensive process which the above scheme attempts to avoid.

Analytical methods for scheduled chemicals and quantification of pollutants

In order to monitor the concentration of the detoxifying solution in a decontaminant, specific analytical methods must be designed to meet the requirements for quantification of pollutants that French regulations make mandatory. These procedures should be used for standard on-line analysis whenever possible.

V. Conclusions

As noted above, safe handling of chemical warfare agents is in contradiction with the basic principle for safe handling of explosives: non-confinement. This is also true of effluents which have been contaminated by explosives. It is impossible to meet the requirements for safe handling of both chemicals and explosives to the same extent. However, a degree of flexibility is possible as regards chemicals as they can be destroyed in both conditions of confinement and non-confinement. Since explosives, on the other hand, can only be destroyed in non-confinement there is a need to explore whether chemical weapons can be destroyed safely using the firing pool technique of controlled explosion in a large quantity of decontamination solution. This may be the only means of efficient, safe and relatively inexpensive destruction which also prevents the release of chemicals into the environment.

Ultimately, the true challenge of dismantling and destroying old and abandoned chemical weapons lies in combining two opposing approaches in order to find a safer means of destruction. It may be impossible to use a method which, though successful in its normal range of application, could prove disastrous when applied to old and abandoned CW. The solution lies in the integration of safety requirements at the design stage of a new process. However, in order to do so, there must be awareness of the safety rules that apply to both chemical weapons and explosives, and such knowledge is not widespread.

12. The German programme for the disposal of old chemical weapons

HERMANN MARTENS

I. Introduction

As a belligerent in both World War I and World War II Germany possesses old chemical weapons (OCW) which date from 1915 to 1945. These weapons continue to be found throughout Germany, particularly at former production sites, filling stations, testing areas and storage facilities. Many old chemical weapons were dumped in the sea at various places, including the Baltic Sea.

Owing to external corrosion, unpredictable ballistic properties and fuse mechanisms, and the extent of decomposition of the chemical fills these weapons are no longer militarily useful as chemical weapons. However, since they are harmful to the environment and a potential hazard Germany took the decision to destroy the old chemical weapons that are located on its territory. The destruction of the OCW takes place at the hazardous waste incineration facility of the German Armed Forces Scientific Institute for Protection Technologies: NBC [nuclear, biological and chemical] Protection at Munster.

The old chemical weapons in Germany are more a 'collection' of various OCW items than an 'OCW stockpile'. Most of the items, including captured old chemical weapons from other countries, were recovered sporadically in the past four decades. Prior to their destruction the condition of each weapon must be assessed by explosive ordnance disposal (EOD) personnel. It is thus not feasible to employ a high degree of automation to speed up the disposal process, although this would reduce costs and improve working conditions. Whenever possible high-risk operations are carried out by remote control.

It is difficult to predict a date for the completion of OCW disposal since there is considerable uncertainty about the total quantity of old chemical weapons that may still be buried in Germany.

Germany is willing to share the experience which it has gained in the disposal of OCW with other countries which face similar problems. However, only some of the experience gained concerning technology, time frames and cost is applicable to the destruction of large CW stockpiles. In addition, Germany possesses special knowledge about handling arsenious compounds.

II. Types of OCW munitions and chemical fills

In Germany there is a great variety of OCW munitions from the 1915–45 period including artillery shells, mortar ammunition, hand and rifle grenades, land mines, bombs and spray canisters, and even storage tanks. A wide spectrum of toxic chemicals was used to fill CW munitions during that period. They include various preparations of sulphur mustard, nitrogen mustard, lewisite, other arsenicals (e.g., adamsite, Clark I and Clark II), phosgene and tabun. In many cases sulphur mustard fills contain arsenicals such as phenyldichloroarsine or 'arsine oil'. Only a small number of shells or bombs filled with tabun have been recovered in Germany, and those have been destroyed.

The old chemical weapons that have been recovered were new or experimental types of chemical munitions at the time they were manufactured. There are numerous such weapons, and the variety is great because many of these weapons were captured from various adversaries in the two world wars. The OCW munitions were dispersed over large parts of Central Europe as a result of both World War I operations and battles (e.g., Flanders), and the moving of chemical weapons to various storage depots in World War II.

Old chemical weapons are believed still to be buried at former test sites and destroyed production or filling sites, like the one at Munster. These chemical weapons include failed batches of spray canisters filled with viscous sulphur mustard.

III. Steps for disposing of old chemical weapons

Reconnaissance and locating old chemical weapons

In Germany, prior to actually searching the terrain for the presence of old chemical weapons a comprehensive examination is made of relevant archives, other available sources and the accounts of eyewitnesses. Aerial infrared photographs provide valuable information, since surface irregularities in areas near former CW production or filling facilities often indicate buried objects such as munitions.

Ground searches are carried out by EOD experts using magnetic probes or other metal detectors. These instruments can locate metal objects to a depth of 6 metres. Locations where a positive detector response is recorded are marked with flags, and precautions are taken to prevent unauthorized access.

Unearthing and identifying old chemical weapons

In order to recover and identify buried objects the EOD experts cautiously remove the soil until the object or objects can be seen. Portable X-ray equipment is used to evaluate the internal structure of the munition. The object is assessed to determine if it is an old chemical weapon, and an evaluation is made of its state of preservation and possible leakage. A decision is taken as to whether or not the object can be moved; some OCW munitions have to be disassembled and decontaminated where they are found. Unidentified items are transported to the nearest dismantling[1] facility and opened for sample-taking and chemical analysis of the contents.

Appropriate safety precautions are taken at all times during the process of unearthing and identification, including the establishment of ample safety zones around the munitions that have been discovered and the utilizing of standard operational procedures (SOPs) for individual protection.

[1] The dismantling of old chemical weapons is the mechanical disassembling of munitions including the removal of the fuses and explosive charges (bursters) and the draining of chemical fills.

Removal and transport of old chemical weapons

Since old chemical weapons are often found scattered over a large area, they have to be collected and transported to an intermediate storage site. Sturdy containers are needed to transport OCW, and these containers must be gas-tight under pressure and coated so that they are able to withstand the use of aggressive decontaminants. In Germany these containers must be approved by the federal authority that regulates the transport of dangerous goods and explosives.

When old chemical weapons are moved, detailed arrangements must be made with the federal state police and environmental agencies about the route to be taken. If significant quantities of OCW are to be transported the police restrict access to the roads on which they will be moved, and personnel trained in detection, decontamination and medical support accompany the transport.

Intermediate storage of old chemical weapons

In Germany old chemical weapons are stored and categorized by type, to the extent possible, in order to facilitate their handling and ultimate destruction, to avoid confusion and to ensure safety. The OCW, which have been removed from various areas, are placed in facilities that provide safe intermediate storage.

The storage bunkers (Stradley type) must meet stringent standards as regards overpressure stability, fragment protection of the walls and doors and gas-tightness in the event of the accidental explosion of munitions with release of toxic chemicals. They must also have chemical agent detection and alarm systems. A ventilation system that filters the air and automatic fire-prevention equipment including sprinklers are needed to ensure worker safety and to protect the environment should an accident occur. In addition, there must be a basin to collect the waste water that would result if decontamination or fire-fighting activities become necessary.

Such a bunker is structured with separate compartments so that the various types of munitions are segregated. The walls are high enough to prevent the spread of any explosions that may occur. The storage facility is subject to frequent inspection in accordance with the safety and security regulations which govern its operation.

Provision must be made for early warning and possible evacuation of those who live near the storage facility. These plans are updated continuously based on the results of risk analyses. Although unlikely, it is impossible to rule out completely the release of toxic chemicals as the result of an explosion, a fire or the crash of an aircraft.

Preparing old chemical weapons for dismantling

The munitions that are excavated are generally coated with dirt and have a rough surface because of corrosion. For this reason, prior to further treatment they have to be cleaned using a high-pressure water jet.

Since the internal structure of the recovered old chemical weapons varies considerably from item to item, it is crucial that each munition be X-rayed individually. The resulting film must then be evaluated and accurate measurements made of all relevant features. Only then can the EOD experts determine how the dismantling of a particular munition should be carried out.

In the final step prior to dismantling the munitions are frozen in order to reduce the internal pressure of volatile liquids or gaseous chemical fills. This reduces or eliminates the spilling of chemical agents from the shells when they are opened.

Dismantling

When old chemical weapons are dismantled they are disassembled, the fuses and explosive charges (bursters) are removed and the chemical fills are drained. In most cases remote-controlled special dismantling equipment—such as defusing tools, machines for drilling, milling and sawing, and chopping benches—needs to be used for these critical and hazardous procedures.

This equipment is operated from a dismantling control bunker. In the event of an explosion or the release of toxic chemicals the personnel are protected from fragments, vapours and aerosols by thick concrete walls and a highly efficient air filtration system. The dismantling process is constantly monitored and videotaped to provide thorough documentation for later evaluation of the operation.

The techniques that have been developed for draining old chemical weapons include: (*a*) draining liquids after piercing or drilling through the wall of the munition; (*b*) expansion of pressurized volatile agents such as phosgene through a pipe system connected to an absorption column; (*c*) extraction of paste fills using special tools such as spatulas; and (*d*) removal of solid fills from opened munitions by melting them using hot steam injection. Currently, none of these draining techniques are used in automated continuous operation. However, a highly automated dismantling line could be used for a large number of sufficiently uniform munitions (typical of current CW stocks), and the United States has shown this to be feasible.[2]

Draining chemical munitions involves a high risk of contamination not only of the dismantling equipment but also of all surfaces inside the dismantling chamber. Equipment which is not part of the actual mechanical operation must therefore be encapsulated. For thorough decontamination of the exposed surfaces inside the chamber a highly effective decontaminant, such as the decontaminant emulsion used by the Federal Armed Forces, must be kept in readiness.

Impermeable protective suits and respirators with filter canisters must be worn by all personnel when they work in the dismantling chamber calibrating, servicing or repairing dismantling equipment. The protective equipment is decontaminated and carefully inspected after any work that might involve exposure to toxic chemicals. In addition, specialized first aid medical assistance and ambulances are available when dismantling operations are conducted.

[2] Presentation to the author at Chemical Agent Munition Disposal System (CAMDS), Tooele, Utah.

IV. Destruction of chemical agents

The German approach to OCW disposal includes incineration of the chemical agent fills at high temperatures as the final step. This results in the complete mineralization of the chemical compounds. The Munster incineration plant has a destruction capacity of approximately 70 tonnes per year, including the contaminated waste and metal scrap from munitions. The flue gases from combustion are scrubbed with sodium hydroxide in an aqueous solution, and emissions from the stack are monitored continuously in order to meet stringent German environmental standards.

Prior to incineration chemical agents, wastes from the dismantling process and empty shells are placed in polyethylene (PE) containers that can be burned together with their combustible contents. The large volume of solid and liquid waste (e.g., protective suits, gloves, boots and exhausted active carbon filters) is disposed of in the same way.

Batch-type incineration technology was chosen to be used primarily because the large quantities of viscous mustard gas discovered in the Munster region exhibit poor solubility in any type of solvent. Munitions filled with phosgene are easily disposed of by neutralization with sodium hydroxide.

In addition to the incineration plant at Munster, another incineration plant is under construction for the disposal of soil contaminated with both arsenicals and other toxic chemicals. Fragments of explosives (TNT) with adherent arsenicals from old pit burning sites must be processed in a special way; presumably a grinding step will be used followed by dilution with inert material before the thermal treatment.

V. The incineration plant

Initial conditions

After approximately five years of planning and construction, the Munster incineration plant began operation in 1980. The type and quantity of OCW materials were evaluated, and a two-stage, batch-type furnace was selected for their incineration. As mentioned above, this choice was made to solve the problem of destroying viscous mustard gas.

The clumps of viscous mustard gas (see figure 12.1) that were found had been in the ground for more than 40 years. They were contained in canisters which had been almost completely destroyed by corrosion. The concentration of sulphur mustard in the clumps ranged from 25 per cent to as little as 0.1 per cent, depending on where measurements were taken. Viscous mustard gas also contains sulphur mustard analogues and various arsenic compounds such as phenyldichloroarsine which is commonly known as 'Pfiffikus'.

Based on these facts a technical solution was chosen that initially employs evaporation of the volatile CW agents and other components. Incineration is followed by gas scrubbing in the flue of the incinerator and arsenic precipitation.

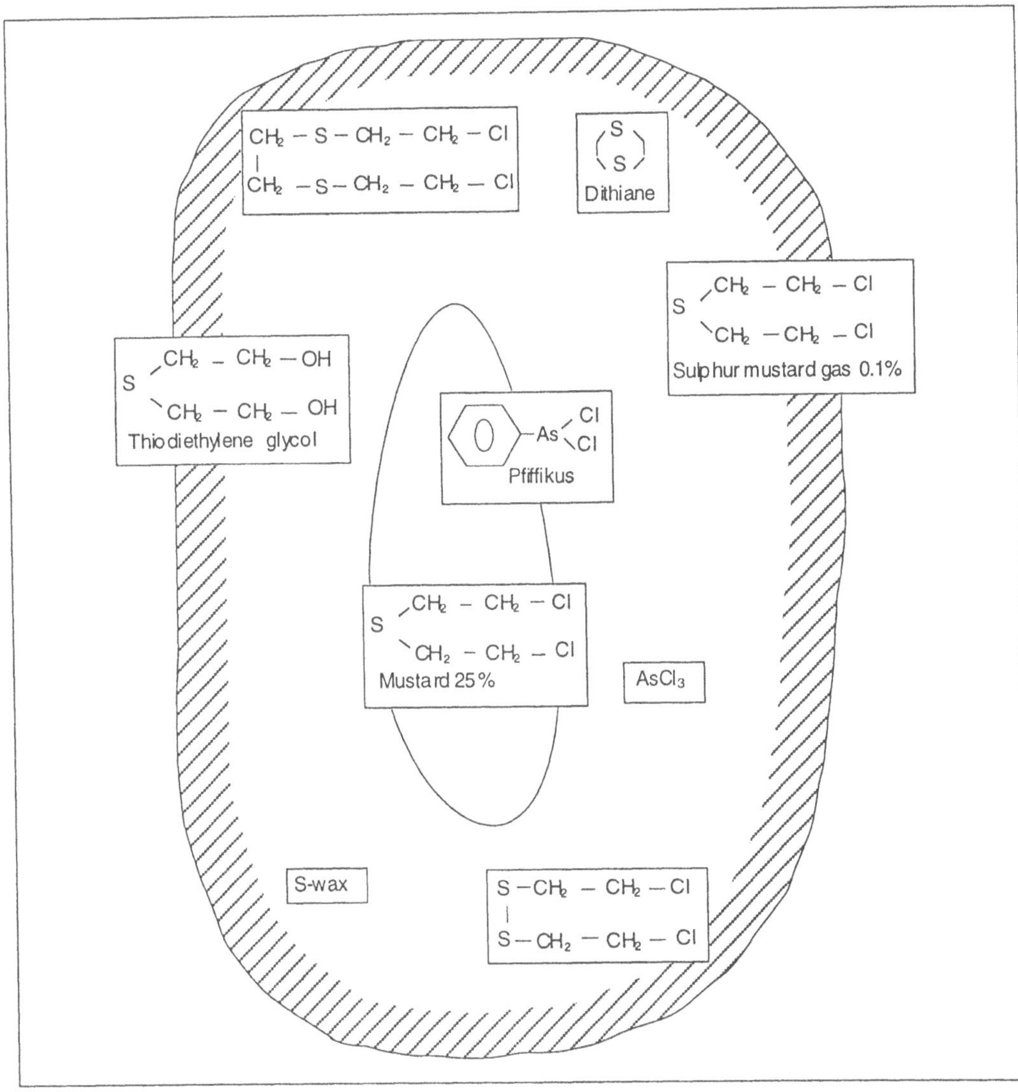

Figure 12.1. Composition of the clumps of viscous mustard gas

Intermediate chemical storage and the charging area

The viscous chemical agents are stored together with their metal casings in 30-litre PE barrels and are kept in an intermediate storage area designed to hold the quantity of chemicals that can be processed by the Munster incineration plant in one week. Contaminated waste is treated in a similar fashion.

Solid chemical agents such as Clark are removed from opened shells by steam injection and are also delivered to the incineration plant in plastic barrels. Prior to incineration the chemicals are analysed for the presence of arsenicals. Batches of chemicals to be incinerated are made up based on the results of this analysis, and the incineration method is chosen which will provide the most effective purification of sewage and effluent air. This procedure is depicted in figure 12.2.

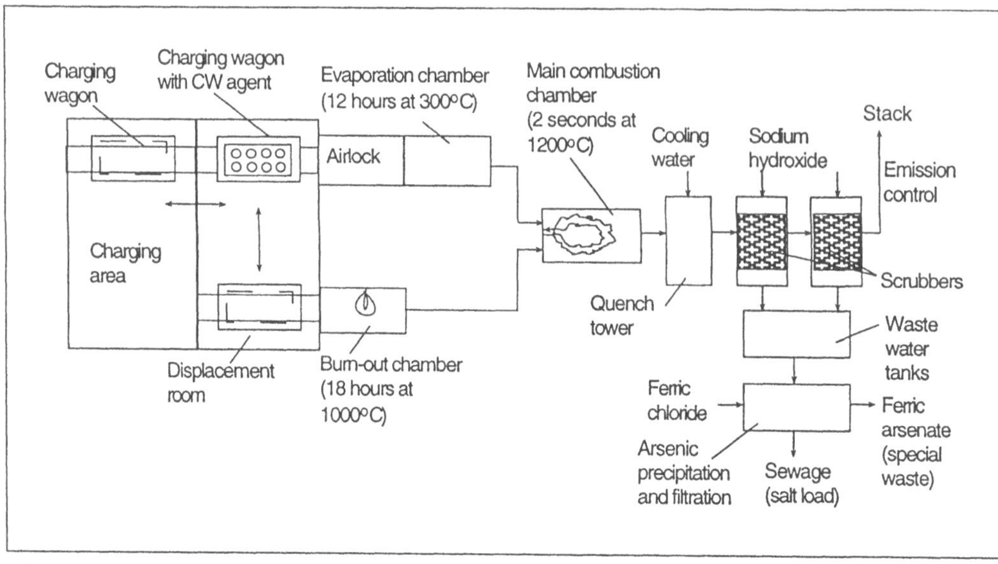

Figure 12.2. Functional diagram of the Munster incineration plant

The first step in the incineration plant is the placing of the PE barrels on a 2.9 x 0.9 metres 'charging wagon'[3] that is covered with fireproof bricks. The lids are removed from the barrels and a cable pulls the wagon into the first chamber of the furnace (the evaporation chamber) through an airlock. Small amounts of liquid CW agents or liquid waste in tin boxes are also placed on the charging wagon. In order to ensure environmental safety and prevent the release of contaminated air, a negative pressure of 0.5–1.0 millibars is maintained in the charging area.

The evaporation and combustion chambers

Inside the evaporation chamber the chemical agents are evaporated and subjected to partial pyrolysis at 300°C in an inert gas environment (nitrogen plus CO_2 and H_2O produced by stoichiometric combustion of fuel oil). This process takes 12 hours for viscous mustard gas owing to the low heat transfer and the need to prevent sudden and violent combustion. Suction from a central fan pulls the evaporated matter into the main combustion chamber (see figure 12.3), which is lined with fireproof bricks. There it is completely incinerated at temperatures of 1000–1200°C; a residence time of two seconds in this high-temperature area is sufficient to achieve total oxidation.

The main combustion chamber is heated with fuel oil using an ultrasonic burner. The charging wagon containing the unevaporated organic chemical and metal residues is moved into the burn-out chamber where metal parts, including cartridge cases, are completely decontaminated at 1000°C. They can subsequently be scrapped and no longer present a chemical hazard. This process requires about the same amount of time as was needed in the first chamber.

When, for example, the CW agent sulphur mustard (HD) is burned, carbon dioxide (CO_2) and water are generated in addition to the environmental pollut-

[3] The charging wagon is a wagon which is loaded with PE or other containers of chemicals that can be incinerated in one batch. The amount depends on the composition of the material.

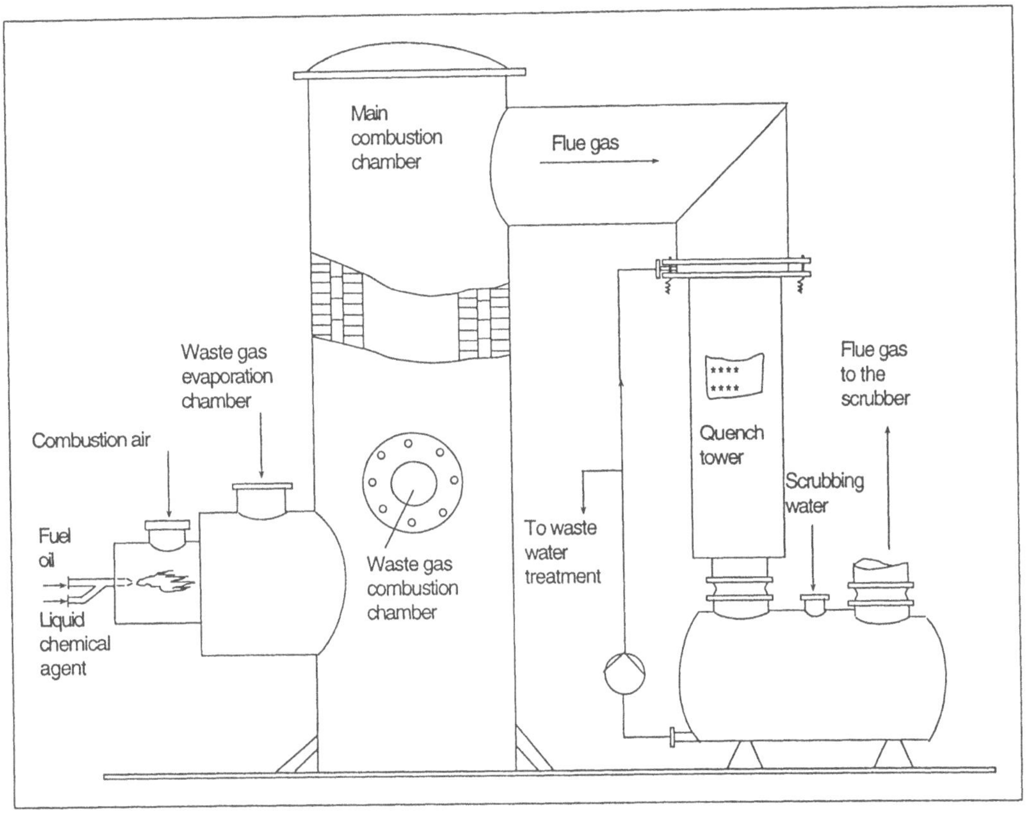

Figure 12.3. The main combustion chamber

ants sulphur dioxide (SO_2) and hydrogen chloride (HCl), which must be eliminated from the flue gas (see figure 12.4). The waste gas produced in the burn-out chamber is fed into the main combustion chamber.

Fluid chemical agents can be fed directly into the main combustion chamber together with the fuel oil through a multi-fuel burner lance. In the process the fluid is pumped from a charging tank through a double-walled heatable pipeline by a mechanical pump.

Cooling and scrubbing the flue gas

The most commonly occurring components of the flue gas are the SO_2 and HCl generated by combustion. In order to remove them, the flue gas is cooled rapidly to 80°C by the injection of cooling water and then washed with alkaline water in two packed scrubbing towers each 6 metres high (see figure 12.5). A caustic soda solution is used to maintain the pH value in the scrubbers at a constant level. This neutralizes both the SO_2 and HCl contaminants, and the SO_2 is oxidized in a reaction tank with hydrogen peroxide (H_2O_2). This process generates sodium sulphate and sodium chloride which are non-toxic but which create salts that can damage the municipal sewage system.

If the initial material does not contain any harmful compounds the water used for scrubbing can be discharged directly into the sewage system. Unfortunately, many of the CW agents produced formerly in Germany contain arsenic compounds and need special treatment (see the discussion in the next subsection).

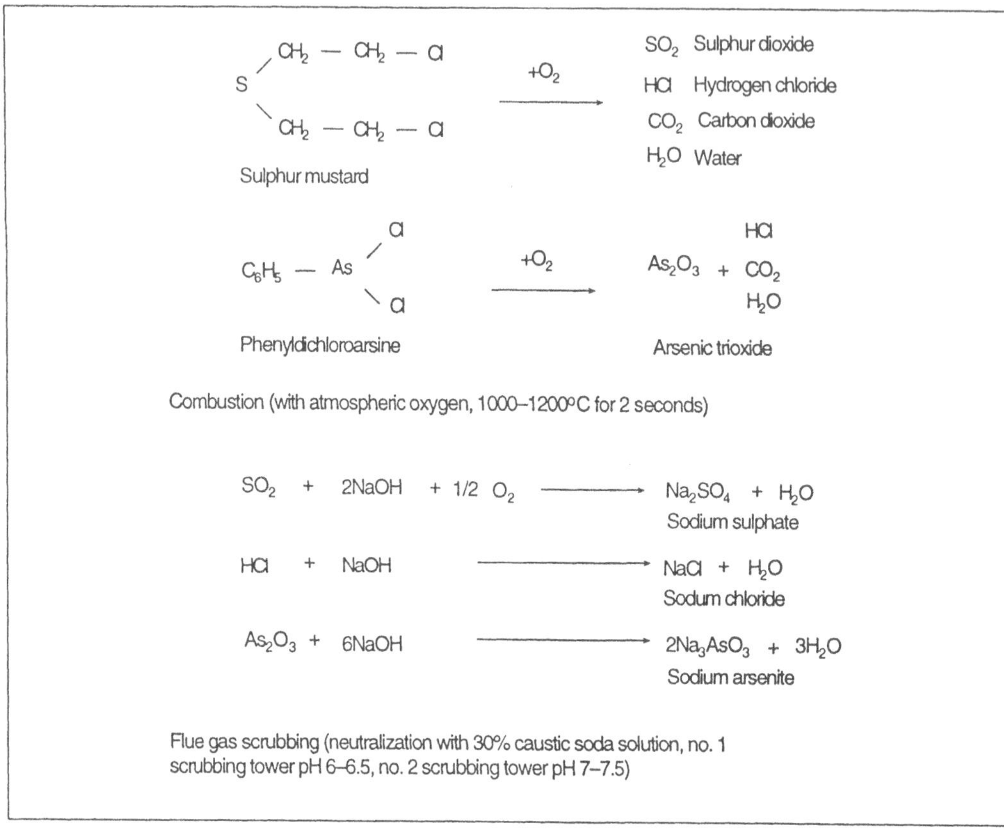

Figure 12.4. Destruction of chemical warfare agents

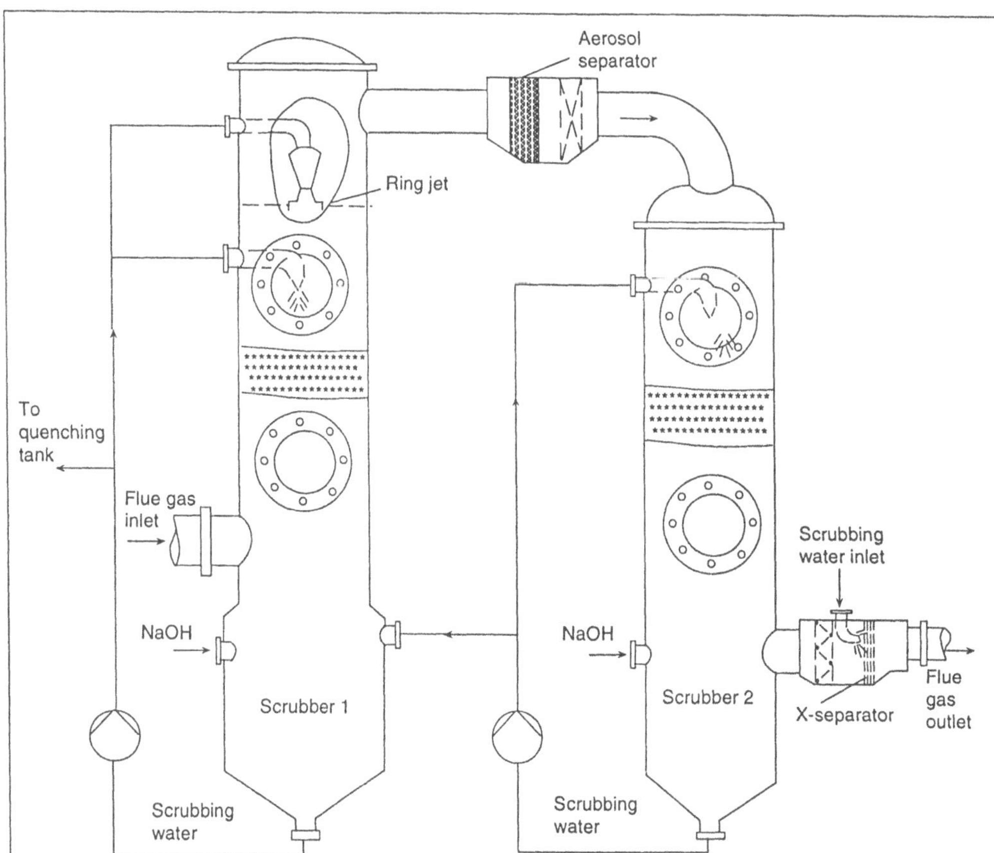

Figure 12.5. Flue gas scrubbing

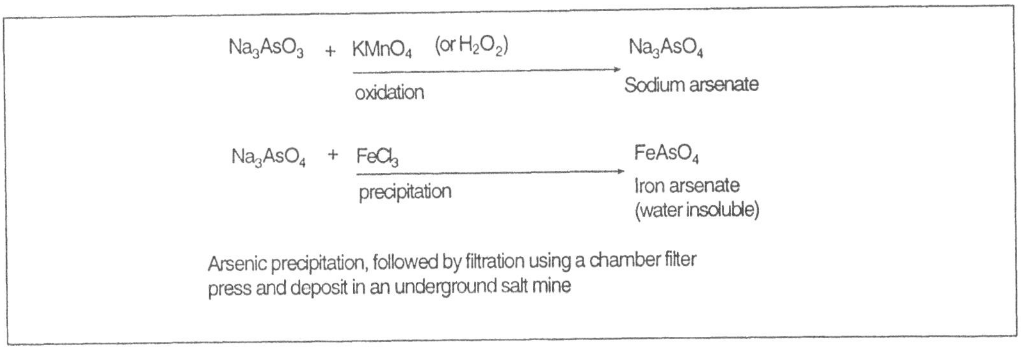

Figure 12.6. Destruction of chemical warfare agents, arsenic precipitation

The flue gas passes through an aerosol separator which removes any remaining liquid or solid particles (e.g., As_2O_3 aerosol) before it is discharged into the open air through the stack. The SO_2, HCl, nitrogen oxides and hydrocarbon emission rates are measured at the stack and monitored continuously in the central operation control room. Dust and arsenic are separated isokinetically from the waste air by a dust probe and are subjected to laboratory analysis by atomic absorption spectroscopy. Rapid cooling of the flue gas (from 1000°C to 80°C) by the introduction of water prevents molecule fragments from recombining to form dioxins or difuranes. The so-called 'dioxin window' between 400°C and 600°C is passed too fast for dioxins to be formed.

Arsenic precipitation

If the combustion material contains arsenic, as do some of the batches of viscous mustard gas that arrive for destruction at the Munster facility, the waste water must be subjected to supplementary treatment (see figure 12.6).

Incineration of the organic arsenic compounds produces arsenic trioxide, which is then washed out as trivalent arsenite in an alkaline environment. Such inorganic arsenic compounds remain strong poisons and potential carcinogens. The conversion of arsenic into a compound that is insoluble in water requires an oxidizing treatment with potassium permanganate (or hydrogen peroxide) in order to obtain pentavalent arsenate. When iron trichloride ($FeCl_3$) is added precipitation of insoluble iron arsenate occurs. The resulting sludge material is dehydrated in a filter press and then filled into 200-litre drums for final deposit in an underground waste dump that was formerly a salt mine. Eventually, the purified filtered water is released into the municipal sewer system. However, because of its high salt concentration a waste water evaporator must be used.

VI. Operational data

The Munster incineration plant operates 14 hours per day in two overlapping 8-hour shifts (see table 12.1). On weekends and holidays the plant is maintained at a lower temperature level since starting up and cooling down the facility are time- and energy-consuming activities. In recent years the plant has been in operation 200–250 days per year and inoperative 50–100 days per year. The throughput of the incineration plant varies considerably depending on the material to be incinerated. Sulphur mustard, for example, can be incinerated at a rate of 20–35 kg/hour.

Table 12.1. Operational data for the Munster incineration plant, as of March 1993

Characteristic	Relevant data
Operating time	
Normal operation	14 hours/day (two 8-hour shifts)
Weekend/holiday operation	Temperature maintained by support firing
Average availability	200–250 days/year
Inoperative periods	50–100 days/year
Throughput	
Sulphur mustard	20–35 kg/hour
Capacity of the charging wagon	24 polyethylene barrels (30-litre barrels)
Energy demand	
Fuel oil	100 litres/hour
Electric power	4 500 kilowatt hours/day
Tap water	Approximately 2m^3/hour
Inert gas	400 newton m^3/hour
Compressed air (6 bars)	60 newton m^3/hour
Total effective thermal power	1.2 megawatts

Source: Munster internal data bank.

Table 12.2. Waste air and waste water data for the Munster incineration plant, as of March 1993

Type of waste	Amount
Waste air (from a 30-metre high exhaust chimney)	
Waste gas quantity	2 800 m^3/hour at 105°C
Waste air emission rates (and allowed levels[a])	
Sulphur dioxide	<5 mg/m^3 (100 mg/m^3)
Hydrogen chloride	<5 mg/m^3 (50 mg/m^3)
Arsenic	Approximately 0.2 mg/m^3 (1 mg/m^3)
Dust	<1 mg/m^3 (30 mg/m^3)
Nitrous oxides	Approximately 200 mg/m^3 (500 mg/m^3)
Waste water	
Waste water quantity	Approximately 0.5 m^3/hour
Salinity	5–40 grams/litre
pH value	Adjusted to approximately 8
Arsenic	Adjusted to a limit of 1 mg/litre

[a] Limits for emission rates under the 1986 TA-Luft regulations.

Source: Munster internal atmospheric emission records.

The energy demand for the combustion chambers is 100 litres/hour of fuel oil, and 4500 kilowatt hours (kWh) of electric power are required each day. The incineration plant has a 258 kilowatt volt ampere (kVA) power generator available for use in an emergency and also an accumulator station with a capacity of 700 ampere hours (Ah). The inert gas used in the evaporation chamber is produced by stoichiometric combustion of fuel oil and equals 400 newton (N) m^3/hour. Sixty Nm^3/hour of compressed air at a pressure of six bars are used to atomize the fuel oil. Approximately two m^3/hour of tap water are used

Table 12.3. Disposal balance for the Munster incineration plant, as of March 1993[a]

Type of waste	Amount
Quantities disposed of since 1980	
Chemical agents	73 tonnes (t)
Contaminated materials[b]	527 t
Total	**600 t**
Quantities in storage and ready for destruction	
At Munster-North Training Area	
Chemical agents	115 t
Contaminated soil	30 000 t
In the states of the Federal Republic of Germany	
Chemical agents[c]	30 t
Contaminated soil	1 000 t

[a] Discovery of additional quantities of World War I and World War II chemical agents is expected.

[b] For example, scrap ammunition and waste materials.

[c] Primarily arsenic agents.

Source: Munster internal disposal balance files.

to cool the flue gas and in the preparation of the scrubbing solution. The total effective thermal power of the incineration plant is 1.2 megawatts.

VII. Control of emissions

Emissions from the incineration plant are monitored continuously to check for SO_2, HCl, residual hydrocarbons (including traces of CW agents) and nitrogen oxides (see table 12.2). Dust is collected at the stack by discontinuous sampling and is subsequently analysed for its arsenic content.

The pH value of the discharged waste water is approximately eight, and the amount of arsenic in the sewage is less than one mg/litre, which is the authorized threshold limit specified in the sewage regulations of the local communities. Comparison of the actual and the allowed levels of emissions indicates that the chemical agent incineration procedures used at the Munster facility cause only minimal pollution.

VIII. The disposal balance of the incineration plant

Since 1980, when the Munster facility began operation, 73 tonnes of pure CW agents have been destroyed (see table 12.3). Much larger quantities of contaminated material have been destroyed. As of March 1993, a total of 600 tonnes had been incinerated.

IX. Cost considerations

An investment of 28 million Deutschmarks (DM) was needed to set up the facilities at the Munster incineration plant as they now exist (see table 12.4). Annual costs are approximately 2.8 million DM, with the interest payments

Table 12.4. Cost for operation of the Munster incineration plant, as of March 1993

Figures are in million Deutschmarks.

Type of cost	Extent of cost
Infrastructure	
Incineration plant and buildings	16.0
Enlargement of the facilities	4.0
Technical improvements	8.0
Total	**28.0**
Current annual cost	
Depreciation	2.8
Operating costs	1.0
Salaries of personnel[a]	0.7
Total annual cost	**4.5**

[a] Personnel include 1 operating manager, 1 chemical engineer, 2 shift leaders, 1 laboratory assistant and 4 mechanics.

Source: Munster internal comptroller's office.

(depreciation costs) of 1.1 million DM the largest cost. Costs for the nine personnel employed at the facility are 0.7 million DM annually. If all costs are taken into account the price for destruction of 1 kg of sulphur mustard CW agent, for example, may be as high as 100 DM. The cost to dispose of sulphur mustard is thus about 10 times more than the cost to produce it.

X. Conclusions

Disposal of old chemical weapons that have been recovered on German territory is a long-term undertaking, and it is difficult to predict how long a period will be needed for their destruction.

Incineration is an appropriate method for the destruction of old chemical weapons (and current chemical weapons) since it causes irreversible and total mineralization of the toxic compounds involved. (In a few exceptional cases, such as phosgene, hydrolysis is a more effective method of destruction.) The double-chambered incineration plant in Munster is likely to continue its operation for a long time to come. Owing to the lack of uniformity and the extraordinary extent of corrosion most of the old chemical weapons need to be handled individually in a very time-consuming and costly process.

Further research and development are needed in order to develop new techniques and procedures to speed up the disposal of OCW. It seems feasible that in the future OCW fills can be incinerated in universal furnaces, such as rotary kilns, and even in a continuous process.

Manual techniques will continue to be employed widely to search for, unearth, transport and dismantle old chemical weapons. Improved physical and chemical methods of identifying the various types of munitions and fills could facilitate and accelerate dismantling and destruction. In any event, considerable effort will continue to be necessary to protect the personnel involved and to ensure decontamination.

13. Postwar destruction of chemical weapons in the former German Democratic Republic

KARLHEINZ LOHS*

I. Introduction

It is well known that during World War II Germany made preparations for chemical warfare and was, indeed, substantially ahead of the Allies in its research and development and in some cases in its production of chemical weapons. Documentation from the 1945–46 Nuremberg War Crime Trials and postwar research in Germany and elsewhere have abundantly substantiated this.[1] Additional evidence was presented at various conferences in 1989–92.[2] All give strong testimony to the complexity of the task of doing away with chemical weapons in the decade immediately following World War II. Too often those responsible for demolition did not have adequate knowledge about the type, the capability or the manufacturing history of the weapons with which they had to deal. In the postwar period the primary weapon disposal problem in Germany was lack of information. The question arose: Where had the German Army (Wehrmacht) and the Nazi élite corps, the Schutzstaffel (SS), destroyed or partially destroyed their supplies of chemical weapons during their retreat? Witnesses could not be found because people were afraid to volunteer information that they feared would lead to accusations of war crime. The psychological climate during the Stalin era and the intense cold war of the 1950s is difficult to imagine today.

This chapter outlines only those disposal operations which were designed to dispose of chemical warfare components and munitions containing chemicals. The author of this chapter directed a team which did not deal with conventional munitions. Such destruction was under the authority of other branches of the police and fire departments of the GDR. Later in the mid-1950s the National People's Army (NVA) had specially trained munition ordnance squads or munition removal units for each region of the country.

[1] Gellermann, G. W., *Der Krieg, der nicht stattfand* [The war which never happened] (Bernard und Graefe Verlag: Koblenz, 1986); Brauch, H. G. and Müller, R. D. (eds), *Chemische Kriegführung, Dokumente und Kommentare* [Chemical warfare: documentation and commentary], vol. 1 (Arno Spitz Verlag: Berlin, 1985); Niedersächsisches, Umweltministerium [Lower Saxony, Ministry for the Environment] *Expertengespräch Rüstungsaltlasten* [Expert talk: military wastes] (Dokumentation: Hanover, 1989); and Oberholz, A., *Tödliche Gefahr aus der Tiefe* [Deadly dangers from the depths] (Kommunalverlag: Düsseldorf, 1991).

[2] Thomé-Kozmiensky, K. J. et al. (eds), *Untersuchung von Rüstungsaltlasten* [Investigation of military wastes], *Verfahren zur Sanierung von Rüstungsaltlasten* [Methods to redevelop military waste sites], *Management zur Sanierung von Rüstungsaltlasten* [Managing the clean-up of old armament wastes], 3 vols (EF Verlag für Energie und Wirtschaft: Berlin 1992); Lohs, K. and Spyra, W., *Chemische Kampfstoffe als Rüstungsaltlasten* [Chemical warfare agents as military wastes] (EF Verlag für Energie und Wirtschaft: Berlin, 1992); and Kiefer, K.-W., Pfaff-Schley, H. and Schimmelpfennig, L. (eds) *Rüstungsaltlasten '91* [Old armament wastes '91] (Erich Schmidt Verlag: Berlin, 1991).

* This chapter is based on the personal experiences and recollections of the author who participated in CW destruction activities in the former German Democratic Republic in 1955–62.

II. The postwar situation in the former German Democratic Republic

In eastern Germany—which was first the Soviet occupied zone and in 1949 became the German Democratic Republic—chemical weapons, depots, production sites and filling stations were disarmed and dismantled or blown up by the Soviet forces according to the provisions of the 1945 Potsdam Declaration and the terms established by the Allied Control Council. The Soviet Union has not released documentation of those activities, but the next few years may witness the opening of the Soviet military archives for that period.

Owing to the number of CW production and storage sites and munition-related activities demolition took considerably longer than was anticipated. In 1952 demolition was made the responsibility of a division of the German People's Police (Hauptverwaltung der Deutschen Volkspolizei). Official documents that should have existed were unavailable or missing, and the police faced an impossible situation.

In 1953 a new police division was established and given the task of disarming, dismantling and removing the former weapon production and storage sites, including those related to chemical warfare. In the autumn of 1955 the Technical Office (Amt für Technik) in Berlin took over these activities. This agency was responsible for both arms procurement and for matters related to the atomic energy industry. Wehrmacht munitions specialists, who had been used as forced labour in the Soviet Union immediately after the war, were employed in the Technical Office under the direction of first, State Secretary Wolf and later, State Secretary Egon Winkelmann. Armament specialists from the former Third Reich worked together with young engineers and scientists who lacked hands-on experience with conventional and chemical munitions because they were trained after the war.

III. Destruction activities at Kapen

Demolition work began in 1952–53 at a former CW filling station at Kapen near Dessau. A small group of chemical technicians, fire-fighters, medical workers and a physician established detoxification and disposal procedures for the chemical shells and bombs which had been located and made preparations for disposal of those which would be found later. The group was also given the task of dealing with unknown substances and residues in the pipelines and sewers of chemical munition plants and sites that had been demolished. Later the Chemie Kapen (VEB Gärungschemie Dessau) company assumed responsibility for the destruction activities.

The Kapen facility was located in a wooded area directly adjacent to an active Soviet ammunition depot about which the German authorities were not given detailed information until 1990. It contained partially destroyed ammunition bunkers, concrete basins that contained tanks full of mustard gas and somewhat usable concrete detoxification pits measuring approximately 3 x 5 metres. The buildings at the site had been demolished.

Kapen was apparently selected as the final CW clean-up site owing to its central location in the middle of the former GDR, near an autobahn. In addition, there was a mustard gas production plant at nearby Ammendorf which contained a considerable amount of mustard gas that was to be transported to Kapen for disposal. There was also a railway line from Kapen to the Ergethan Factory in Stassfurt where arsinöl,[3] a component of the CW agent Winter-Lost, had been produced.

IV. Special problems

The postwar situation in Germany was chaotic and can hardly be imagined today. The records and information that were supplied by Soviet officials when the Kapen facility was turned over to the GDR were incomplete and unreliable, and information from former production workers was often contradictory. The relevant professional literature was laboriously assembled from the available war records and from private sources to the extent possible. Understandably, there was reluctance to admit possession of documents that were branded as 'forbidden Nazi literature' in the GDR.

The task of destruction was complicated by discoveries of munitions whose locations had not been marked on maps during the German retreat. These munitions had been haphazardly hidden, buried and abandoned in forested areas. In order to disarm them before moving them to Kapen for destruction the German People's Police transported them to special transfer stations at facilities which had belonged to the Wehrmacht. Soviet authorization was required for this. There were two transfer stations: one near the Hermsdorfer autobahn intersection for the southern part of the country, and one at Mellenthiner Heide on Usedom Island for the northern part of the country.

At Kapen, together with Herbert Telemann and Kurt Thiesen, the first plant manager, Stangel, constructed special facilities for high-temperature incineration and processing basins for chemical detoxification of non-combustibles. (In the autumn of 1954 Kurt Bischoff became plant manager.) The incineration was primitive but effective. Mustard gas was combined with diesel oil in a ratio of 1:3–1:4 by means of a mixing nozzle and injected into a metre-long burner. Incineration temperatures were approximately 600°C. The combustion products (sulphur dioxide, hydrochloric acid, partially burned diesel oil and carbon dioxide) were released into the air through stacks without any means of gas scrubbing.

The concept of environmental protection as it is known today did not exist at that time, and the 'success' of a combustion method was measured solely by the volume of materials that could be completely burned by the process. In addition to the usual laboratory testing of air samples, measurements were made using Dräger gas detection tubes. The same principle was used to destroy sulphur mustard gas using chlorinated lime in the detoxification basins.

[3] Arsinöl was a crude product containing 50% phenyldichloroarsine, 35% diphenylchloroarsine (DA), 5% triphenylarsine and 5% arsenic trichloride. It was used either as a freezing point depressant for mustard gas or as a raw material for DA. See SIPRI, *The Problem of Chemical and Biological Warfare*, vol. 1, *The Rise of CB Weapons* (Almqvist & Wiksell: Stockholm, 1971), table 4.4, p. 282.

V. Other munition sites and activities

There were, of course, munitions that could not be safely moved to Kapen. For example, in 1955 more than 100 heavily rusted tanks of phosgene from World War I and II were discovered at Schönau (near Eilenburg). They were placed in three flooded quarries, shot by sharpshooters and were sunk immediately thereafter. The escaping clouds of phosgene were neutralized by ammonia gas from tanks installed around the edges of the quarries for this purpose. By today's standards this method seems almost criminal. However, at the time it was the only destruction method available, and it did in fact function safely.

In 1956–57 investigations were made of claims by the residents of Finsterwalde that munitions had been present in the region in 1945, but none were found.

In 1954–61 it became apparent that the entire territory of the GDR needed to be more thoroughly searched for war materials than had been thought necessary in 1952–53. The Soviet Union did not cooperate in supplying information despite efforts by Colonel Günther Ludwig, then chief of the NVA Chemical Services, Colonel Dr Ludwig Mecklinger, Chief of Medical Services (later professor and Minister of Health of the GDR) and his successor in the NVA, Colonel Dr Kurt Geiger. There were numerous Red Army military advisors at the National Defence Ministry in Strausberg near Berlin, but they provided no assistance with these destruction activities.

In 1960 approximately 50 phosgene bombs were discovered by chance in an area used for swimming at Wiehe in the vicinity of Artern. The bombs were recovered with great difficulty by the Kapen salvage crew, transported to Kapen for temporary storage and then taken to the Baltic Sea and sunk with other munitions. The same procedure was used for a badly rusted mustard gas bomb found in a forest near Artern. A search was conducted for remnants of munitions that had been loaded onto railway cars at Frankenberg and Lossa in March–April 1945, but no munitions were found. The former GDR Technical Office and the Ministry of the Interior discontinued the operation and notified the German People's Police of the Erfurt region that no more searches would be conducted.

In 1960 a large area that was highly contaminated by arsinöl and its decomposed compounds was discovered on the grounds of the former Ergethan factory in Stassfurt, a World War II site for the production of arsenic chemical warfare agents. Arsinöl was used as an additive to Winter-Lost, which was produced at the Orgacid factory at Ammendorf. The contaminated soil at the Ergethan factory also contained adamsite and Clark I. The soil was removed, loaded into metal containers, sealed 'for eternity' and placed in salt mine tunnels near Stassfurt. The same procedure was used for a quantity of arsenic trioxide and for 4 tonnes of chloracetophenone (CAP) which were buried in the VEB Kalikombinatt Werra pit of the Alexandershall salt mine located near Eisenach and what was then the border between the Federal Republic of Germany and the GDR.

VI. The status of destruction in 1962

The experience and knowledge acquired in these disposal activities were later recorded in books and journals and used as training materials for police technical personnel, fire fighters and the army.[4] The time-consuming search, removal and demolition programme lasted until the end of 1962. The Government of the former GDR then adopted the position that all chemical weapons, ammunition components, related products and World War II facilities and structures had been found. Imminent danger was said to have been eliminated, but bombs and shells continued to be found during building-site excavations. These finds became the responsibility of the NVA Chemical Services. Government responsibility for munition finds and their eventual demolition remained with the Ministry of the Interior.

A special disposal problem was caused by a 1960 incident in the Baltic Sea near Wieck and Greifswald when fisherman unwittingly hauled a mustard gas bomb onboard their boat. The bomb discharged some of its contents and totally contaminated the vessel.[5] Two fishermen suffered gas burns and had to be taken to the Greifswald University Clinic. The lengthy and complicated detoxification and clean-up of the afflicted vessel were carried out by a crew from the Kapen disposal facility under the direction of Dr Albert Mathes, then assistant director.

The total amount of munitions of all types sunk in the Baltic Sea by Allied forces is unknown. Eyewitness reports of the events following the two world wars confirm the primitive nature of the dumping. As late as 1962 the German People's Army sank a vessel loaded with explosives and ammunition (including barrels containing unknown contents) north-east of the Danish island Bornholm. It was not until 1992–93 that a federal–state commission under the leadership of the Hydrographical Nautical Services in Hamburg looked into the magnitude and impact of postwar dumping activities.[6]

By 1962 the demolition operations carried out in the former GDR had processed approximately 40 000 tonnes of ammunition and warfare chemicals encased in drums, barrels, canisters and the like. Approximately 60–75 per cent of the chemical weapons belonged to the mustard gas group: sulphur mustard, nitrogen mustard and so-called 'mixed mustard'. Arsenic-organic warfare agents, mainly Clark I and Clark II and adamsite, which made up 5–10 per cent of the materials, had already decomposed in the ground where they were

[4] Lohs, K. *Synthetische Gifte* [Synthetic poisons], 1st edn (Verlag des Ministeriums für Nationale Verteidigung: Berlin, 1958), 4th edn (Militärverlag der DDR: Berlin, 1974); Lohs, K. and Martinetz, D., *Entgiftung und Vernichtung chemischer Kampfstoffe* [Decontamination and destruction of chemical warfare agents] (Militärverlag der DDR: Berlin, 1983); Stöhr, R., *Die chemischen Kampfstoffe* [Chemical warfare agents] (Deutscher Militärverlag: Berlin, 1961); and *Chemische Kampfstoffe und Schutz vor chemischen Kampfstoffen* [Chemical warfare agents and protection against chemical warfare agents] (Militärverlag der DDR: Berlin, 1985).

[5] Rößner, H., Stiewe, M. and Zimmermann, R., 'Yperitschäden bei Hochseefischern' [Yperite injuries of deep-sea fishermen], *Zeitschrift Militärmedizin*, vol. 16, no. 4 (1975); and Seidel, R. and Westpahl, H. J., 'Behandlung und Verlauf schwerer Yperitschäden an der Haut' [Treatment and trends of Yperite injuries to the skin], *Deutsches Gesundheitswesen*, vol. 28, no. 5 (1973), pp. 2473–76.

[6] *Chemische Kampfstoffmunition in der südlichen und westlichen Ostsee: Bestandsaufnahme, Bewertung und Empfehlungen*, Bericht der Bund/Länder-Arbeitsgruppe Chemische Kampfstoffe in der Ostsee [Chemical munitions in the southern and western Baltic Sea: compilation, assessment and recommendations, Report by the Federal/Länder Government Working Group 'Chemical Munitions in the Baltic Sea'] (Federal Maritime and Hydrographic Agency: Hamburg, May 1993).

buried. Approximately 10 per cent of the chemicals were irritants, primarily chloroacetophenone; 5–10 per cent were from World War I: 'old wastes' such as rusted steel drums containing chlorine and phosgene, hydrogen cyanide shells with 'polymerized' contents and old white cross, yellow cross and blue cross shells. The details of what was destroyed are no longer available owing to the fact that documents in the archives of the Ministry of the Interior and the Defence Ministry were unable to be located after the collapse of the GDR.

VII. Nerve agents

The GDR did not dispose of large quantities of phosphor-organic chemical weapons of the tabun and sarin group in its original weapon disposal operations. However, in 1964 divers retrieved 400, 10.5-cm tabun-filled shells, found severely decomposed in the harbour basin of Wolgast Usedom. On 14 August 1964 the shells were encased in 36 cement blocks, loaded aboard the coastal vessel *Eisvogel* and sunk to a depth of approximately 3000 metres somewhere between Iceland and Norway in the North Atlantic.[7]

By the end of the war in 1945, 12 000 tonnes of tabun-filled munitions and 400 tonnes of sarin-filled munitions had been secured by the Allied Forces and disposed of in a manner that was not disclosed to the German authorities. The 1964 find of tabun-filled shells in Wolgaster Harbour is evidence that the Soviet Army transported large quantities of these munitions from their German zone. The main production facility for phosphor-organic materials was at Dyhernfurth near Breslau. The production site was said to have been partially blown up by German troops during their retreat. French sources indicate that some of the tabun-filled munitions were found at a depot in the Black Forest in 1945.[8] Other sources indicate finds in Lossa and on Elbe River barges near Torgau.[9]

Russia has provided no information on the sarin II facility at Falkenhagen.[10] However, it seems clear that sarin was not being produced in the Falkenhagen plant at the end of World War II because the German Waffen-SS needed all silver metal equipment[11] for its own plant (part of the Falkenhagen plant) to produce chlorine trifluoride (N-Stoff). Little is known about the disposal activities of the Western Allies in the immediate postwar period, but there is an abundance of information about munition dumps and depots for conventional explosives, armaments and other military chemicals.[12]

[7] Persönliche Mitteilung des damaligen technischen Direktors (Dr Albert Mathes) an den Verfasser [Personal communication from the former technical director (Dr Albert Mathes) with the author].

[8] Capitaine Collomp, 'Les Trilons' [Nerve gases], *Bulletin d'Information Technique et Scientifique*, 23/G, 28 (Ministère de la Guerre, Section Technique de l'Armée: Paris, 1949).

[9] Defa–Dokumentarfilm, 'Es zog eine schwarze Wolke' [A dark cloud has passed], Berlin–Babelsberg. Die im Film benannte Aktion basierte u.a. auf der geheimen Komandosache des Generalfeldmarschall Keitel, Chef des OKW [The action in the film is based *inter alia* on the secret command from Field Marshal Keitel, Chief of the Supreme Command of the Army], 30 Mar. 1945–5 Apr. 1945.

[10] Lohs, K., Schriftliche Mitteilung an den Rat des Kreises Seelow vom 14 Nov. 1992 [Letter to the Council of the Seelow District of 14 Nov. 1992].

[11] Silver is an essential component of the equipment used to produce sarin.

[12] See *Expertengespräch Rüstungsaltlasten* (note 1); Thomé-Kozmiensky *et al.* (note 2); Lohs and Spyra (note 2); and Preuss, J. and Wiegandt, C.-C. 'Rüstungsstandorte des Deutschen Reiches auf dem Gebiet der Bundesrepublik Deutschland' [Armament production sites of the Third Reich on the territory of the Federal Republic of Germany], *Geographische Rundschau*, no. 3 (1992), pp. 175–78.

There were production facilities for the nerve agent lead tetraethyl (anti-knock) at Frose (Saxon-Anhalt) and Gapel–Döberitz near Premnitz (Land Brandenburg). These facilities were included in the munition disposal activities in the GDR owing to the fact that, in the 1930s, the British considered lead tetraethyl and other lead-organic compounds to be potential chemical warfare agents. The USSR demolished these two production plants in a crude fashion, leaving behind destroyed facilities contaminated with extremely toxic lead tetraethyl and equally toxic by-products,[13] but detailed information of these activities is not available. A 300–400 tonnes sludge residue, consisting of lead oxide and 1–5 per cent lead tetraethyl, is estimated to have remained.

The difficult and dangerous work of detoxification and removal was carried out by specially trained workers. Chemical breakdown of the contaminated matter in the drainage conduits, pipes, settling tanks and soil was achieved with a 1 : 1 mixture of sulphuryl chloride and kerosene. The resulting pasty sludges of lead oxide, lead sulphide and lead chloride were deposited in a landfill. One of these former production sites is again in operation producing leaded fuel compounds for civilian use.

VIII. Smoke-producing agents and obscurants

Smoke-producing agents were rarely encountered by the disposal team led by the present author, and then only in relatively small quantities of a few tonnes. Those found were primarily smoke acids (i.e., chlorosulphinated acid or highly concentrated sulphuric acid, a mixture of H_2SO_4 and SO_3, stored in special steel containers.) Smoke-producing powders of zinc chloride—hexachlorocyclo-hexane (HCCH) or varieties of so-called 'Berger mixtures', sometimes mixed with dyes and used as smoke signals—were discovered in only small quantities. These were burned on site or taken to the Kapen disposal facility. A small amount was sent to the German People's Army to be used for training purposes. If the condition of the containers made it possible to transport them, the smoke-producing acids were sunk in the Baltic Sea, as mentioned above. The leaking weapons found by the present author's team were disposed of by specially trained personnel, following the safety precautions in use at the time. The materials were poured directly on the ground at the military training areas or in large open fields far from inhabited areas. For many years these regions were off limits to unauthorized persons.

The disposal of other military chemicals was not considered significant and was not of concern for the disposal team led by the author of this chapter. Training materials, primarily simulation chemicals, were simply burned, and old supplies of neutralizing agents such as chlorinated lime and chloramine were used for the destruction of chemical weapons. Later, excess amounts of neutralizing agents (made surplus by the arrival of new shipments for the army) were given to farmers, who made good use of these agents for pest control.

[13] Müller, K., 'Zur Kasuistik der Bleitetraethylvergiftung' [On the cause of lead tetraethyl poisoning], *Zentralblatt für Arbeitsmedizin und Arbeitsschutz*, vol. 3 (1974), p. 9. See also Lohs (note 4), pp. 179–84.

IX. The status of destruction: 1962–89

In 1962 the plan for the destruction of all CW production and filling sites was revised, and a decision was taken that in the future only situations of obvious and significant danger would be addressed. Sites such as the one in Löcknitz near the Polish border had been destroyed and dismantled by the USSR, but not yet entirely cleaned up. The Orgacid factory in Ammendorf and other 'suspected' sites such as the former Wehrmacht and SS training areas had not been completely decontaminated by the NVA Chemical Services.[14] In order to deal with specific problems, such as the incident involving the *Eisvogel* described above, Lieutenant-General Werner Fleissner, the NVA officer responsible for technology and weapons, occasionally rendered technical assistance in civilian demolition operations.[15] Civilian demolition specialists worked in the R&D facilities of the German People's Army at Altenburg–Thüringen. After R&D work was discontinued there, personnel and equipment were moved to the Military Technical Institute at Königswusterhausen near Berlin, under the direction of General Karl-Heinz Müller.

The extent of munition disposal activities that were carried out from 1962 until the end of the 1970s is unknown. A top secret Ministry of the Interior document[16] included reference to questions prepared by a working group of the Social Democratic Party (SPD) of the FRG under the direction of Egon Bahr that was sent to Hermann Axen, a Central Commission (ZK) member of the German Socialist Unity Party (SED) of the GDR. The questions refer to mustard gas and phosgene bombs stored at World War II munition sites at Hermsdorf–Thüringen and Jessenitz near Schwerin. Axen's answer provided information about dumping but nothing about the extent of disposal activities. There is also no information on reports submitted to civilian authorities or governmental bodies such as the Ministry for the Chemical Industry or the Ministry of Health.

X. Conclusions

The author of this chapter was a supervising participant in all of the disposal activities discussed here which took place in 1955–62; his knowledge of operations before and after this period is incomplete.[17] The sketchy accounts of CW disposal activities in 1989–90 reflect the fragmentary information available at the time. The access to records and supplemental technical literature which has become possible since the collapse of the GDR may eventually fill the gaps in knowledge about that period. However, there is currently a lack of funds for research, for travel to and assessment of the sites in question and for addressing the disposal problems that may exist.

[14] See note 7.

[15] See note 7.

[16] Confidential Classified Matter VVS 1/079 295, vol. 2, Berlin, 25 May 1988, p. 2.

[17] See *Expertengespräch Rüstungsaltlasten* (note 1) and Kiefer, Pfaff-Schley and Schimmelpfennig (note 2).

14. The storage of old and abandoned chemical weapons in Austria

JOSEF STOLZ

I. Old and abandoned CW munitions in Austria

With the signing of the Chemical Weapons Convention[1] in Paris on 13 January 1993 a new chapter opened in the efforts to rid the world of chemical weapons. Austria was one of many original signatory states. Austria does not produce chemical weapons, but as a consequence of World War II it has a CW storage site on its territory. It is therefore particularly interested in the provisions of the Convention which address old and abandoned CW munitions as it faces the problem of the handling and eventual destruction of such munitions.

Since the late 1940s and the early 1950s efforts have been made to cope with the problem. In 1974–76 a 'long-term storage facility under safe conditions' was built for old and abandoned CW munitions.[2] The type of 'long-term storage' envisaged was designed to be permanent and tailored to the particular CW munitions in question.

The old and abandoned chemical munitions were stored in the vicinity of Gross-Mittel, Austria, at a munition plant which had belonged to the German Wehrmacht, and which was located about 40 km south of Vienna. The munitions were probably placed in storage in 1942–43 and are in the so-called 'chemical district' of the plant at its site no. 67. There is no documentation of the storage activities, the kind and amount of CW munitions or their origin.

Site no. 67 was destroyed completely by an 11 November 1944 Allied air raid, and large quantities of CW munitions were partly destroyed, damaged or detonated. Many grenades and rocket projectiles were dispersed over an area of approximately one square kilometre.

The Federal Ministry of the Interior, which is responsible for the problems associated with military items from World War II that remain in Austria, has no information which indicates that there are other locations where CW munitions were left behind by the German Wehrmacht.

The type and calibre of the abandoned CW munitions

The CW munitions stored at site no. 67 consisted of unfused 7.5-cm grenades, fused 10.5-cm grenades and fused 15-cm rocket projectiles.[3]

The 7.5-cm grenades probably originated in France. The chemical warfare agent in them is presumably phosgene (carbonyl chloride), but no data are available to confirm this. The 10.5-cm grenades were equipped with a small smoke impact fuse, technically known as the kl.A.Z.23 Nb, with a small deton-

[1] Conference on Disarmament document CD/1173, 3 Sep. 1992.

[2] Austria, Old chemical weapons: description of a long-term storage facility under safe conditions, Conference on Disarmament document CD/CW/WP.397, 5 May 1992.

[3] Massak, A., *Disposal of Second World War Grenades Containing a Chemical Agent, with Special Attention to Environmental Protection* (Federal Ministry of the Interior: Vienna, 1975).

ation charge (C/32 Nitropenta) and a large central tube charge. The chemical warfare agent was Nitrogen-Lost (tris(2-chloroethyl)amine, also known as HN-3), and the results of repeated analyses are available to confirm this. The 15-cm rocket projectiles were equipped with a bottom fuse (BdZ DOV) and a central tube charge. The chemical warfare agent was identified as Nitrogen-Lost, but no data are available to substantiate this.

The grenades and rocket projectiles had only fragmentary colour and letter coding, which were insufficient for reliable identification. A few of the 10.5-cm grenades were labelled 'Nb'. Military experts assume that the munitions contained a mixture of chemical weapons and smoke in order to produce highly visible explosions. A 1964 report noted: 'After questioning the former responsible persons in the supreme command, it can safely be assumed that Lost was dispersed in the vicinity of Vienna. If there were only ammunition trains, then there are also smoke-grenades with chlorosulphonic acid or with Berger-mixture. It will be appropriate to examine each grenade or its content after opening'.[4]

Officers of the German Wehrmacht with specialized knowledge and training were responsible for these stored CW munitions, as well as for conventional munitions.

The handling of CW munitions found after World War II

From November 1949 to March 1950 approximately 30 000 grenades and rocket projectiles were buried provisionally in the vicinity of site no. 67. The site was located in the Soviet occupation zone, and the initial operation to secure the area took place under Soviet supervision.

A pit (12 metres x 12 metres x approximately 6 metres) was dug, and the CW munitions was sorted by calibre and stacked in the pit. The height of the stacks ranged from 1.8 metres to 2 metres, and the munitions were covered with a layer of calcium hypochlorite that was 0.5–0.6 metre thick. The layer of calcium hypochlorite was then covered with a 4-metre layer of soil. No accidents involving chemical warfare agents are reported to have occurred.

Opening the temporary dumping site

Evaluating the munitions

In 1964 the Austrian Government ordered the pit opened to recover and destroy the CW munitions so that possible pollution of the ground water from the release of chemical warfare agents could be prevented. The Mine Clearing Service of the Federal Ministry of the Interior carried out the operation with the assistance of the Austrian Army, and 100, 7.5-cm and 10.5-cm grenades were removed.[5] The pit was again closed and covered with the excavated soil. The decision to open the pit was motivated by a decision by Vienna to add a third pipeline to its water system in the area known as the Mitterndorfer Senke,

[4] Scholz, S. in *Chemical Warfare Agent Analysis and Proposals for the Disposal of CW Munitions*, quoted in Massak (note 3).

[5] Grössinger, H., 'Sanitation measures in the recovery of mustard gas grenades', *Österreichische Ärztezeitung*, vol. 21, no. 23 (10 Dec. 1966), pp. 2623–26.

where the CW munitions were located, as well as by the desire of the government to protect the ground water.[6]

The opening of the pit allowed the CW munitions to be inspected for corrosion of the fuses and the shells, for identification of the chemical warfare agents and for evaluation of their status. With this knowledge the experts could then plan the destruction of one or more of the chemical warfare agents.

A detection powder was used during the recovery operation to test the soil for the presence of chemical warfare agents, and the air was tested with Draeger tubes and gas detector pumps. These tests revealed that no toxic agents were present in sufficient concentration to be detectable.

In addition, 15 of the 10.5-cm grenades were spot checked by the use of a stationary drill, and samples of the charges were taken for chemical analysis. One grenade had an 'overpressure',[7] and during the drilling there was a partial release of chemical warfare agent. Two people were contaminated and sustained injuries to their skin. Analyses of samples from the 10.5-cm grenades proved that only Nitrogen-Lost was present.

Investigation of the shell casings proved that there had not been significant corrosion since 1949–50. The fuses exhibited varying degrees of corrosion related to the type of metal used in their construction. Those made of aluminium were in relatively good shape, while those made of an alloy of aluminium and zinc had disintegrated partially. For both types, the safety mechanisms were corroded and ineffective and no longer prevented activation of the firing pin. Consequently, the munitions were classified unsafe and could not be transported.

No 7.5-cm grenades were opened, but evaluation of the cloud which resulted from a test detonation of 7.5-cm grenades made it possible to identify the chemical warfare agent as phosgene.

Proposals for further action

Following the investigation of the munitions various experts made proposals for further action, such as: (*a*) unearthing the grenades and placing them in concrete containers, then filling the containers with concrete; (*b*) digging up the grenades, then dumping them in an abandoned mine; (*c*) excavating the grenades, placing them in iron barrels, filling the barrels with concrete and dumping them at sea; and (*d*) dismantling the grenades, then removing and destroying the chemical warfare agent.[8]

Option (*c*) was a method that was often used in the 1950s and 1960s for the disposal of CW munitions, while option (*d*) is the only option which involved true destruction of the chemical warfare agent. The destruction could have been accomplished by: (*a*) mixing 1 part chemical warfare agent and 7 parts diesel oil together and burning the mixture; or (*b*) chemical neutralization of the chemical warfare agent with a solution of 40 parts ethanolamine and 100 parts

[6] Leeb, A., *Documentation: The Isolation of Grenades Containing a Chemical Agent* (Federal Ministry of the Interior: Vienna, 1976). This plan had been under serious consideration since 1960.

[7] Decomposition of the original agent will create products which can increase the internal pressure, resulting in 'overpressure'.

[8] See Scholz (note 4); Massak (note 3); and Meindl, J. in *Chemical Warfare Agent Analysis and Proposals for the Disposal of CW Munitions*, quoted in Massak (note 3).

methyl alcohol (the solvent methyl alcohol could then be redistilled from the reaction mixture).[9]

Elaboration of a draft proposal for destruction of the CW munitions

In 1965 the Austrian Government decided to adopt the approach outlined above in option (*d*). Additional testing was done to select the best method for the neutralization of the chemical warfare agents and for the dismantling of the CW munitions. During the chemical neutralization process the chemical warfare agents must be flushed out. The following methods for opening the grenades were discussed or tested: (*a*) opening by use of a cylinder milling cutter, parallel to the central tube, with the shell in a vertical position; (*b*) opening the shell with a milling cutter; (*c*) opening the shell at the ogive (pointed end) with a vertical turret boring machine, with the shell in a vertical position; and (*d*) drilling the shell at two points opposite each other.

Chemical warfare munitions are generally unsafe munitions. In addition, to the toxicity of their contents, there is danger from the fuses which may be of primitive technical design. When turning lathes are used in the dismantling process care must be taken if the ogive is deformed; such munitions cannot be dismantled using a lathe.

When the shells were opened it became apparent that the consistency of the chemical warfare agents varied with part in a semi-liquid state, part in a jelly-like state and part in droplets. It was not possible merely to pour out the chemical warfare agent, and complicated removal and inspection operations had to be carried out, including close examination of the interior of the shells.

The following chemicals were tested for use in the chemical neutralization process: tris(2-hydroxyethyl)amine and water; chlorinated lime and water; sodium hydrogen sulphate in an alcohol solution; ethanolamine and isobutyl alcohol; and sodium sulphide in an alcohol solution.[10] If tris(2-hydroxyethyl) amine and water, sodium hydrogen sulphate in an alcohol solution, or ethanolamine and isobutyl alcohol are used the anions that are released in the hydrolysis of Nitrogen-Lost will bond chemically and significantly influence the equilibrium of the reaction (hydrolysis). Chlorinated lime and water will react with Nitrogen-Lost in a complicated reaction that is only partially understood and for which no data are available.[11]

These experiments led to several conclusions. In the laboratory the procedures gave partially satisfactory results, but in the field some problems are to be expected. The chemicals needed for these procedures (e.g., hypochlorite and chlorinated hydrocarbons) are hazardous to the environment to varying degrees. Studies conducted over a period of several years revealed certain facts about the destruction of chemical warfare munitions:

1. Currently, it is not possible to dismantle these old CW munitions because the danger to those involved is too great.

[9] Zorn, *Chemical Warfare Agent Analysis and Proposals for the Disposal of CW Munitions* (Vienna Technical University, Institute for Chemical Technology of Organic Substances: Vienna, 1964, 1968).

[10] Zorn (note 9).

[11] Franke, S., *Military Chemistry* (Military Publishing House: Berlin, 1977), pp. 295–305.

2. The chemical neutralization of Nitrogen-Lost cannot be accomplished satisfactorily in the field.

3. No suitable incinerators are available for the destruction of chemical warfare agents.

II. Protecting the environment from the CW munitions

In 1970 the Minister of the Interior ordered that a proposal be drafted for securing the CW munitions so that they could not endanger the environment. It was suggested that the munitions be removed immediately from the temporary dumping site. A report from the Hirtenberger Patronenfabrik supported this opinion. It was also determined that the primer caps of the aluminium fuses were still functional, but that the primer caps of the aluminium and zinc alloy fuses were not.

Reports on the problem were obtained from the Technical University of Vienna and from Federal Army Testing Station no. 53 at Munster, Germany (now the German Armed Forces Scientific Institute for Protection Technologies: NBC [nuclear, biological and chemical] Protection). The Testing Station at Munster prepared a report on the disposal of Nitrogen-Lost by combustion, chemical neutralization and isolation.[12] It was based on the assumption that the stability of the bituminous materials, plastics and concrete had not been altered by the Nitrogen-Lost. The report from the Technical University addressed the general problem of isolating the chemical warfare agents and the stability of bitumen, concrete, aluminium and plastics which had come in contact with Nitrogen-Lost.[13] The proposals of the Technical University were as follows:

1. In isolating the CW munitions from the environment any activity which might raise the temperature of the grenades should be avoided.

2. Concrete containers of various dimensions should be designed for the munitions.

3. The walls of the concrete containers should be covered with aluminium or a layer of plastic fortified with glass fibre.

4. Each grenade should be placed in an aluminium tube.

5. Kieselguhr[14] should be poured over the munitions to fill the containers; it would then serve as an absorbent and also absorb shock. (Sand could also be used to absorb shock and would provide an additional safety mechanism during handling of the filled aluminium tubes.)

6. The filled aluminium tubes should be closed with aluminium caps and sealed with a 2–5-cm layer of casting resin.

[12] Martens, H., 'Concerning the elimination of CW munitions in the vicinity of the Gross-Mittel, Austria munition plant of the former German Wehrmacht', Federal Army Testing Station no. 53, Munster, FRG, report and comments, 1971.

[13] Prey, V. and Schindlbauer, H., *Elimination of Grenades Containing Chemical Warfare Agent Stored at the Gross-Mittel Plant* (Vienna Technical University, Institute for Chemical Technology of Organic Substances: Vienna, 1972).

[14] Kieselguhr is a natural sedimentary rock composed of silicon dioxide (70–90%), water (3–12%) and inorganic by-products. It has a large internal surface which provides excellent adsorption properties.

Preparation for isolating the CW munitions

Based on the above recommendations ferro-concrete (reinforced concrete) containers were constructed to the dimensions outlined in the regulations for the protection of so-called 'type S9' buildings. Reinforced B225-type concrete was used since it is the least porous kind of concrete.[15] The inside dimensions of the containers were 7.5 metres x 5.0 metres x 1.8 metres. The walls of the container were 0.8 metre thick, and the thickness of the saddle-shaped lid was 0.8–1.15 metres. Owing to the fact that concrete can be penetrated by chemical warfare agents it was necessary to cover the interior walls. The examination of different types of bitumen and plastics revealed that bitumen and sealing compounds react with chemical warfare agents by partially or completely dissolving, while plastics became contaminated to varying degrees, with some exhibiting swelling. Aluminium remained stable. However, plates of 2.5-mm stainless steel (NIROSTA type) were used to cover the interior walls, and they were seamlessly welded together on a grid.

Aluminium was chosen for the tubes for several reasons including its: (*a*) stability with Nitrogen-Lost; (*b*) resistance to corrosion; (*c*) mechanical properties; and (*d*) low specific gravity. Two sizes of aluminium tubes were used to accommodate the various sizes of the CW munitions. The filled aluminium tubes were transported in square modules with compartments for the munitions. Each module could accommodate 16 aluminium tubes.

The absorbent kieselguhr has an absorptive capacity of 1 part Nitrogen-Lost to 1 part kieselguhr. The 10.5-cm grenades required that 2.5 kg of kieselguhr be used to fill each aluminium tube (e.g., for chemical warfare agents weighing approximately 0.8 kg twice as much kieselguhr was needed.)

Isolating the CW munitions

Following technical and administrative preparation the work to isolate the CW munitions began in 1974. A detailed approach for dealing with the munitions was elaborated, and work was carried out in six periods from 1974 to 1976 (a total of 282 days). The following steps were used:

1. The temporary dumping pit was opened by mechanical and manual means.
2. The grenades were recovered and cleaned manually.
3. The grenades were examined manually.
4. The grenades were placed manually in the aluminium tubes.
5. The aluminium tubes were filled with kieselguhr and sealed manually.
6. The aluminium tubes were placed manually in the transport modules.
7. The transport modules were placed in a ferro-concrete container by a mobile crane.
8. The transport modules were stored in the concrete containers.
9. The concrete containers were filled with kieselguhr.
10. The concrete lids were placed on top of the concrete containers, which were then buried by mechanical means to a depth of approximately 4 metres using excavated soil.

[15] Dworzak, W., *Ferro-concrete Container for CW Grenades* (Federal Ministry of the Interior: Vienna, 1972).

The munitions were stored in five concrete containers. The total amount of munitions stored was: 4346, 7.5-cm unfused grenades; 23 240, 10.5-cm fused grenades; and 1108, 15-cm fused rocket projectiles.[16] Fewer munitions were stored in the five concrete containers than there had been in the temporary dumping pit. This discrepancy resulted from the fact that fragments of grenades and empty grenades were destroyed at a military proving ground.

III. Recently discovered CW munitions

As of 1993 an additional 121 CW munitions had been discovered in Austria. These grenades are stored above ground in a concrete bunker. The concrete bunker is surrounded on three sides by supporting walls and an embankment, and the entrance is closed by a heavy safety door. Inside the bunker there are two seamlessly welded aluminium containers with lids and 5-mm thick walls. The two containers are filled with grenades stacked in layers that alternate with layers of absorbent material, ending with a top layer of absorbent material. The bunker can be entered and air measurements can be taken.[17]

In 1989 three of the 10.5-cm grenades were opened and their contents scrutinized by the Technical University of Vienna,[18] on orders from the Ministry of the Interior and the Army. The grenades were filled with a liquid and a white solid substance with a weight ratio of approximately 1:1 to 1:1.4. The liquid was pure Nitrogen-Lost, and the white solid was dimer[19] N,N,N',N'-tetrakis-(2-chloroethyl)piperazonium dichloride and a small quantity of tris(2-chloroethyl)ammonium chloride. Both the solid and the liquid convert to tris(2-chloroethyl)ammonium chloride, and subsequently to tris(2-hydroxyethyl) amine (triethanolamine) and a small quantity of 4-morpholine ethanol.

Analysis of the hydrolysis of Nitrogen-Lost by the Technical University gave the results presented in figure 14.1. It was not possible to prove analytically that the compounds listed as III and V in figure 14.1 would result. The final products of hydrolysis under test conditions proved to be triethanolamine—tris(2-hydroxyethyl)amine—and Nitrogen-Lost—tris(2-chloroethyl)amine—in large quantities and a small quantity of 4-morpholine ethanol. If Nitrogen-Lost is increasingly diluted it will hydrolyse rapidly.

A report was also submitted by the Federal Institute for Water Quality of Vienna on the toxicity of the white solid substance in water.[20] Hydrolytic changes in the water used in testing revealed some methodological problems, but these were resolved when a hydrolytic balance was achieved. The conversion products were classified as mildly hazardous (Water Endangering Class 1).

[16] See Leeb (note 6).

[17] Neuninger, H., *Recommendations Concerning the Disposal of Grenades Containing a Chemical Agent* (Federal Ministry of the Interior: Vienna, 1989); and Neuninger, H., *Report on Outdoor Testing* (Federal Ministry of the Interior: Vienna, 1989). For a discussion of Dräger-Stitz probes see, *Testing Old Dangerous Waste Using Dräger Tubes*, Dräger brochure no. 343, 2/1989 (Dräger Werk: Lübeck, 1989).

[18] Schindlbauer, H., *Analysis of the Reaction Behaviour of N-Lost in Old CW Grenades* (Vienna Technical University, Research Institute for the Chemistry and Technology of Petroleum Products: Vienna, 1990).

[19] Dimer is an association of two identical molecules linked together.

[20] Kohl, W. and Rodinger, W., *The Aquatic Toxicity of N-Lost Reaction Products* (Federal Institute for Water Quality: Vienna, 1990).

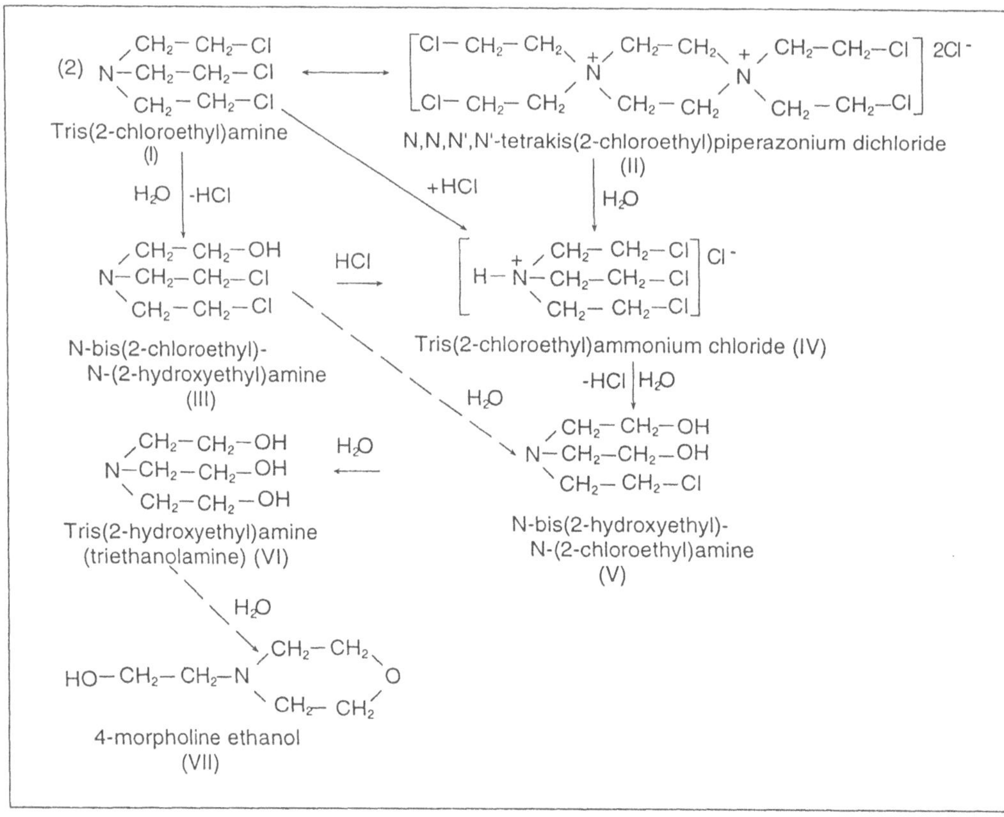

Figure 14.1. The hydrolysis of Nitrogen-Lost

Source: Schindlbauer, H., *Analysis of the Reaction Behaviour of N-Lost in Old CW Grenades* (Vienna Technical University, Research Institute for the Chemistry and Technology of Petroleum Products: Vienna, 1990).

The results of the tests by the Technical University proved that the Nitrogen-Lost which was analysed was a chemical warfare agent with high purity and one that was still effective after 50 years.

IV. The recovery and disposal of CW munitions in Austria

Prior to 1938 the following production sites for explosives, propellants and munitions were among those known to exist in Austria: the Dynamit Nobel explosives facility at St Lambrecht, the state-owned explosive facility at Blumau,[21] the state-owned propellant facility at Trofaiach, the state-owned munitions facility at Enzesfeld, the Hirtenberger ammunition facility, the ammunition facility at Lichtenwörth, the metalware facility at Berndorf, and the Böhlerwerk facility for the manufacture of arms and munitions at Kapfenberg.[22]

Most of these production facilities were established in the latter half of the 19th century and are located in the area south of Vienna. This was probably owing to its strategic location in the Austro-Hungarian empire, the proximity of military facilities and proving grounds, and the small population of the area.

[21] Mötz, J., 'The explosive facility in Blumau', *German Weapons Journal (DJW)*, no. 3 (1987).

[22] Schausberger, N., *Armament in Austria 1938–1945: A Study Relating to the Interaction between Industry, Politics and Strategy* (Brothers Hollinek: Vienna, 1970).

There are no official documents listing sites, locations or facilities for the manufacture of explosives, propellants and chemical warfare agents in Austria in World War II, and no documents are available about the clearing of such sites. Most such production facilities, sites and plants were destroyed in World War II by aerial attack, during the fighting, by deliberate destruction to avoid their falling into enemy hands or by dismantling and plundering during the period of occupation. It is known that among other facilities during World War II a propellant facility had been planned to be built at Ulmerfeld, and there was an explosives facility at St Lambrecht and an aircraft factory at Wiener Neustadt.

Currently, no high explosives or propellants are produced in Austria, and they must be imported if needed. Austria does produce explosives for civilian use in blasting activities as well as blasting caps, igniters, detonators, ignition switches, circuit testers and the like.

The Ministry of the Interior is responsible for the recovery and disposal of conventional munitions and other material remaining in Austria as a result of World War II. The Mine Clearing Service is obligated to see that such activities are carried out. (Munitions that are used by the Austrian Army are exempt from these regulations, and the Federal Ministry of Defence is solely responsible for the recovery and disposal of such munitions.) Conventional munitions are disposed of by the Mine Clearing Service by the following methods:

1. Old munitions can be exploded at designated proving grounds. In most cases this is done together with the Federal Ministry of Defence. This method is used for munitions which are larger than 2 cm.
2. Munitions smaller than 2 cm are incinerated in a combustion furnace with external heating. The core of the furnace is a perforated steel vessel in which the infantry munitions are heated to the deflagration point. The fused metal which remains in the steel vessel is disposed of after the furnace has cooled.
3. Large bombs (those larger than approximately 250 kg) which cannot be exploded are cut up after the explosive charge has been deactivated. Propellants are used to ignite the bombs which are then burned.

If technical investigations of munitions need to be done this is carried out by a laboratory for criminal investigation which reports to the Federal Ministry of the Interior.

V. Conclusions

This chapter summarizes the manner in which the old and abandoned CW munitions found in Austria were stored in 1974–76. The practical and technical experience gained from this operation provides information about how such munitions may be handled.

In the future the CWC will not allow this type of storage, which is termed a 'special kind of a deposition' in the Convention. Austria will thus need to make other arrangements for the disposition of any new munitions that are discovered. Under Articles III and IV of the Convention each State Party must declare and destroy any chemical weapons anywhere on its territory. There is an exception in Article III, paragraph 2, and Article IV, paragraph 17 which leaves it to the discretion of each State Party to apply the provisions of the articles and of the relevant provisions of Part IV of the Verifications Annex concerning chemical weapons which were buried on its territory before 1 January 1977 and which remain buried, or which were dumped at sea before 1 January 1985.[23]

It is the opinion of the author of this chapter that—under Article III, paragraph 2 and Article IV, paragraph 17 of the CWC—the old and abandoned chemical weapons in Austria which have been stored safely in five concrete containers can remain as they now are. However, the abandoned chemical weapons which were discovered subsequently and stored in a bunker above the ground will need to be addressed according to the provisions of the CWC. Article III of the CWC necessitates that these munitions be declared. They will also need to be dealt with according to the procedures laid down in Article IV and the Verification Annex—specifically paragraphs 13–15 of the section entitled 'C. Regime for abandoned CW' of Part IV (B).

Under the CWC Austria is: (*a*) obliged to declare any CW munitions that are discovered after its initial declaration of chemical weapons; (*b*) responsible for the destruction of subsequently discovered CW munitions; (*c*) authorized to make an agreement with the abandoning State Party about destruction of the abandoned chemical weapons; and (*d*) required to offer appropriate cooperation for such destruction.

[23] See Conference on Disarmament document CD/1173 (note 1).

15. The destruction of old chemical munitions in Belgium

JEAN PASCAL ZANDERS

I. Chemical warfare in Belgium in World War I

During World War I the northernmost part of the front line cut through the Belgian province of West Vlaanderen, running roughly from the coastal town of Nieuwpoort on the Yzer estuary over Diksmuide and Ypres to the French border.[1] Both ends of the front line were alternately occupied by British and French troops, with Belgian forces holding the centre. In February 1918 the area controlled by Belgian troops extended to the North Sea, and by June Belgian forces held most of the Ypres salient. The front was relatively static until the final series of Allied offensives late in 1918. After the First Battle of Ypres (autumn 1914), which frustrated German hopes of capturing the French Channel ports, the Belgian front remained calm although interrupted by some violent fighting, particularly in the Ypres salient (e.g., the Second Battle of Ypres in the spring of 1915, the Third Battle of Ypres in the summer of 1917 and the Allied breakout in 1918). In addition, Belgium assisted Britain and France in their major offensives in France. That assistance consisted of limited actions, such as raids or artillery duels, to occupy German troops.[2] However, the flooding of the Yzer River to halt the German advance in 1914 meant that the area was not suitable for offensive operations.

The relative quiet of the battlefront and the resulting hope of surprise probably explain why experiments with new toxic substances were carried out in Flanders. The Ypres salient was one of few sectors where the prevailing winds were not south-westerly. The presence elsewhere of disadvantageous winds from the south-west contributed to Germany's adoption of the use of shells for its chemical attacks, while the Allies, notably Britain, employed cloud gas attacks until the final days of the war. Appendix 15A summarizes the documented chemical warfare operations which took place on Belgian territory.

Modern chemical warfare is regarded as having begun on 22 April 1915. On that date German troops opened approximately 6000 cylinders along a 7-km line opposite the French position and released 150–168 tonnes of chlorine gas. Tear-gas (T) shells were also fired into the cloud and at the northern flank, the boundary between French and Belgian troops. Between 24 April and 24 May Germany launched eight more chlorine attacks. However, chemical warfare had not been assimilated into military doctrine, and German troops

[1] Belgium is a multilingual country, and the relative importance of Flemish and French has shifted dramatically in the past, resulting in various spellings for names of places. The spellings used here are those of the current official language for the respective places unless they have an English name. In 1900–30 French was the dominant language for all administrative and legal purposes. This is reflected in names of governmental bodies and military units. The names used in official documents are also retained.

[2] Lheureux, V. (2nd Lt.), 'L'utilisation des gaz de combat sur le front belge pendant la guerre 1914–1918' [Use of gas on the Belgian front in the 1914–18 war], and its annexes, Travail de fin d'études présenté pour l'obtention du titre de licencié en Sciences Sociales et Militaires, Ecole Royale Militaire, 126e Promotion Toutes Armes, 1989–90 [Dissertation submitted to obtain the degree of master in social and military sciences, Royal Military Academy, 126th Promotion All Arms, 1989–90], p. 46.

failed to exploit their strategic surprise. Chemical weapon attacks in following weeks were fundamentally different as they supported local offensives and thus served tactical purposes. In each case the amount of gas released was much smaller than that employed on 22 April, and crude individual protection against gas enabled Allied soldiers to hold the lines.

Prior to the April 1915 use of a chlorine cloud, gas shells filled with T-stoff (xylyl bromide or benzyl bromide) or a mixture of T-stoff and B-stoff (bromoacetone) had been employed. In addition, as early as 14 February 1915 (i.e., approximately the same period as CW trials on the Eastern front) two soldiers of the Belgian 6th Division had reported ill after a T-shell attack.[3] In March 1915 French troops at Nieuwpoort were shelled with a mixture of T- and B-stoff (T-stoff alone had proved unsatisfactory). In response to the British capture of Hill 60 (approximately 5 km south-east of Ypres), German artillery counter-attacked with T-shells on 18 April and the following days.[4] In the hours before the chlorine attack on 22 April the 45th Algerian Division experienced heavy shelling with high explosive (HE) and T-stoff.

Such attacks continued throughout the Second Battle of Ypres.[5] Although Germany overestimated the impact of T-shells, on 24 April their persistent nature appears to have been exploited for the first time for tactical purposes. Near Lizerne (approximately 10 km north of Ypres) German troops fired 1200 rounds in a wall of gas (Gaswand) behind Belgian lines to prevent reinforcements from reaching the front. The park of Boezinge Castle, where Allied troops were concentrated, was attacked in a similar manner.[6]

In 1915 only two additional gas attacks are known to have occurred in Belgium after the Second Battle of Ypres. In the first of these T-shells were used against a Belgian unit on 16 August; in the second attack, near Ypres on 19 December, a phosgene cloud was used for the first time. This marked the 'return' of German gas operations from the Eastern front. In 1916 the first British cloud gas attacks from trenches took place in Belgium, but gas was used infrequently in Flanders in 1916.

However, 1917 marked a turning point for three reasons. First, after 23 April the number of gas attacks in Belgium increased significantly. Shells and mortar bombs began to be used instead of cylinders, and this contributed to the assimilation of gas tactics into the planning of operations. Second, during the night of 12–13 July German forces used mustard gas for the first time when 50 000 shells were fired into the Ypres salient. Two days earlier, at Nieuwpoort, Germany had used blue cross (diphenylchloroarsine) shells for the first time in an attempt to counteract the measures being employed to protect troops against gas.[7] Third, with shells in general use the flooded areas, which had previously

[3] Decuyper, Dr, *Journal de Campagne*, Brugs Genootschap voor Geschiedenis, 1968 [Diary of the campaign, Bruges Society for History], in Baccarne, R. and Steen, J., *Boezinge 1914–1918: Gasaanval 2° Slag om Ieper* [Boezinge 1914–18: gas attack at the Second Battle of Ypres] (Uitgegeven onder de auspiciën van de N. V. Bank van Roeselare en West Vlaanderen [Published under the auspices of the N. V. Bank van Roeselare en West Vlaanderen], 1979), p. 135 (in Dutch).

[4] Foulkes, C. H., *'Gas!': The Story of the Special Brigade* (William Blackwood & Sons: Edinburgh and London, 1936, 2nd reprint), p. 34.

[5] Haber, L. F., *The Poisonous Cloud: Chemical Warfare in the First World War* (Clarendon Press: Oxford, 1986), pp. 25, 33–34. See also the detailed chronicle by Baccarne and Steen (note 3), pp. 128–88.

[6] Baccarne and Steen (note 3), p. 171.

[7] For a discussion of the cross codes, see the subsection 'Colour coding using cross codes' in section II of chapter 6 in this volume.

shielded Belgian troops from cloud gas attacks, no longer functioned as a protective barrier. The use of CW was now common on the front lines. This led to the little-known Belgian retaliatory gas shell attacks in the autumn and to the general use of gas shells for retaliation, harassment and counter-battery fire throughout 1918.[8]

Belgium was occupied for most of the war and was unable to produce chemical munitions indigenously; instead, shells were purchased primarily from France. The shells that were purchased included: hydrogen cyanide (shell no. 4), phosgene (shell no. 5) and—for the final offensive in 1918—mustard gas (shell no. 20). Belgium also bought chloropicrin shells (NC) from Britain. Although it is not known how they were procured, Belgium obviously possessed large quantities of German T-stoff shells since field manuals which refer to captured munitions extensively detail their use. The Belgian forces fired approximately 260 000 chemical rounds (25 per cent of all shells), and 55 per cent of these rounds were consumed in the period from the start of the final offensive on 28 September until the end of the war.[9]

There were four phases of chemical warfare in Flanders. In the first phase, from April 1915 until December 1915, cloud gas attacks dominated, and there was some use of T-shells. Nieuwpoort and the Ypres salient were the main combat areas. Waterways or flooded fields separated the battlelines in the central sector, and this probably explains the absence of cylinder attacks there.

In the second phase, December 1915–March 1917, there were few gas operation on the Belgian front. Only approximately 10 such operations are on record, and the majority of these were carried out in the Ypres salient. In this phase new chemical agents such as phosgene were introduced, the first British counter-attacks took place and the use of cylinders was gradually replaced by the use of shells.

The third phase, April 1917–September 1918, was marked by a dramatic shift in the nature of chemical warfare activities. Gas tactics had been assimilated into the military strategies of the major belligerents to varying degrees and chemical agents began to be used liberally for various purposes. In addition, new compounds, such as mustard gas, were introduced which changed the nature and perception of chemical warfare. In the Third Battle of Ypres large quantities of artillery shells, many of which contained gas, were used by all participants.

In Flanders two particular events influenced developments. On 23 April 1917, at Nieuwpoort, an unexpected wind diverted a massive chlorine cloud which German forces had released against French troops and which drifted across the entire length of the front held by Belgian forces. Belgian commanders and troops, who had previously not been concerned about gas warfare because of the relative safety offered by the flooded areas and waterways, were

[8] Only few people, including military and local historians, are aware of Belgian gas operations and defence measures. There are three main reasons for this. First, most literature on gas operations is of British, French, German or US origin and deals with the activities of each country's own troops, giving the erroneous impression that the sector held by Belgian troops was hardly attacked with CW. Second, the literature focuses mainly on operations by special troops or special events. By the time Belgian troops began using gas, chemical artillery bombardments had become routine and thus hardly worth mentioning. Third, Belgian official histories studiously avoided direct references to Allied chemical attacks and retaliation, and no mention was made of Belgian involvement. The only known analysis based on original documents is the unpublished dissertation by Lheureux (note 2).

[9] Lheureux (note 2), pp. 92–102, and documents reproduced in its appendices V and X.

now forced to organize gas defence and to conduct training. Three months later, during the Third Battle of Ypres and after several German gas shell attacks against Belgian positions, the Belgian High Command took the decision to stock chemical munitions. However, for reasons which are not known the first Belgian use of gas did not take place until the end of October or early November following several weeks of gas attacks by German forces.[10]

From April 1917 the entire front in Flanders was the site of general chemical warfare. Mustard gas could be used effectively in Flanders because it was less affected by the 'microclimate'. (The inundated land was no longer a factor that prevented widespread use of cloud attacks.) Its introduction directly affected the Belgian approach to CW. Nevertheless, offensive operations were only possible south of Diksmuide. When Belgian troops began extending their lines towards the Ypres salient in 1918 greater involvement in gas operations became inevitable. However, until the end of the third phase and even during the German spring offensive of March 1918, official Belgian policy was one of in-kind retaliation only.

At the beginning of the fourth phase, September 1918 to the armistice, the Belgian High Command issued directives for the use of gas as a component of offensive operations. The availability of yperite shells (to be supplied by the French) was even a precondition for Belgian participation in the final offensive, which began at the end of September 1918. In this final phase of chemical warfare in Flanders the Belgian, British and French forces used gas shells extensively in the preparatory attack stages, targeting German artillery positions in particular.[11] In addition, units of the British Special Brigade supported tactical engagements with portable mortars. German forces, on the other hand, used gas as part of their defensive strategy. Civilians behind German lines had not received gas masks, and there were numerous casualties when villages were subjected to gas attacks by both sides. Gas was still being used in early November 1918, and Belgian communiqués for 2–11 November reported 'intense activity'.

Several chemical warfare 'milestones' took place in Flanders: chlorine, phosgene, blue cross and mustard gas were all first used on the Belgian front, and yperite derived its name from its use in Belgium. More important perhaps than the introduction of these gases was the rapid adoption of a strategy of general and intense use of CW in West Vlaanderen which began to be practised some months before the Third Battle of Ypres. This change in strategy coincided with the first use of the vesicant yellow cross and affected both the military and civilians for the remainder of the war. In addition, improved artillery made easy targets of areas in the rear that were occupied by civilians. Belgian armed forces are estimated to have suffered 40 per cent of all war casualties and 80 per cent of all gas casualties in 1918, the latter figure represents 16 per cent of all casualties for that year.[12] The psychological effects of chemical warfare lingered long after the armistice not least because of the many accidents involving unexploded munitions and the slow and painful recovery of many of those who had been gassed.

[10] Lheureux (note 2), p. 84. Official communiqués issued in Sep. 1917 mention retaliation, but without reference to the use of gas. The first explicit mention is in the communiqué of 3–9 Nov. 1917.

[11] Lheureux (note 2), pp. 93–94.

[12] Lheureux (note 2), p. 108.

II. Early disposal of old chemical munitions

Immediately after the armistice on 11 November 1918 the removal of munitions from the former battlefields became a task of the utmost urgency. The problem was twofold. First, some 30 per cent of the ammunition fired between 1914 and 1918 failed to detonate, and unexploded munitions littered the zone where the front line had remained stagnant for four years. Second, retreating German troops had left many munition dumps. Civilians began searching for shells in order to obtain the valuable metals contained in them. Crude methods which did not distinguish between explosive and toxic munitions were used to remove the explosive so that the iron and copper in the shell casing or driving band could be sold. The weapon components which could not be sold were buried or dumped in the rivers. Inevitably, there were numerous casualties. Collecting scrap iron and copper was so lucrative that a subsidiary of a British firm, F. N. Pickett et Fils, was established to take advantage of the situation, and a number of small Belgian companies were also set up. Men, women and children collected the iron and copper for independent iron mongers (not Pickett) but for such low wages that in 1920 the authorities intervened and ended the practice.[13]

In 1920 the Service de Destruction des Munitions (SDM, service for the destruction of munitions) was created as part of the Commission Centrale de la Récupération (CCR, central commission for recovery) to deal with the munition problem in addition to its other responsibilities. Initially, the SDM had sections attached to each Provincial Service of the CCR. In West Vlaanderen the SDM was assisted by another branch of the CCR, the Service des Régions Dévastées (SRD, service of the devastated regions), which operated only in the former front-line area.[14] The principal task of the SDM was to retrieve and destroy munitions, and it operated its own destruction sites. However, manpower levels were soon drastically reduced, in part as a result of the removal of German munition dumps from the area between the front line and the border between Belgium and Germany. Some military and civil officials expressed optimism that West Vlaanderen would be completely cleared by the end of 1922. The problem was reappraised after huge German munition dumps were uncovered, and it was learned that municipal officials had not reported the existence of such dumps because they had failed to appreciate the danger to the local population.[15] The Haut Commissariat de la Côte (high commission of the coast), which was responsible for the sector between Ostend and the French border, had its own Service de la Côte (service of the coast), clearly as a matter of prestige, to collect and destroy all munitions found in the area. It answered directly to the CCR, rather than to the SDM.[16]

[13] Craenen, P. (2nd Lt.), 'Historiek van de Belgische Ontmijningsdienst' [History of the Belgian Bomb Disposal Service], Afstudeerwerk voorgelegd tot het bekomen van de titel van licentiaat in de Militaire- en Sociale Wetenschappen. Koninklijk Militaire School, 125e Promotie Alle Wapens, Brussels, 1989 [Dissertation submitted to obtain the degree of master in social and military sciences, Royal Military Academy, 125th Promotion All Arms, 1989], pp. 1–2 (in Dutch); and Baccarne, R. and Steen, J., *Poelkapelle 1914–1918* (Drukkerij-Uitgeverij Almar: Wervik, Belgium, 1968, 2nd edn), pp. 170–72 (in Dutch).

[14] Documents and literature often confuse these agencies when dealing with former front-line areas.

[15] Craenen (note 13), chapter 1.

[16] 'Rapport relatif à l'organisation et à l'activité du service à la Côte' [Report on the organization and the activities of the service of the coast], Annexe au Document no. 48818/A/2730 de la Commission Cen-

There is little information on the munition disposal activities of these organizations for 1920–22 and less for the period before World War II. Data on the disposal of CW ammunition are almost non-existent.[17] It is not clear whether the SDM systematically destroyed chemical shells at its depots, but the records which exist suggest that until at least the end of 1922 the SDM delivered toxic munitions to F. N. Pickett et Fils at Poelkapelle (see below). Most documents provide information on units operating in West Vlaanderen. Reference is made, however, to the destruction of toxic munitions which could not be transported to West Vlaanderen by the service at the Morhet depot (approximately 6 km south-west of Bastogne) in the south-eastern province of Luxembourg.[18]

The only location where the Belgian Army is known to have destroyed toxic munitions on a regular basis was Lombardsijde beach (just north of Nieuwpoort). Information about the destruction activities conducted there is detailed in a series of reports on a 2 February 1922 accident in which six labourers were killed and two injured. There are no reports about activities at the site after that date. The Service de la Côte operated the destruction site. It processed only shells which had been fired and which usually had been found in fields or during reconstruction work along the coast between Ostend and Nieuwpoort.[19] The shells were blown up, and by February 1922, 1556 tonnes of munitions had been destroyed, including 1463 tonnes of explosives (94 per cent) and 93 tonnes of gas shells (6 per cent). The rate of destruction in the four weeks prior to the accident was 11.5 tonnes of explosive and 3.5 tonnes of toxic munitions per week.[20] However, the annual rate of destruction was relatively low since most work had to be carried out during the off-season in bad weather to give the local inhabitants the opportunity to access the dunes (where high explosives were being destroyed) and the beach. Chemical shells were exploded in pits between the low- and high-water marks, and the work was limited by both the tide and the wind since destruction could occur only when the wind blew offshore.[21]

The activities of F.N. Pickett et Fils in the Houthulst–Ypres area remain obscure. The company had its headquarters in Wimereux, France, with offices in Brussels, New York, Paris and Warsaw, and the firm's letterhead describes it as Démonteurs de Munitions Toxiques et Explosives (dismantlers of toxic and explosive munitions).[22] A postal and telegraph address on another letter indicates

trale de la Récupération, Brussels, 6 Feb. 1922, Dossier MDN [Ministère de la Défense Nationale], Cabinet, Commission Centrale de la Récupération, Centre de Documentation Historique, Evere (in French).

[17] The archives of the Dienst voor Historische Documentatie (CDH, Service for Historical Documentation) contain only one folder on the CCR with a number of documents relating to 1921–22. All other documents were reportedly lost following the German invasion in May 1940.

[18] Craenen (note 13), p. 5.

[19] 'Rapport relatif à l'organisation et à l'activité du service à la Côte' (note 16); and 'Rapport sur l'explosion de Lombartzyde du 2 février 1922 par le Lieutenant-Général Tollen, Président CCR' [Report on the explosion at Lombardsijde on 2 Feb. 1922 according to Lt.-Gen. Tollen, President of the CCR], Annexe au Document no. 48818/A/2730 de la Commission Centrale de la Récupération, Brussels, 6 Feb. 1922, Dossier MDN, Cabinet, Commission Centrale de la Récupération, Centre de Documentation Historique, Evere (in French).

[20] It was only possible to destroy explosive munitions during a few weeks each year, so work at Lombardsijde must have started soon after the armistice if the weekly rates given in the report are typical.

[21] The prevailing winds are south-westerly (i.e., onshore).

[22] 'Letter from F. N. Pickett et Fils to Lt.-Gen. Tollen, President of the CCR', Brussels, 18 May 1922, Dossier MDN, Cabinet, Commission Centrale de la Récupération, Centre de Documentation Historique, Evere (in French).

that a decision-making centre was located in London.[23] The company set up its bomb disposal unit in concrete shelters in Vrijbos at Poelkapelle soon after the armistice.[24] The unit was on the Ypres–Kortemark railway line, adjacent to a British military camp and just south of the SDM munition destruction site.

On 24 April 1921, Belgian authorities and Pickett signed a contract to eliminate munitions of all calibres, their components and gas apparatus for use in trenches.[25] Article I of the contract required the Belgian state to furnish Pickett with at least 1000 tonnes of ammunition per month and a total of 15 000–16 000 tonnes in 10 months. The first 15 000 tonnes compensated Pickett for constructing a railway line into the compound where the munitions were stored. The firm was required to pay 13.5 francs per tonne (exclusive packing) for the remaining munitions (article III). The material which resulted from the dismantling and destruction process became the property of the company, but its shipment abroad was subject to export regulations and licensing in some instances (articles III and XIII). The contract provided for a permanent on-site representative of the Belgian Government, and CCR officials had access at all times. There was also an agreement which detailed safety precautions.

Article IX of the contract dealt extensively with the destruction of toxic shells. It focused on safety precautions related to ventilation, equipment and clothing and on emergency procedures and safety standards for dismantling munitions. The permitted methods of destruction were also enumerated: 'The liquid, corrosive and toxic compounds must be destroyed on-site, removed, transformed into harmless products, or disposed of in a borehole with a depth of 2.50 metres after they have been neutralized. The holes will be filled and closed by the Pickett company, at its expense, to the complete satisfaction of the Belgian authorities'.[26]

The manner in which Pickett proceeded is not entirely clear. It is claimed that even today nothing will grow in some parts of the former compound because toxic chemicals were disposed of by simply pouring them into the soil. If the Belgian Government's permanent representative did not act in complicity with the company to conceal what took place, there remain three possible explanations for the barren areas. First, something other than chemical agents may have caused the areas which lack vegetation. Second, if these areas are identical to the bore-holes, the chemicals which were placed in them may not have been adequately neutralized, or a mixture of the end products of decontamination may have been placed in each hole, resulting in a chemical reaction that poisoned the soil. Third, Pickett's standards for decontamination may have been lower prior to the intervention of Belgian authorities or when the first (known) decontamination activities were carried out under the 31 May 1920 contract.

23 'Letter to Lt.-Gen. Tollen, President of the CCR', London, 14 Sep. 1922, Dossier MDN, Cabinet, Commission Centrale de la Récupération, Centre de Documentation Historique, Evere (in French). The location is given as '22, Queen Anne's Gate, London SW'.

24 The villages of Houthulst, Langemark–Poelkapelle and Westrozebeke border on each other just south of Houthulst Wood, which creates some confusion about the location of the installations. Although most official documents referred to Westrozebeke as the village where Pickett was located, presumably because of the proximity of Westrozebeke barracks, both the firm and the barracks were in Poelkapelle.

25 'Contrat entre l'Etat Belge et la firme 'Pickett' (F. N. Pickett et Fils à Wimereux, France)' [Contract between the state of Belgium and the firm Pickett (F. N. Pickett and Sons, Wimereux, France)] Brussels, 24 Apr. 1921, Dossier MDN, Cabinet, Commission Centrale de Récupération, Centre de Documentation Historique, Evere, (in French). Article XVII of the contract notes that it replaced an earlier 31 May 1920 contract.

26 Quotes translated from French.

Article IX of that contract provides support for this supposition: 'The dismantling of gas shells will be done by following the methods approved by the President of the CCR to the letter'.

A 23 February 1922 letter from Lieutenant-General Tollen, President of the CCR, to the Minister of Defence[27] provides evidence that the installation initially processed green and yellow cross shells and that its work was satisfactory. However, a 15 September 1922 report stated that successful decontamination was achieved only by trial and error. The destruction processes for phosgene[28] and mustard gas each posed particular problems which led to delays, although in the end the decontamination was successful.[29] Blue cross shells appear to have posed significant problems during the start-up phase since, on 23 February 1922, Tollen asked to visit Pickett's special installation at Dannes, France, in order 'to avoid all miscalculations'.

The only surviving document which outlines Pickett's disposal methods in Poelkapelle is the report of a Belgian delegation. Based on the report, it appears that during the period of the May 1920 contract chemical compounds were burned in specially built furnaces rather than neutralized and then discharged into the soil. The report stated that although the processes for destroying different types of chemical warfare agents varied, the differences were relatively small, and an installation for the destruction of one gas could be used to destroy other gases as well. The installation in question was described as:[30]

> essentially a furnace in refractory materials with a horizontal bottom, the furnace continued in: (1) a slightly inclined chimney consisting of a sandstone flue in which there is a continuous flow of water, (2) a horizontal chimney in refractory bricks with a length of 5 or 6 metres, (3) as a continuation of the first, a vertical chimney with a height of about 10 metres.
>
> The flow of water circulating through the first enters the conduit through openings bored in the higher part and leaves it through an opening located at the lower end next to the point where the angled chimney and the horizontal chimney join.

The report also dealt with the methods of destruction for the main CW gases: chlorine, phosgene, yperite and the various contents of green cross shells. It noted that the processes in use were the correct ones and that the results were satisfactory. Although no ill effects were observed for the employees, the local

27 'Lettre du Lieutenant-Général Tollen, Président de la CCR au Ministre de la Défense Nationale' [Letter from Lt.-Gen. Tollen, President of the CCR to the Ministry of Defence], Brussels, 23 Feb. 1922, document no. 49386/A/2950, Dossier MDN, Cabinet, Commission Centrale de la Récupération, Centre de Documentation Historique, Evere (in French).

28 The use of the terms 'phosgene' and 'green cross' in the documents is apparently based on the country of origin of the compound, not on chemical properties. For a discussion of the cross codes, see the subsection 'Colour coding using cross codes' in section II of chapter 6 in this volume.

29 'Lettre du Lieutenant-Général Tollen, Président de la CCR, au Ministre de la Défense Nationale' [Letter from Lt.-Gen. Tollen, President of the CCR, to the Ministry of Defence], Brussels, 15 Sep. 1922, document no. 52705/A/2950, Dossier MDN, Cabinet, Commission Centrale de la Récupération, Centre de Documentation Historique, Evere (in French), 3 pp. and annex.

30 Institut d'Hygiène de Biologie et Chimie de l'Armée, 'Rapport par Paul Erculisse, Pharmacien de 1e Classe, Chef du Laboratoire de Chimie Toxicologique, Dr Sillevants, Médecin de Régiment, Chef du Laboratoire de Physiologie Toxicologique et le Capitaine-Commandant Vanden Berghe' [Report of Paul Erculisse, Pharmacist, Head of the Laboratory of Chemical Toxicology, Dr Sillevants, Regimental Physician, Head of the Laboratory of Physical Toxicology, and Captain-Commandant Vanden Berghe], document no. C/28, Brussels, 3 Mar. 1922, Dossier MDN, Cabinet, Commission Centrale de la Récupération, Centre de Documentation Historique, Evere (in French).

Table 15.1. Pickett et Fils production rates (explosive and chemical munitions)

	Deliveries to Pickett (tonnes)	Destroyed by Pickett (tonnes)	Percentage destroyed
First contract (31 May 1920–21 Apr. 1921)	10 200	1 800	*17.64*
Second contract (21 Apr. 1921–1 Sep. 1922)	22 230	26 130[a]	*117.54*
Total	**32 430**	**27 930**	***86.12***[b]

[a] Included munitions not destroyed under the first contract.

[b] Percentage of destruction of total delivered.

Source: 'Lettre du Lieutenant-Général Tollen, Président de la CCR, au Ministre de la Défense Nationale' [Letter from Lt.-Gen. Tollen, President of the CCR, to the Ministry of Defence], Brussels, 15 Sep. 1922, document no. 52705/A/2950, Dossier MDN, Cabinet, Commission Centrale de la Récupération, Centre de Documentation Historique, Evere (in French), 3 pp. and annex.

population or vegetation, it was recommended that the furnace operators wear full protective clothing at all times. Depending on the tasks they were performing employees who did not work in the immediate vicinity of the furnace were to use gas masks or to have them ready to use (i.e., hanging from their necks). These precautions were deemed necessary since there was likely to be leaking from some of the 30–40 gas shell fillings which could be emptied at a single facility. It was feared that the accumulation of gas vapours during the lengthy destruction process could lead to serious chemical intoxication. The report noted that no chlorine or hydrochloric acid were present in the gas from the chimney and that there was therefore no danger to the area around the destruction facility.

In September 1922 Tollen wrote an extensive progress report on the work being carried out by Pickett.[31] It is the only known document which contains figures and charts (summarized in table 15.1). Tollen noted that Pickett had improved its destruction rates considerably from a monthly average of 189.5 tonnes, under the first contract, to 1686 tonnes, under the second contract. However, the monthly output varied considerably. During the start-up phase Pickett experienced serious difficulties, and seasonal and weather conditions also affected the destruction process.

Tonnage is not a reliable indicator of work volume. When work concentrated on munitions which required complex or multiple types of handling the output figures dropped significantly. Similarly, lower tonnage figures were reported when small munitions, such as grenades, were processed. By 1 September 1922 Pickett had received 7400 tonnes of gas shells from the Belgian authorities (22.82 per cent of the total of all munitions received) and had destroyed 510 tonnes of yellow cross shells, 4350 tonnes of blue cross shells and 2000 76-mm green cross shells. The report added that the SDM had 2300 tonnes of munitions in its stocks in addition to the munitions awaiting destruction at Pickett (approximately 2500 tonnes). Three-quarters of the total of approxi-

[31] 'Lettre du Lieutenant-Général Tollen, Président de la CCR, au Ministre de la Défense Nationale' (note 29).

mately 4800 tonnes of the munitions remaining were green cross shells, and one-quarter of the shells were yellow cross.

Tollen's report indicated that Pickett was constructing a second furnace, similar to the first, but even under optimal conditions the destruction of green cross shells would require another seven months (i.e., until March 1923). A 14 September 1922 letter from Pickett to Tollen hinted that the work might be completed sooner, after which Pickett wished to close the facility.[32]

Other private firms also destroyed munitions. There are no detailed records of these activities, and the name of only one firm, Emile Deckers of Schoten (near Antwerp), is known. It cannot be determined whether these firms worked independently or under government contract or if they treated toxic munitions. In addition, in 1918 the British military had set up a camp for the destruction of explosive and chemical munitions on the Ypres–Kortemark railway at Poelkapelle. Approximately 300 workers carried out destruction activities in two sheds, removing the toxic components from shells and then burning them. The camp closed in 1924.[33] Both Belgian and British troops also dumped large quantities of explosive and chemical munitions in the North Sea off the Belgian coast.

There is a gap in the history of Belgian munition disposal units until just before World War II. It is not known whether the units were disbanded or continued to exist under other names. Toxic munitions, however, continued to create problems for decades after the armistice, particularly for farmers and cattle breeders. An article in a veterinary journal described an incident in October 1935 in which a number of pigs were killed by phosphorus poisoning.[34] After months of investigation the cause of the poisoning was discovered when the owner of the pigs witnessed the spontaneous explosion of a phosphorus shell in a field where the pigs often grazed. In another incident a farmer obtained a large English shell which he requested a smith to convert to a roller. When holes were drilled in the shell to pass a shaft through it, the smith was nearly overcome by the mustard gas which was released.[35]

Toxic gases from unexploded World War I munitions were also discharged in Belgium in World War II during the May 1940 campaign. Cattle died when a bomb containing a mixture of phosgene and sulphur-containing phosphorus leaked, and investigations in August–December 1942 revealed that exploding shells had breached the soil in 1940 allowing toxic gases to contaminate the grass. Before World War II the area had been considered safe.[36]

[32] 'Lettre de Pickemetal au Lieutenant-Général Tollen, Président de la CCR' [Letter from Pickemetal to Lt.-Gen. Tollen, President of the CCR], Londres, 14 Sep. 1922, Dossier MDN, Cabinet, Commission Centrale de la Récupération, Centre de Documentation Historique, Evere (in French).

[33] Craenen (note 13), p. 14.

[34] Grains of solid phosphorus were found in the grass which the pigs ate. Before the cause was ascertained, there was speculation that someone had intentionally poisoned the animals. De Jonckheere, J., 'Vergiftiging van varkens door fosfor: Overdruk van het' [Poisoning of pigs by phosphorus], *Vlaamsch Diergeneeskundig Tijdschrift* [Flemish veterinary journal], vol. 6, no. 1 (Jan. 1937), pp. 8–11 (in Dutch).

[35] 'Een stikgasbom veroorzaakt bijna een erg ongeval' [A poison-gas bomb almost causes serious accident], *Het Ypersche*, 14 May 1938 (in Dutch).

[36] Samyn, E. (Cdr.), 'Brief aan den Heer Caenepeel, Rijksveearts te Ieper, betreft "Verslag nopens schade aan hoornvee op het Zwaanhof te Boesinge", Commissariaat-Generaal voor's Lands Wederopbouw, Dienst voor Ruiming van Springstoffen' [Letter to Mr Caenepeel, State Veterinary in Ypres, concerning 'Report regarding damage to horned cattle of Zwaanhof [farm] at Boesinge', Commissioner-General for the Reconstruction of the Country, Service for the Removal of Explosives], Direktie. ref. S/3110, Westroozebeke, 7 Dec. 1942 (in Dutch).

III. Belgian chemical weapon production

Even less is known about the production of chemical warfare agents in Belgium in the 1930s than about Belgian use of chemical weapons. It is generally believed in political and military circles that Belgium never produced chemical munitions or prepared for offensive chemical warfare. However, a historical study of a fortress at Steendorp (a village near Temse, south-west of Antwerp) revealed that the Belgian military engaged not only in anti-CW research, but also that mustard gas and adamsite (DM) were produced just before World War II.[37] A letter by the commander of the SDM, Captain-Commander E. Samyn, refers to a contaminated area on the fortress's retaining walls, but no other documentary evidence of CW-related activities has survived or been publicly released.[38]

Experiences in World War I and debate after the war led Belgian authorities to consider chemical weapons as potentially decisive in a future war. Initially, the interest in these weapons may have been defensive, and international political developments may have influenced the decision to build up a small retaliatory capability—thereby concretizing Belgium's reservations of in-kind retaliation to the 1925 Geneva Protocol. Throughout the 1920s anti-gas personnel, sometimes conscripts with specialized degrees, were included in Belgian Army units. The Service de Protection contre les Gaz (SPG, service for protection against gases) was created in the mid-1920s and reported directly to the Minister of War. In addition to protection against CW, the SPG studied chemical warfare agents. Its headquarters were in Brussels. Laboratories were established at Vilvoorde, and a factory was eventually set up at Steendorp Fortress—the Ateliers d'essais et de contrôle du matériel anti-gaz (AMAG, workshop for the testing and control of anti-gas equipment). Steendorp Fortress was obsolete and had been declassified in 1924, after being partially destroyed in World War I. Nevertheless, the authorities denied a request by a civilian firm to build a factory within its perimeter because of the explosives and gas shells which were still present.

The construction of AMAG began in the latter part of 1928. Its main activities were the production and testing of anti-gas equipment, including gas masks, canisters and air filters for large areas. However, the manufacture of gas appears to have been one of the original purposes of the facility since the first buildings housed both a gas tank and a scrubber. In addition, there was a conduit made of varnished or glazed ceramic pipes, which was specially built to discharge liquid waste directly into the Scheldt River rather than into the existing sewerage system, and two large yellow-tiled concrete precipitation tanks for decantation—known locally as 'acid pits'. The facility system was not entirely ready until early 1934, and it can therefore be assumed that there was no significant production before that date. Part of the facility still exists.

[37] Colaes, R. and Gils, R., 'Fort Steendorp: een vestingbouwkundige, heemkundige en ecologische benadering' [Fortress Steendorp: an architectural, sociological and ecological approach], *Jaarboek 1991* (Gemeentemuseum: Temse, 1992), pp. 285–311 (in Dutch).

[38] Only three people, a workman, a maintenance worker and a person with general knowledge of the work, are known to be alive; they still keep the pledge of secrecy they were required to make. The authors collated the information on the basis of extensive interviews on a broad range of topics related to the history of the fortress. Gils, R. (Lt. Col., ret.), private communication with the author, 16 Apr. 1993. The following summary is done with his permission.

AMAG produced mustard gas and adamsite, and chloropicrin was produced once. Mustard gas was initially synthesized using ethene and sulphur dichloride. Production at AMAG was not continuous, usually consisting of one-week runs during which several 200-litre barrels were filled. These were either transported to Zwijndrecht Fortress (near Antwerp) for bulk storage or used for on-site filling of ammunition. Later, a German process was adopted which used thiodiglycol and hydrochloric acid. The thiodiglycol was purchased in France. Adamsite was synthesized on-site from diphenylamine and chlorarsine. Some open-air testing of adamsite shells was carried out at Steendorp Fortress. Red and white phosphorus were also manufactured and were tested at the Beverlo military camp. Production ceased in 1940.

When German forces invaded Belgium, AMAG stocks were transferred to the local tramway, which was built into the fortress, and then transported by train to a French gas facility in the Pyrenees. No other information is available about the operation.

IV. Sea dumping of old chemical munitions

Chemical munitions from World War I were dumped into the sea on three occasions: first, just after the war when Allied troops disposed of huge amounts of conventional and toxic ammunition off the Belgian coast near Zeebrugge; second, in 1954–72 when Belgium regularly dumped 'problem munitions' (i.e., those unable to be determined conventional munitions and therefore presumed to be CW) encased in concrete into the Gulf of Biscay; and third, in 1980 when there was a single, large dumping of problem munitions into the Gulf of Biscay.

As mentioned above, information about the disposal of ammunition between the two wars is sketchy as few, if any, records seem to have survived. This is particularly true of the dumping site near Zeebrugge. An official account of dumping there, prepared for a meeting of a technical group of the Oslo Commission, stated that in 1920 Belgian Armed Forces dumped an estimated 35 000 tonnes of munitions retrieved from the Flanders battlefields into the North Sea.[39] Shortly afterwards indications of the dump disappeared from sea charts. The dump was rediscovered in 1972 and is now marked on charts as the Paardenmarkt (horse market). Fishing and anchoring in the area have been banned. Measurements and surveys have been conducted to determine the size and potential risk of the dump.

The Paardenmarkt is located 1.5–3.5 km from the eastern wall of the external port of Zeebrugge and 1–2 km from the shoreline of Knokke, a major seaside resort. It covers 3.6 square km. At low tide, the seabed is approximately 4–5 metres deep. The munitions are known to be of German origin, and, based

[39] Ministère de la Santé Publique et de L'Environnement, Institut d'Hygiène et d'Epidémiologie, Unité de Gestion du Modèle Mathématique de la Mer du Nord et de l'Estuaire de l'Escaut, 'Munitions de gaz de guerre immergées au large de la côte belge' [Gas munitions from the war submerged along the Belgian coast], référence OSCOM 12/14/1, para 7.29, document présenté à la réunion du groupe technique 'SACSA' de la Commission d'Oslo [Document presented at the meeting of the SACSA technical group of the Oslo Commission], Copenhagen, 11–15 Mar. 1991, p. 1 (in French). The estimate was made based on the accounts of witnesses who stated that 300 tonnes were dumped daily in a 6-month period. Ministerie van Volksgezondheid en Leefmilieu, Instituut voor Hygiëne en Epidemiologie, Beheerseenheid Mathematisch Model Noordzee en Schelde-estuarium, 'Paardenmarkt: Interdepartementale evaluatievergadering, Verslag [The Paardenmarkt: an interdepartmental evaluation meeting, report], Brussels, 17 July 1990, p. 3 (in Dutch).

on German Army supply data, it is believed that chemical munitions (primarily 77-mm shells filled with mustard gas) comprise as much as one-third of the dump (11 600 tonnes). The shells, some of which remain in their original crates, are in good condition.[40] Most are now covered by at least one metre of silt, thus placing them in an anoxic environment and protecting them from the worst effects of erosion.

Salvaging companies, the Flemish section of Greenpeace and the municipal authorities of Knokke have lobbied to have the munitions removed. However, technical experts from various governmental departments concur that recovering the shells would pose a greater hazard to the marine environment and the coastal resorts than letting the munitions remain where they are. If the munitions were exposed to air, corrosion would occur, and an explosion during handling would endanger the lives of the salvagers. A report by a commission with representatives from the Defence, Public Works, and the Environment departments therefore concluded that the Paardenmarkt does not pose acute danger if anchoring and fishing prohibitions are respected and that the danger is less than in other areas, such as the Baltic Sea. Moreover, there have been no recorded accidents in the zone.[41] The commission recommended against recovery and suggested instead that the situation be monitored. Its conclusions were transmitted to the March 1991 meeting of the Standing Advisory Committee for Scientific Advices (SACSA) of the Oslo Commission, which adopted the recommendations in June 1991.[42]

Press reports in April 1989 of a survey of shipwrecks along the Belgian coast by the Ministry of Public Works caused political controversy over the Paardenmarkt. The Flemish daily *Het Volk* questioned official claims that the site posed no risk to Zeebrugge.[43] *Het Volk* claimed that the dump was a threat to tourists, and that this was compounded by the presence (and planned expansion) of a large liquid natural gas terminal at the end of the sea wall and the projected construction of a nearby underwater pipeline for Norwegian natural gas. Concern was also expressed about plans to build a marina next to the eastern mole. The spectre of a serious accident with a ship adrift—a scenario reminiscent of the capsizing of the British ferry *Herald of Free Enterprise* in March 1987 just outside Zeebrugge—was also raised. *Het Volk* accused the government of underestimating the potential danger in order to avoid a delay in the construction of three gas storage tanks. The Inspector General of the Ministry of Public Works and other officials claimed that the sea wall and liquid natural gas terminal had been built to resist both earthquake and explosion and that the risks were minimal.[44] (Zeebrugge is also a major NATO munition supply port.) Sus-

[40] Strubbe, J., 'Telefaxbericht ter attentie van de heer Oosthoek' [Telefax to the attention of Mr Oosthoek], Ministerie van Openbare Werken, Bestuur der Waterwegen, Brussels, 11 July 1989 (in Dutch), private archive; and Vraag no. 496 van de heer Loones van 8 december 1989, 'Noordzee–Gasberg–Onderzoeken–Opruiming' [North Sea–gas mountain–investigations–disposal], Belgische Kamer van Volksvertegenwoordigers, *Vragen en Antwoorden*, no. 97, Gewone Zitting 1991–1992 [Bulletin of written parliamentary questions], 13 Feb. 1990, pp. 7574–75 (in Dutch).

[41] Ministerie van Volksgezondheid en Leefmilieu (note 39), pp. 3–4.

[42] Vraag no. 163 van de heer Maertens d.d. 21 Jan. 1993, 'Gifgasmunitiestort "Paardenmarkt": Stand van zaken' [Poison gas munition dump 'Paardenmarkt': situation], Senaat, *Vragen en Antwoorden*, no. 51 (6 Apr. 1993), pp. 2413–14 (in Dutch).

[43] Vandenbussche, F., 'Bom onder Zeebrugge' [Bomb under Zeebrugge], *Het Volk*, 8 Apr. 1989, pp. 1, 2 (in Dutch).

[44] 'Tijdbom op de zeebodem' [Time bomb on the sea bottom], *Het Laatste Nieuws*, 10 Apr. 1989, p. 5; T. H., 'Openbare Werken onderzoekt munitiedepot in Noordzee' [Public works investigates munition

picion increased when a spokesman for the Distrigaz company, the owner of the terminal, declared that the dumped munitions had not been taken into account in risk analyses for the terminal because government officials had failed to inform the company about the dump.[45] The issue resurfaced in written parliamentary questions when the Minister of Public Works promised to conduct a thorough study. The study was apparently never completed, although the Ministry of Defence has conducted various other studies.

The only known Belgian policy on sea dumping dates from the period after World War II. No documents are known to exist about other sea-dumping operations between the wars, and there is little information about the handling of dangerous munitions in the years just after World War II. There are strong indications that between 1945 and the early 1950s toxic ammunition was disposed of along the Belgian coast.[46] After 1954 three navy frigates—the *Godetia*, the *Zinnia* and, in particular, the *Kamina*—carried out sea dumping, although Belgian authorities are reported to have initially chartered civilian vessels.[47] The dumping continued until 15 February 1972 when Belgium acceded to the 1972 Oslo Convention for the Prevention of Marine Pollution by Dumping from Ships and Aircraft (known as the Oslo Convention).[48]

During the 18 years between 1954 and 1972 Belgium dumped 810 tonnes of old dangerous munitions encased in concrete blocks into the Gulf of Biscay. The procedure was relatively straightforward. In Poelkapelle each munition was placed in a metal case; the cases were then grouped in concrete blocks. It is believed that the concrete casings were not strong or heavy and that munitions made up the larger part of the gross weight. The blocks were transferred by rail from Poelkapelle to Zeebrugge where they were loaded onto the frigates. Delays in Zeebrugge were usually minimal, and no special precautions were taken other than the standard safety procedures for any munition transport. As soon as they were loaded the ships departed for the Gulf of Biscay. Each shipment had a modest total weight.[49]

After signing the Oslo Convention Belgium immediately ceased its dumping operations and the munitions were stored outdoors in Poelkapelle. The stock-

depot in the North Sea], *Gazet van Antwerpen*, 10 Apr. 1989, p. 9; and 'Rode bolletjes in kaart' [Red dots mapped], *Brugsch Handelsblad*, 14 Apr. 1989, pp. 1, 3 (all in Dutch). See also Strubbe (note 41).

[45] Noterman, J., 'Knokke allait construire une marina sur un lit de bombes' [Knokke will construct a marina on a bed of bombs], *Le Soir Illustré*, 20 Apr. 1989, pp. 10–12 (in French).

[46] Private communication, Apr. 1993. This deduction follows from the fact that before 1954 the Belgian Navy did not dump toxic munitions in the Gulf of Biscay and the Army did not stock or destroy them on land. It is apparently corroborated by a draft answer to a written parliamentary question, prepared by the General Staff of the Navy, stating that 'the positions of dumping zones of World War I munitions were well known'. The plural forms, used in both the French and Flemish draft, are significant. Letter from the Chief of Staff of the Navy to the Minister of Defence, reference ZS 404061, 16 May 1989 (in Dutch), private archive. The official reply made no mention of other dumping zones and stated that 'the position of sunken warships is well known'. Both documents state that there are no other 'similar' [to the one at Zeebrugge] dumping zones off the Belgian coast. Vraag no. 331 van de heer Van Dienderen van 28 Apr. 1989, 'Zeemacht: Berging van gevaarlijke tuigen voor de kust' [Navy: salvaging of dangerous contraptions off the coast], Belgische Kamer van Volksvertegenwoordigers, *Vragen en Antwoorden*, 1989 (in Dutch).

[47] Ontmijningsdienst van de Landmacht, 'Bezoek Minister van Landsverdediging' [Bomb Disposal Service of the Army, Visit by the Minister of Defence], Poelkapelle, 15 May 1992, annex II-B, p. 1 (in Dutch).

[48] Ratified on 8 Feb. 1978 and published in *Belgisch Staatsblad*, 4 May 1978. The Royal Decree of 7 Nov. 1983 implementing the ratification act was published in *Belgisch Staatsblad*, 25 Jan. 1984. The Oslo Convention is also discussed in chapter 18 in this volume.

[49] Interview with Maj. De Vuyst, Commander of the Bomb Disposal Service, 6 Apr. 1993.

pile grew each year at an average of 20 tonnes. In the absence of other destruction procedures the Ministry of Defence reconsidered sea dumping. The government appointed an ad hoc technical committee composed of representatives of the ministries of Defence, Foreign Affairs and Transport, who recommended sea disposal. On 14 May 1980 the Minister for the Environment and Public Health invoked the 'emergency' provision of Article 9 of the Oslo Convention and issued a licence for dumping the accumulated stock of 225 tonnes of toxic munitions in the Gulf of Biscay. Strong international protests followed because the Belgian authorities informed the other contracting parties after the fact rather than consulting them prior to the decision as required by the Oslo Convention.[50] Finland, Ireland, the Netherlands, Norway and Sweden did not oppose the dumping operation as such but urged Belgian authorities to seek alternative methods of disposal. France made a reservation regarding the arsenic compounds, the proposed dumping site and the possible migration of the sunken munitions to the continental shelf. Only the Federal Republic of Germany formally opposed the entire operation.[51]

In contrast to the approach taken two decades earlier great attention was paid to environmental hazards, and the munitions were encased in 850–1250-kg reinforced concrete blocks which had been subjected to pressure tests. The munitions constituted 8.3 per cent of the gross weight of approximately 2700 tonnes. The Belgian authorities decided on a single operation, and a Danish ship was chartered because the frigates did not have adequate cargo capacity.[52] The dumping was carried out in November 1980. Since then the Army has stockpiled munitions under the open sky, more recently under roofing.

V. Description of the stock of old chemical munitions[53]

Each year the Bomb Disposal Service of the Army (Ontmijningsdienst van de Landmacht, ODLM) responds to 3000–3500 calls to remove old munitions and retrieves 200–250 tonnes of ammunition, the largest part in West Vlaanderen. Approximately 90 per cent of these old munitions can be positively identified as conventional munitions, and the ODLM regularly destroys them by underground explosion at their base in Poelkapelle. Large-calibre munitions, such as bombs from World War II, are detonated on the high sea with the assistance of the Navy in order to avoid structural damage to nearby private property from the resulting shock wave. The remaining munitions are considered 'problem munitions' since the exact nature of a particular artillery shell cannot be determined. Most are definitely toxic. Some are incendiary, smoke or tracer devices; some may be explosive but cannot be identified as such because of corrosion and decay.

[50] WM, 'Opruiming granaten Houthulst komt pas begin 1995 op gang' [Disposal of grenades at Houthulst will only start beginning 1995], *De Standaard*, 12 Jan. 1993, p. 8.

[51] Ontmijningsdienst van de Landmacht (note 47), annex II-B, p. 2.

[52] Interview with Maj. De Vuyst (note 49).

[53] This chapter is based on information and charts provided by the ODLM and interviews with Maj. De Vuyst (note 49) and Cdr. Alfons Vander Mast (14 Apr. 1993) of the ODLM, unless otherwise noted.

Table 15.2. The evolution of the stocks of toxic munitions, 1980–96

Year	Annual tonnage	Cumulative tonnage
1980	2	2
1981	18	20
1982	16	36
1983	18	54
1984	20	74
1985	19	93
1986	20	113
1987	16	129
1988	26	155
1989	15	170
1990	18	188
1991	9	197
1992	13	210
1993	18	228
1994	16	244
1995	17	261
1996	28	289

Source: Bomb Disposal Service, 1993; and private communication with Cdr. Alfons Vander Mast of the Bomb Disposal Service of the Army (ODLM), 30 Jan. 1997.

Table 15.2 shows that an annual average of slightly more than 17 tonnes of such munitions was retrieved in 1981–96.[54] The ODLM believes that the annual figures may begin to drop, although such a trend could also be the result of the introduction of automatic cleaning equipment that enabled better identification of individual munitions.

Official statements quote an average annual figure of 20 tonnes of munitions to account for unexpected large finds, but the discovery of a single dump could upset the statistical prognosis. In November 1988 a large German munition dump was unearthed, consisting of 165 mustard gas shells from the Battle of Berlare on the Schelde River. This single discovery accounts for the peak value that year. More recently, on 25–26 July 1993 heavy rains uncovered 28, 77-mm German chemical shells, each weighing approximately 7 kg, near the Flemish town of Ninove.[55] In August 1993 concern was expressed that roadwork in Monceau-sur-Sambre (just outside Charleroi in the Walloons) would disturb a site where the British are believed to have buried approximately 40 tonnes of German shells, presumably filled with mustard gas, in a seven-metre deep ditch which they subsequently covered with slate.[56] Confirmation of the dump's exis-

[54] The figure for 1980 is low because all recovered munitions for the period which presented a problem were dumped in the Gulf of Biscay.

[55] DSG, '28 gasbommen opgegraven' [28 gas bombs unearthed], *Het Laatste Nieuws*, 26 July 1993. The article suggests that the German shells are from a World War II munition dump. According to Cdr. Alfons Vander Mast (Private communication, 5 Aug. 1993) the shells are similar to those from World War I which are stored at Poelkapelle. The shells were discovered on a hill that dominates the region, and which was used by German forces as a gun emplacement in World War II. The shells are corroded and decayed, and it is no longer possible to determine whether they were fired. However, the quantity which remains strongly suggests that they were not fired. Ninove was still several km behind the German front line at the time of the armistice.

[56] BW, 'Mosterdgas houdt Monceau-sur-Sambre in de ban' [Mustard gas keeps Monceau-sur-Sambre in spell], *De Morgen*, 18 Aug. 1993, p. 4. (in Dutch).

tence and its subsequent removal to Poelkapelle would alter the statistical prognosis. When work on the Ypres–Veurne motorway resumes, it can be expected that large quantities of World War I munitions will be unearthed since the proposed route will essentially follow the former front line. Some major finds of dumps with toxic munitions were made in 1995 and 1996.

Storage methods for old munitions have evolved considerably since World War II. Initially, munitions were piled in long rows of shells placed on top of each other in open wooded areas. In the early 1980s, when long-term storage of toxic munitions became necessary, eight concrete platforms were constructed—each with a surface greater than 90 square metres and separated by earthen blast walls—to limit damage in the event of accidental detonation. Smaller 'pyramids' of a maximum of 25–50 shells (depending on calibre) were stored in U-shaped metal frames. This enhanced security, but a review of storage procedures and methods was necessitated by an accident on 7 May 1986 in which eight gas shells exploded killing four persons.[57] This led to the introduction of box pallets in December 1988 (a storage method similar to that used for wine bottles). Wooden partitions prevent physical contact between the shells, thus sharply reducing the risk of 'sympathetic explosion'. The shells are placed so that the detonators do not face each other. Box pallets also reduce handling. If a box pallet is placed on the back of a lorry, personnel can place retrieved shells in it. Once back at the military base a fork-lift truck can unload the box pallet and its contents. This method facilitates the relocation of munitions and their eventual transfer to a dismantling facility. Currently, the pyramidal storage method is in use on only one platform. Shells are also placed under moveable roofs to reduce the effects of sun and rain.

Leaking munitions, especially those filled with persistent mustard gas, pose a particular hazard. Shells which can be identified and easily retrieved are placed in airtight containers filled with active carbon or are encased in plaster. Such shells represent fewer than 0.34 per cent of the gross weight of stored munitions. In most cases a faint odour reveals the presence of a leaking shell. Reports are prepared about such observations, but no further action is taken. Searching for the leaking shells would expose personnel to unnecessary risk since the three storage methods—dumps, pyramids and box pallets—would make it necessary for potentially unstable munitions to be handled.

All casualties since World War II have been caused by explosions, not gas. An explosive ordnance disposal (EOD) robot, was acquired in April 1991 but is not often used. It can be employed to place munitions in box pallets and for other handling during dismantling. In rare cases a shell may detonate or release its contents spontaneously as a result of decay—when a seeping chemical starts a reaction—or because of the action of temperature and humidity. Such an incident took place on 21 May 1992, when a 150-mm shell ruptured, releasing a white cloud. There were some personnel in the vicinity, but there were no casualties. Immediate investigation of the area by a person wearing nuclear, biological and chemical (NBC) protective clothing revealed no traces of a CW agent,

[57] The subsequent investigation found that the 4 soldiers died from the explosion and had not been affected by the gas.

Table 15.3. Summary of munitions at Poelkapelle

Munition type and calibre	Nationality	Number stocked
Grenades	Various	143
Cylinders	German	12
77-mm shells[a]	German	15 493
105-mm shells	German	3 973
150-mm shells	German	875
170-mm shells	German	76
210-mm shells	German	61
60-pounder shells	British	739
4-inch Stokes mortars	British	33
Livens mortars	British	136
4.5-inch shells	British	2 296
6-inch shells	British	113
Total		**23 950**

[a] This category also includes 75-mm and 76-mm shells. According to detailed statistics based on 1991 data prepared for the consulting agency Coppée–Courtois, there were only 200, 75-mm and 76-mm shells and mines compared with 10 508, 77-mm shells. Less than 5% of the ammunition in this class is of French origin. British shells amount to only a handful. Vander Mast, A. (Cdr.), private communication, 5 Aug. 1993.

Source: Bomb Disposal Service, 1 Apr. 1993; and private communication with Cdr. Alfons Vander Mast of the Bomb Disposal Service of the Army (ODLM), 30 Jan. 1997.

nor did a more thorough technical investigation some hours later by a member of the Technical Service of the Army (based in Vilvoorde, outside Brussels).[58]

There are also toxic munitions in two additional dumps and on five of the eight platforms. The munitions are distributed on the platforms by calibre and sometimes by country of origin, and several platforms will gradually reach full storage capacity simultaneously. At the end of 1994 a sixth platform was filled, leaving only two platforms for temporary stockpiling of explosive munitions. The depot at Poelkapelle is rapidly reaching its maximum storage capacity, and the government has therefore been forced to act to open the dismantling facility as soon as possible. It would be unsafe to construct additional platforms since there must be adequate distance between the platforms and between the platforms and the perimeter road, the administrative buildings and the nearby Army munition depot. Each platform has a storage limit in terms of net explosive weight. In the case of toxic shells, however, the net weight of the burster charge is much lower than that of explosive munitions. Only 10 per cent of the permitted net explosive weight per platform is attained with the current storage method, and storage capacity is limited instead by volume.

In January 1997 the two dumps and five platforms held a total of 23 950 shells, grenades and other devices that were either filled with toxic substances or which could not be positively identified as explosive munitions (see

[58] 'Duitse gifbom ontploft' [German poison bomb explodes], *Het Laatste Nieuws*, 23 May 1992, p. 1; and NT, 'Gasbom barst open op DOVO-domein' [Gas bomb bursts open at DOVO site], *De Standaard*, 23 May 1992, p. 13 (in Dutch). A local action group claimed that several of the military personnel had to be treated in hospital for the effects of mustard gas and that the remainder of the liquid was hosed away, contaminating a nearby small river. The Ministry of Defence and an ODLM spokesperson denied the charges. For a review of these claims, see Driesmans, R., 'Bomincident: "sneller ontmantelen"' [Bomb incident: 'accelerate dismantling'], *Het Laatste Nieuws*, 25 May 1992 (in Dutch).

table 15.3). The vast majority (85.5 per cent) are of German origin, and 77-mm artillery shells alone make up more than 65 per cent of the total. The detonators in these shells used the state-of-the-art technology of that time. However, World War I did not allow sufficient time for thorough field testing, and an estimated 30 per cent of the shells are 'duds'. The Allied economic embargo forced Germany to use inferior materials and further reduced the reliability of the detonators. The detonators of these old German shells often rot and fall off, but part of the fuse may remain inside a shell and can cause it to explode.

The ODLM estimates the minimum net volume of toxic compounds at 5000 litres and the maximum at 25 000 litres. The divergence can be attributed to allowing for variations in the amount of CW agents in the shells, losses due to leaking munitions and a margin for explosive munitions which currently cannot be identified as such. The cumulative net weight of CW agents is believed not to exceed 10 tonnes.[59] The consultancy company Coppée-Courtoy identified 48 different product combinations which may be present in the munitions on the basis of an analysis of the relevant literature; they include: CW agents in the narrow sense as well as irritants and harassing agents, smoke devices and incendiaries. These compounds constitute a subset of all agents or combinations used in World War I. Documented use on the Belgian sector of the front appears to have been a major determinant in narrowing down the list of possible toxic compounds.

VI. The dismantling facility: a brief overview of the policy-making process

After the final dumping operation in November 1980, the Ministry of Defence began looking into alternative ways to dispose of dangerous munitions. In 1984 experts visited chemical demilitarization facilities in the Federal Republic of Germany, France and the United States. The visit to Kampfmittelbeseitigungsanlage der Bundeswehr (chemical munition disposal facility of the German Armed Forces) in Munster led to further negotiations with the West German Ministry of Defence which resulted in a draft agreement on ad hoc technical assistance in 1987.[60]

Initially, technological obstacles and lack of experience with CW destruction precluded a rapid solution to the problem. The political decision-making process and debate about departmental responsibilities further postponed its resolution. In the latter half of the 1980s, the proposed solutions ranged from returning the shells to the countries of origin to having them destroyed at the Munster facility. The former proposal was impractical because in many cases it was impossible to identify the individual munitions and because it would have led to diplomatic difficulties. The latter idea was abandoned because transport across Belgium and part of Germany was considered too hazardous. The elimi-

[59] Ontmijningsdienst van de Landmacht (note 47), annex III-D, p. 1.

[60] The document was apparently never formally signed. Ontmijningsdienst van de Landmacht (note 47), annex II-B, p. 2. The same wording was used in the brochure issued for the ceremony to mark the beginning of construction of the dismantling facility. Ontmijningsdienst van de Landmacht, 'Ontmantelingsinstallatie voor de Toxische Chemische Munitie' [Dismantling facility for toxic chemical munitions], Poelkapelle, 21 June 1993, p. 3 (in Dutch).

nation of these alternatives led the government in January 1989 to its only possible option—construction of a dismantling installation at Poelkapelle.[61]

A complex set of factors affected the decision-making process. At the end of the 1980s the unstoppable growth of Belgium's public debt caused political concern and forced ministers to cut spending in their departments. After the demise of the Warsaw Pact there were many demands on the Ministry of Defence as various quarters tried to collect the so-called peace dividend. Meanwhile the estimated cost of the destruction facility rose from an estimated 150 million Belgian francs in 1989[62] to 500 million francs three years later. Operational costs, including personnel costs, were estimated at 65 million francs for the first year and at 55 million francs per year thereafter.[63] The Ministry of Defence declared that it was unable to shoulder the burden of investment alone and sought to share the cost with other ministries. A proposal to seek European Community (EC) funds on grounds that the problem arose from a European war involving several current major EC members failed to materialize.

A second important factor was the federalization process in which specific executive powers were transferred from the federal government to the regions and linguistic communities. As a consequence, it was unclear for some time whether a federal or a regional ministry would have to share the financial burden. The old chemical munitions at Houthulst and in the North Sea are located on Flemish territory. Technically, the question of the *Paardenmarkt* was resolved by acceptance of the recommendation that it was safest to let the old munitions remain in the sea sediment. Politically, it was settled by the position taken by the Council of State on a bill to prohibit storage of chemical weapons on Belgian territory.[64] The Council of State asserted that the region's competence regarding public works was limited to those areas necessary for the execution of responsibilities under the constitution transferred to the regions; thus the problem of old toxic munitions had to be resolved at the federal level. Presumably, the federal government shared the opinion of the Council of State, and the Ministry of Defence conducted further negotiations with the federal Ministry of the Environment rather than with the regional Flemish departments of the environment or public works.

[61] Reply by Defence Minister Guy Coëme to questions in the Senate, reported in 'Overheid bouwt installatie voor ontmanteling gasbommen' [Authority builds installation for dismantling gas bombs], *Financieel Economische Tijd*, 17 Jan. 1989, p. 24; ESN and MDW, 'Plannen ontmantelingsinstallatie oude munitie over half jaar klaar' [Plans for dismantling facility for old munitions ready in half a year], *De Standaard*, 17 Jan. 1989; ESN, MDW and VKV, 'Gasbommen worden ontmanteld' [Gas bombs to be dismantled], *Het Nieuwsblad*, 17 Jan. 1989; and EW, 'Coëme belooft gifgasbommen te ontmantelen' [Coëme promises to dismantle gas bombs], *De Morgen*, 17 Jan. 1989, p. 6.

[62] Vraag no. 361 van de heer Van Dienderen van 9 juni 1989, 'Houthulst, Uitschakeling gasbommen' [Houthulst: elimination of gas bombs], Kamer, *Vragen en Antwoorden*, no. 69, 18 July 1989, p. 5577 (in Dutch).

[63] Vraag no. 34 van de heer Van Hooland van 3 juni 1992, 'Vernietiging van mosterdgas in Houthulst' [Destruction of mustard gas in Houthulst], Senaat, *Vragen en Antwoorden*, no. 15, 21 July 1992, p. 631 (in Dutch).

[64] Eerdekens, C., Collart, J. and Onkelinx, L., 'Proposition de Loi relative à l'interdiction de la fabrication, du dépôt et de l'utilisation, par la Belgique, d'armes chimiques, Avis du Conseil d'Etat' [Bill pertaining to the prohibition of the production, storage and use of chemical arms by Belgium, comment by the State Council], Chambre des Représentants de Belgique, Session ordinaire 1988–89, no. 739/3, 8 Jan. 1990, p. 6 (in French).

The political debate accelerated after the January 1989 Paris Conference on the Prohibition of Chemical Weapons.[65] At the conference a Belgian military expert expressed concern about the danger the CW stocks posed to people and the environment.[66] Subsequent press reports and television pictures of decaying shells at Poelkapelle led to the creation of a local action group that began lobbying efforts with local, provincial and national politicians. Their campaign resulted in the establishment of emergency plans for use in the event of an accident and extension of the existing air exclusion zone over the storage site. Perhaps more importantly, the group managed to place the issue high on the national political agenda and local MPs began questioning the Minister of Defence at regular intervals. The ODLM and the action group, supported by the peace movement and local politicians, became allies in a common cause to lobby the government from both within and outside the establishment.[67]

On 2 June 1989 the government took the decision to build a dismantling facility at Poelkapelle, and the Ministry of Defence budgeted an additional 150 million francs for 1989–92 to cover the cost, one-tenth of which was earmarked for a technical study.[68] A protocol was subsequently signed with the Department of the Environment to determine the respective responsibilities for the dismantling and destruction of the munitions and the budgetary implications.[69] Soon afterwards replies to written parliamentary questions indicated that both ministries had launched preparatory scientific and technical programmes in their respective areas in anticipation of construction of the dismantling facility. The protocol also outlined basic responsibilities: the Ministry of Defence was to take care of the dismantling of shells, and the Ministry of the Environment was to actually destroy the toxic compounds. The Minister of Defence declined to respond to questions about the destruction process.[70]

Following various preliminary studies by consulting firms, the Brussels-based firm Coppée-Courtoy was awarded a contract on 4 November 1991 to prepare a detailed report.[71] Burning the CW agents was the preferred option,

[65] For a discussion of the conference, see Lundin, S. J., 'Chemical and biological warfare: developments in 1988', *SIPRI Yearbook 1989: World Armaments and Disarmament* (Oxford University Press: Oxford, 1989), chapter 4, pp. 116, 129–30; and Lundin, S. J., 'Multilateral and bilateral talks on chemical and biological weapons', *SIPRI Yearbook 1990: World Armaments and Disarmament* (Oxford University Press: Oxford, 1990), chapter 14, pp. 534–35.

[66] Doornaert, M., 'België betaalt hoge prijs voor afbraak chemische wapens' [Belgium pays a high price for demolition of old chemical weapons], *De Standaard*, 11 Jan. 1989, p. 1 (in Dutch). The article misquoted the expert, reporting that the stockpile was 160 000 tonnes instead of 160 000 kg.

[67] The action group did not respond to a request for an interview or provide written information. It is not known if the group still exists or what it now has as its agenda.

[68] 'Ontmantelingsinstallatie gasgranaten in Poelkapelle' [Dismantling facility for gas shells in Poelkapelle], *Het Nieuwsblad*, 3 June 1989, p. 10 (in Dutch); and '"Feu vert" à un centre de destruction des armes chimiques près d'Ypres' [Green light for a centre for the destruction of chemical arms at Ypres], *La Meuse*, 8 June 1989, p. 10 (in French).

[69] Vraag no. 361 (note 62).

[70] Vraag no. 553 van de heer Van Steenkiste van 6 februari 1990, 'Houthulst–Chemische munitie–Eindrapport' [Houthulst–chemical munitions–final report], Kamer, *Vragen en Antwoorden*, no. 104, 3 Apr. 1990, p. 8284; Vraag no. 164 van de heer Ghesqière van 28 september 1990, 'Chemische wapens-Ontmantelngsinstallatie te Houthulst' [Chemical weapons-dismantling facility at Houthulst], Senaat, *Vragen en Antwoorden*, no. 9, 4 Dec. 1990, p. 363; Vraag no. 36 van de heer Kuijpers van 12 december 1990, 'Verbrandingsinstallatie voor gifgassen te Houthulst-Mosterdgas' [Incineration facility for poisonous gases at Houthulst: mustard gas], *Bulletin van Vragen en Antwoorden-Senaat*, no. 14, 15 Jan. 1991, p. 578; and Vraag no. 37 van de heer Kuijpers van 12 december 1990, 'Ontmanteling van chemische wapens' [Dismantling of chemical weapons], Senaat, *Vragen en Antwoorden*, no. 17, 5 Feb. 1991, pp. 727–28 (in Dutch).

[71] Ministerie van Volksgezondheid en Leefmilieu, Dienst Sanitaire Bouwkunde, 'Studieovereenkomst inzake de vernietiging van toxische chemische afval opgesteld in de installaties van de Ontmijningsdienst

and the Ministry of the Environment requested that Coppée-Courtoy propose the best incineration methods for various compounds and contaminated equipment as well as methods for disposing of chemicals which could not be burned. Special consideration had to be given to the effect on the environment of the alternative methods, and Coppée-Courtoy was asked to design containers for CW agents which could be burned without endangering the environment and which were inexpensive. The study also addressed transport and general costs, and Coppée-Courtoy was required to submit specifications and other information to the Ministry of Defence about the construction and operation of the dismantling facility.

The contract was an important policy-making document for two reasons. First, the Ministry of the Environment defined its options clearly. Second, progress by the Ministry of Defence on its construction project depended on the recommendations of the feasibility study. Its conclusions also determined the final apportionment of costs between the two departments. The comprehensiveness of the contract masked the disagreement between the two ministries. It can also be assumed that the linguistic factor played a role,[72] since the departments had been headed by ministers from different linguistic groups in the past two governmental coalitions.[73] The contract did contain passages—on impending debates on cost-sharing and the dependence of the Ministry of Defence on the results of the feasibility study—that appeared likely to rekindle debate. The Ministry of Defence denied its responsibility for missed deadlines by referring to its dependence on the Coppée-Courtoy study. In a reply to a written parliamentary question, the Minister of Defence declared: 'It is regrettable that the instruction to realize the study of the Environment's tasks, the conclusions of which are necessary for the start of the realization of the work of the Ministry of Defence, was only given by that department early in November 1991. The study's results are expected in August 1992'.[74]

Other elements apparently also played a role. Coppée-Courtoy reportedly delayed signing the contract for almost 10 months,[75] and the study revealed several technical problems related to the proposed destruction process that had to

van de Landmacht te Poelkapelle' [Study contract pertaining to the destruction of toxic chemical waste stored at the facility of the Bomb Disposal Service of the Army at Poelkapelle], Onderhandse Overeenkomst, GSB/ENV/5109. Brussels, 4 Nov. 1991, 8 pp. and annexe.

[72] Neervoort, F., 'Als de wereld van zijn gifgas af wil, wil België het wel voor iedereen opslaan' [If the world wants to get rid of its poison gas, Belgium is willing to store it for everybody], *Vrij Nederland, vol.* 50 (14 Jan. 1989) (in Dutch). This Dutch journalist linked the debate on cost to the appropriation of 40 billion Belgian francs for the purchase of several weapon systems, contracts which would benefit the ailing Walloon defence industry.

[73] Until early 1992 the Defence Minister was Guy Coëme (francophone Socialist), and the Secretary of State of the Environment attached to the Prime Minister was Miet Smet (Flemish Christian Democrat). Leo Delcroix (Flemish Christian Democrat) was Defence Minister. Laurette Onkelinx (francophone Socialist) was Minister of the Environment and replaced in 1993 by fellow party member Magda De Galan.

[74] Vraag no. 1 van de heer Loones van 17 maart 1992, 'Chemische munitie: Ontmantelingsinstallatie' [Chemical munitions: dismantling facility], Senaat, *Vragen en Antwoorden*, no. 5, 12 May 1992, pp. 222–23 (in Dutch).

[75] Driesmans, R., 'Ontmanteling gasbommen kost 200 miljoen fr. duurder dan voorzien was' [Dismantling gas bombs costs 200 million francs more than anticipated], *Het Laatste Nieuws*, 27 Feb. 1992, p. 33 (in Dutch). This accusation was made by Senator Maria Tyberghien-Vandenbussche (Flemish Christian Democrat) defending the record of her fellow party member, the Secretary of State of the Environment. See also 'Senator Maria Vandenbussche zet haar kruistocht tegen toxische oorlogsbommen verder' [Senator Maria Vandenbussche continues her crusade against toxic war bombs], *Het Wekelijks Nieuws*, 1 May 1992 (in Dutch).

be addressed before further progress could be made. In November 1992 the ministries of defence and the environment, headed by new ministers, signed a second protocol, replacing the first, which detailed their responsibilities regarding preparations and execution of the dismantling and destruction processes and their respective budgetary commitments for the construction and operation of the dismantling facility. On 27 November 1992 the Ministerial Council approved the awarding of the building contract to the Roegiers-Soberi company. In early May 1993 Parliament approved supplementary budgetary allocations for both ministries: 240 million francs for the Ministry of the Environment and 264 million francs for the Ministry of Defence. Construction of the facility was completed by the end of 1994.[76]

VII. The dismantling facility: a brief overview of planned procedures[77]

In 1989 Flemish public television broadcast pictures of the storage site at Poelkapelle. Memories of World War I gas casualties and pictures of the chemical warfare victims of the 1980–88 Iraq–Iran War may have contributed to the demand by those living near the facility for the immediate and safe destruction of the old chemical munitions stored there.

On 21 June 1993 the ministers of Defence and the Environment laid the first stone of the dismantling installation. The complex has three major components: the identification, the dismantling and the decontamination buildings. The facility has been designed to provide maximum security for the personnel handling the toxic munitions, the workers in an adjacent facility and the people living in the surrounding villages.

The purpose of the identification building (figure 15.1) is to determine the nature of the munitions. After initial cleansing, a munition is placed on a conveyor belt by a 'loader', who describes its external characteristics to an operator in the control room via an intercom. The rest of the process is highly automated. The shell is automatically weighed and transported to a corundum blaster for cleansing and then automatically transported to the radioscopy chamber for inspection of its interior. The use of X-rays allows for unambiguous identification of the munition and choosing of the most appropriate dismantling method. The shell is then marked with symbols for this information and placed on a pallet for interim storage. The data which has been gained, including the X-rays, is stored on an optical disc, which accompanies the munition during its dismantling.

The second phase, dismantling, is the most complex (figure 15.2). Here, too, most operations are automated. The procedure to be used is chosen based on whether the toxic compound is liquid or solid. A munition with liquid contents

[76] Wetsontwerp houdende aanpassing van de algemene uitgavenbegroting van het begrotingsjaar 1992-secties ministerie van Volksgezondheid en Leefmilieu en ministerie van Landsverdediging [Bill pertaining to correction of the overall expenditure budget for the budgetary year 1992: sections Ministry of Public Health and the Environment and Ministry of Defence], Kamer van Volksvertegenwoordigers van België, 48e Zittingsperiode-1992–93 GZ-Handelingen-PLEN, 5 May 1993, pp. 224–26.

[77] This section is based on information and charts provided by the ODLM and interviews with Maj. De Vuyst (6 Apr. 1993) and Cdr. Alfons Vander Mast (14 Apr. 1993) of the ODLM, unless otherwise noted.

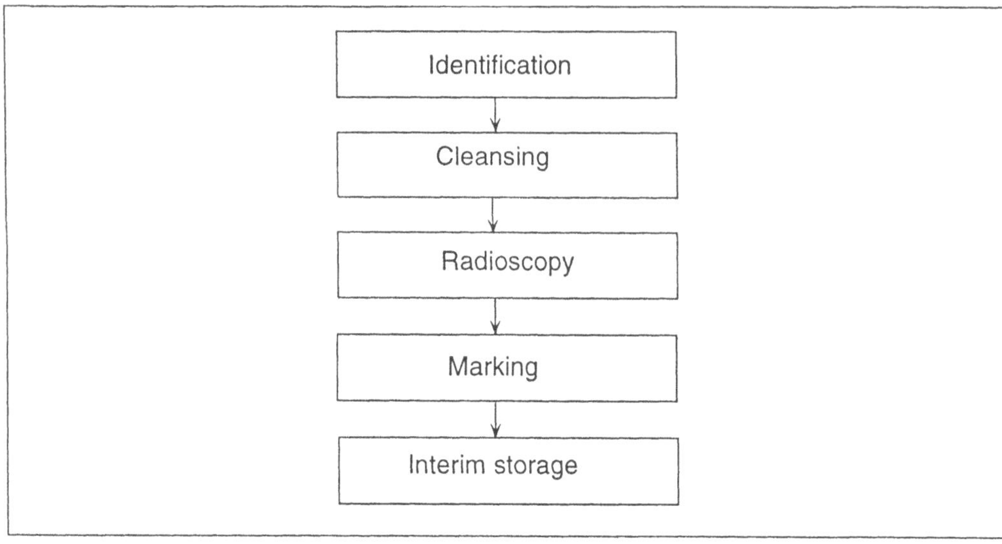

Figure 15.1. Procedures in the identification building

is placed on the conveyor belt by a loader wearing full protective clothing, under the guidance of an operator in the control room. The shell is secured and placed under a gas-tight bell, and a hole is made in it by an automatic drill. A sample of the contents is analysed in the laboratory to determine the exact composition of the toxic agent. Based on this analysis the contents are pumped to the appropriate container using the pressure–underpressure principle. Phosgene is the only CW agent which can be completely neutralized in the dismantling facility. Its high toxicity makes it dangerous in a confined space, and it must be processed by a phosgene scrubber. The shell casing is subsequently rinsed and chemically decontaminated. It is then placed on a pallet with other explosive ammunition to await destruction.

Shells with solid contents are placed in a milling cutter by a loader wearing full protective clothing. The X-ray film on the optical disc helps the technicians determine the exact point which separates the detonator and explosive part of the shell from the rest of the munition. After milling, the detonator is automatically removed for decontamination and is later destroyed together with other explosive ammunition. The remaining part of the shell is handled manually in order to separate the toxic components from any remaining explosive substances. The components are then packed and stored in special containers pending removal or destruction.

Treatment of liquid and solid agents takes place in separate sections of the dismantling facility, and both operations can proceed simultaneously. Shells with both liquid and solid contents are processed using a combination of the above procedures, which is determined by analysis of the munitions.

The third major component of the complex is the decontamination building (figure 15.3). Personnel pass through it when entering and leaving the dismantling building. When entering, they undress completely and put on underclothing and full protective clothing with an autonomous breathing apparatus. They are then driven to the dismantling building.

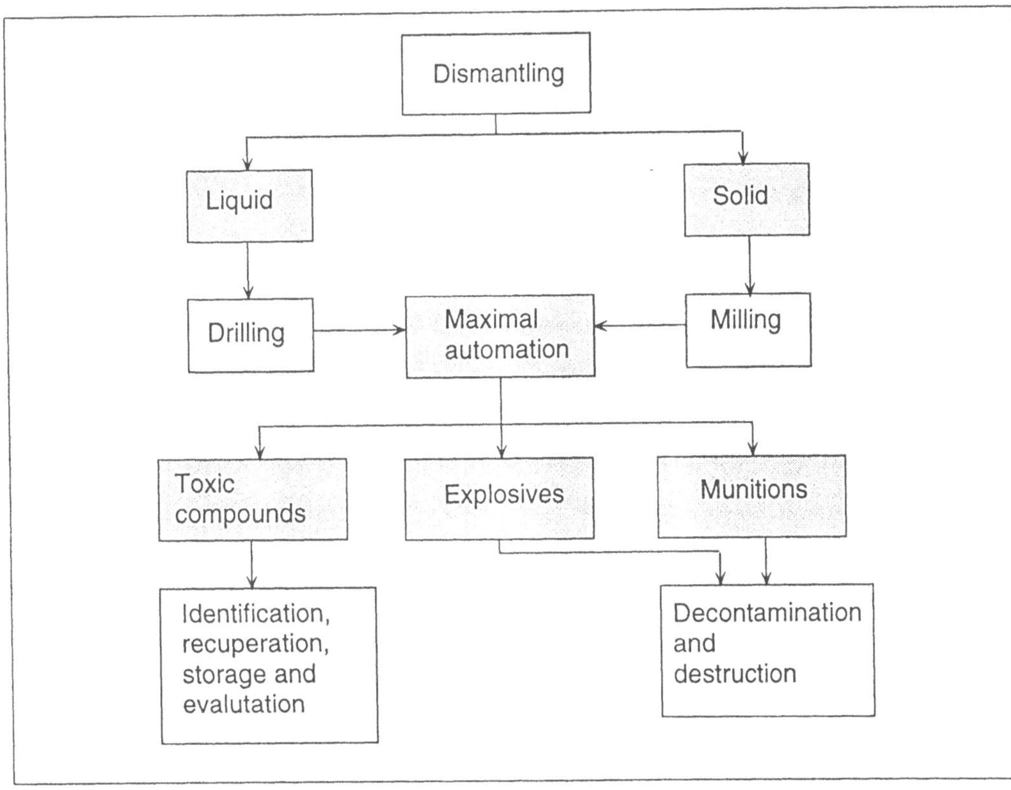

Figure 15.2. Procedures in the dismantling building

At the end of the workday, the personnel are again driven to the decontamination area. They enter immediately and undress completely, following a procedure to avoid contaminating the inside of their protective suits. The clothing is then transferred to a separate room to be decontaminated and cleansed after which it remains in a drying chamber for 24 hours. It is checked, tested, repaired (if necessary) and then stored in the stockroom. The breathing apparatus is also decontaminated and the underclothing washed. The personnel first walk through a 'rinsing' area and then proceed into a shower, where they wash thoroughly. They proceed to the dressing room where their collect their personal belongings and then leave the building through an airlock.

When it is working at full capacity the dismantling facility can process 8–10 shells per workday and approximately 2000 munitions annually. In 1993 it was estimated that the then existing stock of approximately 18 000 munitions could be disposed of in less than 9 years. (Coppée-Courtoy had calculated that 7.5 years would be needed.) Some of the munitions are of the explosive type, and these will not need to be dismantled. They can be destroyed at Poelkapelle. More chemical shells will be found, and if they continue to be located at the current rate, it will take approximately 15 years to dismantle all of them.

Interim storage of the CW agents and toxic waste is ultimately the responsibility of the Ministry of Defence. Under its research contract with the Ministry of the Environment,[78] Coppée-Courtoy was required to propose a design for a

[78] The Ministry of the Environment did not respond to a request for an interview. The information summarized here is compiled from unclassified charts and working documents provided by the Ministry of Defence and the ODLM and from comments and explanations given in interviews. This section of the

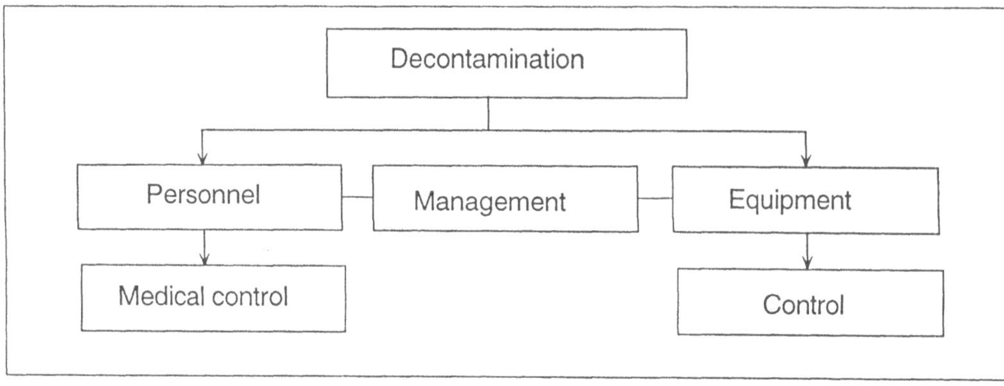

Figure 15.3. Procedures in the decontamination building

storage container that can resist the corrosive effects of CW agents and be ecologically destroyed together with its contents. A 20-litre polyvinyl chloride flask may meet these requirements. The contents of one shell would be emptied into each such flask in order to avoid unexpected chemical reactions that could be caused by mixing two different agents or agents tainted by chemical reactions within a munition. Exceptions could be made if chemical analysis proved that the contents of two shells had exactly the same composition. The storage problem is further complicated by the diversity of dangerous chemical substances, which can be irritating and toxic and also simultaneously explosive or flammable. Moreover, some materials such as solvents are toxic, although they may have been used for other purposes. Unexpected chemical compounds may be present as a consequence of reactions with shell casings or because a CW agent has seeped into the burster charge. It may be possible to dispose of such munitions by a controlled explosion, but it is not clear if this would have other unwanted effects, such as the release of noxious fumes.

The actual destruction process will be carried out by the Antwerp-based Indaver company, the only company in Belgium able to process the kind of toxic waste that results from dismantling chemical munitions. Indaver will not accept some compounds for processing, and these substances would be hazardous to store or transport. The compounds in question are highly flammable white phosphorus and combinations of white phosphorus with celluloid, molasses, carbon disulphide and carbon disulphide and tar oil. The only alternative is to destroy them by explosion at Poelkapelle, but this generates toxic fumes in the form of phosphorus pentoxide (P_2O_5) and phosphoric acid (H_3PO_4).

The various steps of the burning process can only be determined after the composition of the CW agents has been determined in the dismantling unit. Some serious problems were identified and the Minister of the Environment commissioned at least two universities to conduct supplementary research, some of it confidential. In late summer 1993 the Minister of the Environment ordered a study on the lethality of World War I CW agents.

chapter is added to indicate the problems involved and cannot be considered as complete or as the policy of the Ministry of the Environment.

Appendix 15A. Chemical attacks on the Belgian front, 1915–18

Date	Location	Type of attack	Comments
1915			
14 Feb.	Belgian sector, north of Ypres salient	Tear-gas (T) shells	2 Belgian soldiers of 6th Division reported ill
Mar.	Nieuwpoort	T- shells	Mixture of T- and B-stoff, 15-cm shells
22 Apr. (day)	Pilkem, north of Ypres	T- shells	Intense shelling of 45th Algerian Division with HE and T-stoff
22 Apr. (5 p.m.)	Bikschote, Poelkapelle, Langemark, Pilkem	Chlorine cloud, T-shells	150–168 tonnes chlorine gas released from almost 6000 cylinders along 700-metre front; T-shells fired with chlorine cloud
22 Apr. (dusk)	Steenstrate, north of Ypres	T- shells	Shelling with T-stoff (15-cm shells) of 6th Belgian Division defending against German attack on northern edge of Ypres salient
22 Apr. (dusk)	Boezinge	? T- shells	Artillery barrage of village, including gas shells
24 Apr. (2 a.m.)	Poelkapelle–St Juliaan, north-east of Ypres	Chlorine cloud	15 tonnes chlorine released on left flank of 2nd Canadian Division
24 Apr.	Lizerne, north of Ypres	T- shells	1200 T-shells fired to create gas wall (Gaswand) behind Belgian 6th Division to prevent reinforcements from reaching the front
24 Apr.	Boezinge	T- shells	Similar attack against troop concentrations in Boezinge Castle park
27 Apr.	Steenstrate–Lizerne	Chlorine cloud	German forces open northernmost gas cylinders, which had remained closed on 22 Apr. because of technical problems
29 Apr.	Steenstrate–Sas	Chlorine cloud	Small cylinder attack fails because of high winds and prompt retaliation by British artillery, which smashes many cylinders
1 May	Hill 60, south-east of Ypres	Chlorine cloud	60 cylinders opened against 1st Dorsets; casualties on both sides because of blowback
2 May	Ypres salient, British sector, ? south-west of Pilkem	Chlorine cloud	Attack against British 12th and 84th brigades[a]
5–6 May (night)	Hill 60	Chlorine cloud	Large number of cylinders released on 1400-metre front south of Hoge
10 May	Hill 60	Chlorine cloud	Small attack, few British casualties, probable release of remaining cylinders from 5–6 May attack

Date	Location	Type of attack	Comments
24 May (2:45 a.m.)	Frezenburg, east of Ypres	Chlorine cloud	Last gas discharge in Second Battle of Ypres from freshly installed cylinders; shells also used
16 Aug.	Belgian sector	B-stoff	Attack using 150-mm shells, presumably filled with bromacetone, against company of 1st Cycle Battalion near Noordschote (*c.* 10 km south of Diksmuide); last known gas attack against Belgian troops in 1915
19 Dec. (5:15 a.m.)	Wieltje, north-east of Ypres	Chlorine and phosgene cloud	First use of phosgene; 180 tonnes in 80 : 20 chlorine : phosgene ratio released from 9300 cylinders against British troops, who suffer relatively light casualties because of ample warning and 'gas discipline'
1916			
30 Apr. (12:35 a.m.)	Wulvergem, south of Ypres	Chlorine and phosgene cloud	Cloud launched on 3.2-km front; 512 British casualties, 89 of whom die
13 June	Ypres salient	Chlorine cloud	First British attack in Belgium; 300 cylinders opened along front of 20th Division
16 June (12:40 a.m.)	Wulvergem	Cloud gas, presumably chlorine	80% of 2000 large and 3000 small cylinders opened from German trenches
8 Aug.	Wieltje	Chlorine and phosgene cloud	Last German cloud gas discharge of the war against British troops; mixture contained higher percentage phosgene (50 : 50) than usual; 370 British troops killed, 434 injured; 174 casualties in 1st Canadian Division
29 Aug.	Steenstraat	. .	Attack on the Belgian front[b]
1 Sep.	Mesen (Messines), south of Ypres	Cloud gas	British discharge 118 cylinders; cloud drifts back into British trenches
5 Oct. (11 p.m.)	Nieuwpoort	White star cloud	British discharge 2050 cylinders; cloud affects civilians as far away as Raversijde, near Ostend
1917			
23 Apr.	Nieuwpoort	Chlorine cloud	Massive German cloud attack against French in Nieuwpoort spreads over entire Belgian front when wind turns unexpectedly; effects felt 43 km away; 1 Belgian killed, numerous casualties
4 May	Nieuwpoort, Diksmuide	. .	Attack on the Belgian front
24 May– 7 June	Mesen–Wijtschate	Gas[c]	British Special Brigade conducts night gas attacks in preparation for attack on Mesen–Wijtschate Ridge on 7 June; additional 4700 cylinders, 4800 projector drums and 4250, 4-inch bombs containing 300 tonnes gas readied but not discharged because of weather

Date	Location	Type of attack	Comments
29 May	Mesen–Wijtschate	Phosgene projector drums, mortar bombs	700 projector drums and 1000, 4-inch bombs fired
1 June	Mesen–Wijtschate	Phosgene projector drums	900 projector drums fired
2 June	Mesen–Wijtschate	Phosgene projector drums	100 projector drums fired
2–5 June	Nieuwpoort	Various agents	Daily attacks on the Belgian front
3 June	Wijtschate Wood, Grand Bois, unnamed wood	Oil drums, mortar bombs	1500 oil drums and 1000 bombs fired from Stokes mortars
4 June	Mesen	Oil drums	1500 oil drums fired at woods on Mesen Ridge
6 June	Ramskapelle, Pervijze	. .	Attack on the Belgian front
7 June	Mesen–Wijtschate	CBR, PS and SK shells, white phosphorus and thermit bombs	30-minute artillery barrage at zero hour; 450 white phosphorus and 300 thermit bombs fired during infantry advance
7 June	Ploegsteert Wood, south of Mesen	Phosgene shells, tear gas	German artillery barrage on 3rd Australian Division while taking assault positions
7 June	Nieuwpoort	Chlorine cloud	Release of chlorine gas cloud against French forces; 120 casualties in 1 hospital[d]
8 June	Boezinge	Gas bombs	50 British 120-kg heavy gas bombs fired on village[e]
20–21 June (night)	Hoge, east of Ypres	Gas	British attack, large number of gas casualties in 46th German Regiment
6–10 July	Nieuwpoort	Green cross, lachrymator, Blue cross (DA)	Heavy artillery bombardment of British troops, high percentage of gas shells; 10 July, 1st use of DA shells
12 July	Oothof	. .	Attack on the Belgian front
12–13 July	St Jan–Potijze, north-east of Ypres	Phosgene, Yellow cross	Heavy pre-emptive artillery bombardment of British 15th and 55th divisions before Third Battle of Ypres; 1st use of mustard gas, 50 000 shells fired
13 July (night)	Ypres area	Projector	British Livens projector attack
15 July	Bellewaarde Lake, near Hoge	Gas	German gas attack on British trenches
15–21 July	Ypres salient	Projector and mortar	British fire 5100 projector drums and 14 000 Stokes mortar bombs (100 tonnes gas) in preparation for Third Battle of Ypres
16 July	Langemark	Gas	British gas attack on 226th Reserve Regiment
20 July (night)	Boezinge	Projector	British Livens projector attack

Date	Location	Type of attack	Comments
20–21 July (night)	Near Ypres, Wijschate	Blue cross	German use of DA (repeated several nights later); British find unexploded blue cross shells near Wijtschate on 28 July
21 July	Pildem, Boezinge	Gas mines	British gas attack kills 150 in German 102nd Reserve Regiment
21 July (2 a.m.)	Langemark	Projector	British gas attack on German 392nd Regiment
21–22 July (night)	Nieuwpoort	Yellow cross	2821 British casualties, including 17 dead
23 July	Nieuwkapelle, Hazewind, Diksmuide	..	Attack on the Belgian front
24 July	Hazewind, Diksmuide	..	Attack on the Belgian front
25 July	Nieuwpoort	..	Attack on the Belgian front
26 July	Nieuwkapelle, Hazewind, Diksmuide	..	Attack on the Belgian front
End July–mid-Aug.	Ypres salient	Yellow cross	Systematic night attacks on British positions, including targets in the rear
28 July (night)	Nieuwpoort	Yellow cross	Heavy German artillery attack
28–30 July	Ypres salient	Phosgene, chloropicrin, stannic chloride	British counter-battery fire using large variety of gas shells; 80 000 shells reportedly fired
29 July	Kouseboom	..	Attack on the Belgian front
31 July (3:50 a.m.)	Ypres salient	Oil drums, thermit	British fire 330 oil drums and 1300 thermit bombs at start of Third Battle of Ypres
5–17 Aug.	Ramskapelle	..	Several attacks on the Belgian front
22 Aug.	Hazewind, Diksmuide	..	Attack on the Belgian front
7 Sep.	Hazewind, Diksmuide	..	Attack on the Belgian front
11 Sep.	Hazewind, Diksmuide	..	Attack on the Belgian front
18 Sep.	Langemark	Projector	Many casualties in all 3 regiments of 36th German Division
25 Sep.	Hazewind, Diksmuide	..	Attack on the Belgian front
27 Sep.	Warneton, south of Ypres, near French border	Phosgene	British drum attack, many German casualties
7 Oct.	Ramskapelle	..	Attack on the Belgian front
9 Oct.	Ypres	..	Attack on the Belgian front
16 Oct.	Poelkapelle, Passendale	..	Attack on the Belgian front
19–20 Oct. (night)	Poelkapelle	Projector and mortar	136 drums fired on Meunier House and 285, 4-inch bombs fired at Poelkapelle Brewery

Date	Location	Type of attack	Comments
26–27 Oct.	Diksmuide	Gas cloud, projectors, mortars, flame projectors	British attack from Belgian trenches in quiet sector to support operations in Ypres salient; integrated attack with gas cloud, 200 CG bombs and 150 thermit bombs fired from Stokes mortars
29 Oct. (night)	Diksmuide	Projector	1480 projector drums fired at targets 0.8 km south of Diksmuide
31 Oct.	Hazewind, Diksmuide	. .	German attack on the Belgian front
End Oct. or early Nov.	Nieuwpoort, Diksmuide	Shells	1st Belgian chemical attacks retaliate against German shelling with gas
1 Nov.	Oostkerke	. .	German attack on the Belgian front
2–5 Nov.	Passendale, Oostkerke, Nieuwpoort, Diksmuide	. .	German attack on the Belgian front
2 Nov.	Diksmuide	SK	400 SK bombs fired at German heavy trench mortar battery in Diksmuide
6 Nov.	Diksmuide	Projector	960 projector drums fired
7 Nov.	Pervijze, Oostkerke, Nieuwkapelle, (? Diksmuide)	Mustard gas shells	1st mustard gas attacks on Belgian positions
9 Nov.	Diksmuide	Mortar	40 mortars fire 1530 bombs on town
12–15 Nov.	Passendale, Oostkerke, Nieuwpoort, Diksmuide	. .	Attack on the Belgian front
20 Nov.	Passendale, Oostkerke, Nieuwpoort, Diksmuide	. .	Attack on the Belgian front
21 Nov.	Oostkerke	. .	Attack on the Belgian front
17 Dec.	Merkem	. .	Attack on the Belgian front
18 Dec.	Merkem	. .	Attack on the Belgian front
1918			
18 Feb.	Nieuwpoort, Boezinge	. .	Attack on the Belgian front
22 Feb.	Nieuwpoort, Boezinge	. .	Attack on the Belgian front
24 Feb.	Nieuwpoort, Boezinge	. .	Attack on the Belgian front
25 Feb.	Nieuwpoort, Boezinge	. .	Attack on the Belgian front
26 Feb.	Nieuwpoort, Boezinge	. .	Attack on the Belgian front
3 Mar.	Nieuwpoort, Boezinge	. .	Attack on the Belgian front
5–22 Mar.	Nieuwpoort, Boezinge	. .	Attack on the Belgian front
11–15 Mar.	Ypres salient	Yellow, green and blue cross	Heavy German shelling with mustard gas; green and blue cross directed at strong points and headquarters
17–23 Mar.	Ramskapelle	. .	Attack on the Belgian front
18 Mar.	Diksmuide, Pervijze	. .	Attack on the Belgian Front
21 Mar.	Pervijze	. .	Attack on the Belgian Front
27 Mar.	Nieuwpoort, Boezinge	. .	Attack on the Belgian Front
28–30 Mar.	Ramskapelle	. .	Attack on the Belgian front
3 Apr.	Ramskapelle	. .	Attack on the Belgian front

Date	Location	Type of attack	Comments
4 Apr.	Ramskapelle	..	Attack on the Belgian front
8–12 Apr.	Ramskapelle Nieuwpoort	..	Attack on the Belgian front
9–25 Apr.	Kemmelberg, south-west of Ypres	Yellow, green and blue cross	Battle for Mount Kemmel
16 Apr.	Ramskapelle	..	Attack on the Belgian front
17 Apr.	Ramskapelle	..	Attack on the Belgian front
24 Apr.	Ramskapelle	..	Attack on the Belgian front
25 Apr.	Ramskapelle	..	Attack on the Belgian front
25 Apr.	Kemmelberg	Yellow, green and blue cross[f]	Barrage intended to isolate French troops during final assault
27 Apr.	Boezinge, Langemark	..	Attack on the Belgian front
3 May	Nieuwpoort, Boezinge, Langemark	..	Attack on the Belgian front
5 May	Boezinge	..	Attack on the Belgian front
9 May	Merkem	..	Attack on the Belgian front
10 May	Boezinge, Langemark	..	Attack on the Belgian front
14–18 May	Nieuwpoort	..	Attack on the Belgian front
2 June	Nieuwpoort	..	Attack on the Belgian front
3 June	Nieuwpoort	..	Attack on the Belgian front
6 June	Nieuwpoort, Ramskapelle, Pervijze	..	Attack on the Belgian front
8 June	Nieuwpoort	..	Attack on the Belgian front
22 June. (night)	Ypres salient	Gas cloud	Beam attack form railway trains against 31st German Division causing 500 casualties
23 June	Oostduinkerke	..	Attack on the Belgian front
12 July	Oothof	..	Attack on the Belgian front
24 July	Ypres salient	Gas cloud	Beam attack from railway trains against 1st Landwehr Division, which was withdrawn because of severe casualties
20 Aug.	Steenstrate, Boezinge	..	Attack on the Belgian front
21 Aug.	Steenstrate, Boezinge	..	Attack on the Belgian front
22 Aug.	Steenstrate, Boezinge	..	Attack on the Belgian front
23 Aug.	Nieuwpoort	..	Attack on the Belgian front
24 Aug.	Steenstrate, Boezinge	..	Attack on the Belgian front
26 Aug.	Ypres salient	Gas cloud	Bean attack from railway trains
26 Aug.	Steenstrate, Boezinge	..	Attack on the Belgian front
30 Aug.	Steenstrate, Boezinge	..	Attack on the Belgian front
11 Sep.	Ypres salient	Projector	British launch 580 projector drums
12 Sep.	Nieuwpoort	..	Attack on the Belgian front
15 Sep.	Nieuwpoort	..	Attack on the Belgian front

Date	Location	Type of attack	Comments
1–14 Oct.	Staden, Nieuwpoort, Pervijze, Moorslede, Diksmuide	. .	Attack on the Belgian front
13–14 Oct.	Wervik	HS shells	More than 1000 HS shells fired; attack during which Adolf Hitler was gassed
17 Oct.	Loppem	. .	Attack on the Belgian front
22 Oct.	Zomergem	. .	Attack on the Belgian front
23 Oct.	Bellem	. .	Attack on the Belgian front
27–28 Oct.	Avelgem	Yellow cross	500 rounds fired, gassing at least 350 civilians
30 Oct.	Zomergem	. .	Attack on the Belgian front
Early Nov.	Hansbeke	Yellow cross	27 civilians reported dead
2 Nov.	Tournai Wood	Gas drum	British fired 50 gas drums
8 Nov.	Bossuit, north of Tournai	Thermit bombs	252 thermit bombs fired in last attack of the war by British Special Brigade

T-stoff = German code-name for liquid lachrymatory, mix of brominated aromatic hydrocarbons, filled in lead canisters; B-stoff = German code-name for bromoacetone; HE = High explosive; CBR = British designation for 50% phosgene and 50% arsenic chloride; PS = British designation for chloropicrin; SK = British designation for ethyl iodoacetate; DA = Diphenylchloroarsine; CG = Phosgene; HS = British code-name for bis(2-chloroethyl) sulphide (i.e., mustard gas). (For a discussion of the cross codes, see the subsection 'Colour coding using cross codes' in section II of chapter 6 in this volume.)

[a] In *The Poisonous Cloud: Chemical Warfare in the First World War* (Clarendon Press: Oxford, 1986), L. F. Haber disputes claims that a German attack occurred on that day, based on lack of documentary evidence in British or German archives. However, the 29 June 1915 *London Gazette*—reproduced in Latter, J. C. (Maj.-Gen.), *The History of the Lancashire Fusiliers, 1914–1918 in Two Volumes* (Gale & Polden: Aldershot, 1949)—reports the awarding of the Victoria Cross to Private J. Lynn of the 2nd Lancashire Fusiliers, who 'handled his machine-gun with very great effect against the enemy' who were 'advancing behind their wave of asphyxiating gas'. Lynn died the following day of gas poisoning.

[b] All references to 'attack on the Belgian front' are from annexe P of Lheureux, V. (2nd Lt.), 'L'utilisation des gaz de combat sur le front belge pendant la guerre 1914–1918' [Use of gas on the Belgian front in the 1914–18 war], and its annexes, Travail de fin d'études présenté pour l'obtention du titre de licencié en Sciences Sociales et Militaires, Ecole Royale Militaire, 126e Promotion Toutes Armes, 1989–90 [Dissertation submitted to obtain the degree of master in social and military sciences, Royal Military Academy, 126th Promotion All Arms, 1989–90]. Lheureux's compilation is based on documents from the Belgian Armed Forces and on weekly communiqués issued by Belgian General Headquarters. The compilation is the only known study of chemical warfare operations against and by Belgian troops which uses source documents; it does not detail the origin or nature of the attacks.

[c] Specifics of the most important attacks are given in the following entries.

[d] Mentioned in Foulkes, C. H., *'Gas!': The Story of the Special Brigade* (William Blackwood & Sons: Edinburgh and London, 1936, 2nd reprint), p. 117, without specifying how the gas clouds were generated.

[e] Baccarne, R. and Steen, J., *Boezinge 1914–1918: Gasaanval 2° Slag om Ieper* [Boezinge 1914–18: gas attack at the Second Battle of Ypres] (Uitgegeven onder de auspiciën van de N. V. Bank van Roeselare en West Vlaanderen [Published under the auspices of the N. V. Bank van Roeselare en West Vlaanderen], 1979), p. 135 (in Dutch), states that the bombs were filled with chlorine and sulphur dioxide.

[f]Yellow cross was used near Ypres, Verbrande Molen (south of Zillebeke Vijver), Hoge (north of Zillebeke Vijver) and on a narrow stretch running south-west by north-east from Scherpenberg (1.5 km north of Loker) and east from Dikkebusch Vijver to approximately 3.5 km south-west of Ypres. Green cross was used west, north and east of Reningelst. Blue cross was used in the attack on 2 French divisions in trenches stretching from south of Kemmelberg to an area between Kemmel and Wijtschate.

Sources: Baccarne, R. and Steen, J., *Boezinge 1914–1918: Gasaanval 2° Slag om Ieper* [Boezinge 1914–18: gas attack at the Second Battle of Ypres] (Uitgegeven onder de auspiciën N. V. Bank van Roeselare enWest Vlaanderen [Published under the auspices of the N. V. Bank van Roeselaere en West Vlaanderen], 1979) (in Dutch); Breton, W., *Les combats de Steenstraat Avril–Mai 1915* [The Battle of Steenstraat, April–May 1915] (Librairie Militaire Berger–Levrault: Paris, 1918); Delfosse, *Les grenadiers à Steenstraat/De grenadiers te Steenstraat* [The grenadiers at Steenstraat] (Ets. L. Collignon: Brussels, 1934); Edmonds, J. E. (Brig.-Gen.) and Becke, A. F. (Maj.), *History of the Great War Based on Official Documents: Military Operations, France and Belgium, 1915*, vol. 2 (MacMillan: London, 1928); Edmonds, J. E. (Brig.-Gen.) and Becke, A. F. (Maj.), *History of the Great War Based on Official Documents: Military Operations, France and Belgium, 1916*, vol. 1 (MacMillan: London, 1932); Edmonds, J. E. (Brig. Gen.) and Litt, D., *History of the Great War Based on Official Documents: Military Operations, France and Belgium, 1917*, vol. 2. (MacMillan: London, 1948); Edmonds, J. E. (Brig.-Gen.) and Litt, D., *History of the Great War Based on Official Documents: Military Operations, France and Belgium, 1918*, vol. 4 (His Majesty's Stationery Office: London, 1947); Edmonds, J. E. (Brig.-Gen.), Litt, D. and Maxwell-Hyslop, R., *History of the Great War Based on Official Documents: Military Operations, France and Belgium, 1918*, vol. 5 (His Majesty's Stationery Office: London, 1947); Edmonds, J. E. (Brig.-Gen.), Wynne, G. C. (Capt.) and Becke, A. F. (Maj.), *History of the Great War Based on Official Documents: Military Operations, France and Belgium, 1915* (MacMillan: London, 1927); Falls, C. and Becke, A. F. (Maj.), *History of the Great War Based on Official Documents: Military Operations, France and Belgium, 1917*, vol. 1 (MacMillan: London, 1940); Foulkes, C. H., *'Gas!': The Story of the Special Brigade* (William Blackwood & Sons: Edinburgh and London, 1936, 2nd reprint); Haber, L. F., *The Poisonous Cloud: Chemical Warfare in the First World War* (Clarendon Press: Oxford, 1986); Hanslian, R. and Bergendorff, F., *Der Chemische Krieg. Gasangriff, Gasabwehr und Raucherzeugung* [Chemical warfare: gas attack, gas defence and smoke generation] (E. S. Mittler & Sohn: Berlin, 1925); Lheureux, V. (2nd Lt.), 'L'utilisation des gaz de combat sur le front belge pendant la guerre 1914–1918' [Use of gas on the Belgian front in the 1914–18 war], and its annexes, Travail de fin d'études présenté pour l'obtention du titre de licencié en Sciences Sociales et Militaires, Ecole Royale Militaire, 126e Promotion Toutes Armes, 1989–90 [Dissertation submitted to obtain the degree of master in social and military sciences, Royal Military Academy, 126th Promotion All Arms, 1989–90]; Miles, W. and Becke, A. F. (Maj.), *History of the Great War Based on Official Documents: Military Operations, France and Belgium, 1916*, vol. 2 (MacMillan: London, 1938); Moore, W., *Gas Attack! Chemical Warfare 1915–18 and Afterwards* (Leo Cooper: London, 1987); and Latter, J. C. (Maj.-Gen.), *The History of the Lancashire Fusiliers 1914–1918 in Two Volumes* (Gale & Polden: Aldershot, 1949).

16. Chemical weapon agent and historic chemical munitions disposal: the British experience

RON G. MANLEY

I. Introduction

The destruction of chemical weapon stocks, CW-filled weapons and the plant associated with the production, bulk storage and weapon filling of CW materials is a difficult and hazardous task. This chapter outlines the approach of the United Kingdom to this problem. It describes the method used during the 1970s for the disposal of the nerve agent GB (sarin), the associated pilot plant and storage facilities, residual stocks of mustard gas (20 tonnes), ethyl iodoacetate—KSK (10 tonnes) and α-bromobenzyl cyanide—BBC (10 tonnes) and techniques currently used for the safe disposal of recovered, 'historic' chemical munitions. However, as it was one of the factors which influenced the disposal methods selected, it is first worth briefly reviewing the World War II and post-war history of CW agent production within the United Kingdom.

During World War II a number of chemical warfare agents were manufactured in the UK. These included mustard gas, phosgene, BBC, lewisite, diphenylamine chlorarsine, methyl dichloroarsine and KSK. However, of these only mustard gas and phosgene were manufactured and filled into munitions on a large scale.[1] At the end of the war manufacture ceased and no CW agents were produced until the early 1950s, when a one tonne per week pilot plant was constructed to undertake process research to establish parameters for the production of GB. This plant was operated during the two-year period 1954–56. The GB produced during this period was stored in bulk containers and used solely for stability studies. No nerve agent was filled into munitions. In 1956 the United Kingdom took a unilateral decision to forgo renewal of its offensive CW capability and confine its future efforts solely to research into effective countermeasures against the CW threat. Since 1956 the preparation of CW agents in the UK has been limited to the small quantities necessary to support this research programme.

A programme for the destruction of the UK's remaining stocks of CW agents and production facilities was drawn up and successfully carried out during the 1960s and 1970s. Small numbers of old chemical munitions, however, continue to be recovered from Ministry of Defence (MOD) ranges and also during the excavation of land previously used for military purposes. In recent years the rate of recovery of these historic munitions has increased, as many former MOD sites are redeveloped for civil purposes.

[1] Past production of chemical warfare agents in the United Kingdom, Conference on Disarmament document CD/856, 11 Aug. 1988.

II. Disposal options

To date those technologies with a proven track record for the destruction of CW agents in a safe and effective manner have been based on either chemical or thermal degradation processes.

The toxicity of most CW agents can either be eliminated or reduced to an insignificant level by chemical treatment. GB, for example, is rapidly hydrolysed by treatment with a 10–20 per cent aqueous solution of sodium or potassium hydroxide. Mustard gas can also be hydrolysed using aqueous/alcoholic solutions of sodium or potassium hydroxide. A 10–20 per cent chlorine bleach solution is known to be equally effective. Phosgene can also be readily hydrolysed; although as it is an industrial intermediate, recovery and recycling may also be a viable option. Most of the arsenical agents can be chemically oxidized and neutralized to yield arsenites. In this instance, however, the problem of disposing of the toxic arsenic salts remains.

Many of the CW agents are flammable (the notable exception being GB) and have acceptable calorific values. They lend themselves therefore to destruction by incineration. Even GB, although non-flammable, can be readily destroyed by this method. It has been demonstrated that, in properly designed and operated plants, very high destruction removal efficiencies can be obtained (>99.99999% removal) for these materials. While direct high-temperature incineration has been shown to be an effective method of destroying G (organophosphorus) agents, their higher vapour pressure and greater inhalation toxicity make the operation more difficult. The problem is not in the actual incineration step, but rather in the handling and feeding of the waste to the furnaces and the protection of the operating staff. One solution to this problem is to hydrolyse the G agent prior to incineration, thus reducing the need for such high levels of containment at the incineration step. However, it can equally be argued that once the agent has been hydrolysed incineration is no longer the most appropriate disposal method for the hydrolysate.

The principal disadvantage of the alkaline hydrolysis of nerve agents is seen to be the large volume of hydrolysate produced. A fivefold increase in the overall volume is not uncommon. However, a number of options are available for the disposal of this relatively non-toxic material. The options available within the UK were: incineration; dilution, followed by disposal to coastal waters; and concentration/solidification, followed by burial in a licensed landfill.

While the organic content of the hydrolysate can be further degraded by high-temperature incineration, where sodium or potassium hydroxide have been used to effect the hydrolysis, the very low calorific value and high salt content of the product make this a demanding and expensive solution. It also leaves the problem of disposing of the inorganic salts recovered from the effluent gas cleaning plant of the incinerator. Providing the hydrolysate can be shown to be free of anticholinesterase activity and the concentration of salts present is low enough to comply with regulatory discharge limits, British regulations permit its disposal to coastal waters. Alternatively, the water component of the hydrolysate can be removed by either forced or natural evaporation. The residual salts are then collected and appropriately packaged before being buried in a properly constructed secure landfill (i.e., designed to contain any leachate).

Experience suggested that no single technology was ideal for all types of agent. While it was possible to apply certain general principles to the overall programme, the actual destruction of each category of CW material and its associated facility had to be tackled on a case-by-case basis. Giving due regard to safety and environmental aspects, the most appropriate technique in each case was chosen.

III. Disposal of bulk CW stocks

Although mustard gas can be destroyed by chemical degradation the process becomes both difficult and messy when, as is normally the case, significant quantities of impurities are present. Incineration, however, is largely unaffected by this factor and therefore was the process adopted during the 1970s for the disposal of the UK's remaining stocks of mustard gas. Over the years considerable experience has been gained in the safe incineration of this material. Small amounts of material recovered from old munitions, which continue to be found occasionally in the UK during construction and excavation work, are still processed by this method in the Chemical & Biological Defence Establishment (CBDE), Porton hazardous waste incineration plant.[2] No major problems have been experienced with this method and the basic process, although gradually modified over the years to comply with changing safety and environmental legislation, remains fundamentally unchanged.

All of the UK's nerve agent was produced and stored at the Chemical Defence Establishment (CDE), Nancekuke, a coastal site in the south-west of the country. On completion of the bulk agent storage trials it was destroyed by alkaline hydrolysis using a 20 per cent sodium hydroxide solution. The destruction operation was carried out, as a batch process, over the two-year period 1967–68. It is important to note that, because the disposal operation was carried out using part of the nerve agent pilot plant facilities (suitably modified), all of the infrastructure necessary for undertaking such an operation in a safe and effective manner was already present. This included: a building specifically designed for work with this agent, filtered ventilation, effluent treatment, indirect heating and cooling systems, changing and decontamination facilities, medical facilities and, most important of all, experienced staff. Without these assets the operation would have been much more difficult and taken far longer time.

As the potential hazard from the products of alkaline hydrolysis of GB were assessed to be very low, further treatment by incineration was considered to be unwarranted. Given the location of the site it was judged that, of the available options for the disposal of this material, dilution followed by sea disposal was the most appropriate. Prior to being diluted and discharged to the ocean, via an undersea outlet, each batch of hydrolysate was tested to confirm that it was free of any anticholinesterase activity. Dilution of the waste ensured that the precipitated salts were redissolved and the fluoride ion concentration was below the limit for discharge. A working party under Professor R. H. S. Thompson, Courtauld Professor of Biochemistry at Middlesex Hospital Medical School, appointed by the Minister of Defence for Equipment in 1969, concluded in its

[2] The disposal of these historic munitions is discussed in more detail in section V.

report: '(*a*) The effluent system at CDE, Nancekuke, is entirely satisfactory to treat effluent arising from all foreseeable eventualities and embraces safeguards not common to normal practice, and (*b*) the effluent from CDE, Nancekuke, causes negligible pollution of the sea or of the atmosphere'.[3] Studies of the marine algae population (diatoms, etc.) in the vicinity of Nancekuke concluded that there was no evidence that effluent from the site at Nancekuke had resulted in any measurable pollution of the surrounding coastal waters.

The few remaining tonnes of KSK were hydrolysed at Nancekuke and the hydrolysate transferred to a commercial chemical company which recovered the iodine for reuse as an industrial chemical. A similar quantity of BBC was hydrolysed in a batch process using sulphuric acid. After neutralization and being certified free of BBC the reaction product was diluted and discharged, via the undersea outlet, to the ocean.

IV. Decontamination and disposal of the GB pilot plant

Once the bulk of the agent had been destroyed the problem remained of decontaminating the pilot plant and its associated support services, which were located in several buildings. When production ceased in 1956 the plant was drained down and as much of the CW agent as practicable removed. Any potential points where liquid might be trapped or held up were identified and dealt with where possible. The plant was then sealed, filled with a 15 per cent sodium hydroxide solution and left to stand for a lengthy period (several weeks). On removal the sodium hydroxide solution was checked and certified free of any anticholinesterase activity before being discharged via the site's effluent system. The plant was then sealed and 'mothballed'.

In 1976 the decision to close the site at Nancekuke was taken and it became necessary to clear it of all chemical activity and to dismantle the GB pilot plant. Although known to be free of gross contamination, it was considered highly probable that traces of agent were still present in the plant. Previous experience had shown that gaskets, seals and glands, for example, were unlikely to have been completely decontaminated when the plant was filled with the sodium hydroxide solution. The assumption was therefore made that the plant was contaminated with agent and the decommissioning procedures planned accordingly.

The first step was to ensure that the key support services were still operational or, where necessary, re-established. An experienced team of workers then began to dismantle the plant cubicle by cubicle. Starting at the top and gradually working down the plant was disassembled, piece by piece. All joints were broken, gaskets removed and potentially contaminated surfaces exposed and decontaminated. Where practicable the individual plant items were submerged in decontaminant for at least 24 hours. The whole operation was monitored using appropriate detection and monitoring equipment. A number of alarms did occur and agent was detected during the dismantling operation. Once the plant itself had been removed all the auxiliary services were similarly treated. The systematic decontamination of the cubicle walls was then under-

[3] *Report of a Working Party on the Transport of Nerve Agents from Nancekuke to Porton and the Disposal of Effluent from Nerve Agent Production* (Her Majesty's Stationery Office: London, 1970).

taken prior to their being removed and checked for residual contamination. Finally, the walls of the building and the drains were treated with decontaminant and washed through with water. All decontamination effluent was collected, tested and certified safe before discharge.

The majority of the components which had been in contact with GB and its organophosphorus intermediates were fabricated from silver. After chemical decontamination and being certified free of agent these components were melted down and the recovered silver cast into ingots. The silver ingots were then recycled via the commercial market. Other components, after being certified free of contamination, were either disposed of as scrap or buried in an approved landfill. Nineteen governments accepted the invitation made to the Committee on Disarmament to send representatives to the UK in 1979 to inspect the dismantling process. The site was finally cleared and vacated in September 1980 and has since reverted to use by the Royal Air Force.

V. Disposal of historic chemical munitions

The routine recovery and disposal of old munitions discovered in the UK is undertaken by the British Army's Explosive Ordnance Disposal (EOD) teams. However, on the discovery of a suspected chemical munition the standard procedure is for the EOD team dealing with the incident to contact the CBDE. Except in the very rare situation where the munition is judged to be too unsafe to be moved, arrangements are made for it to be suitably over-packed (i.e., placed in a sealed steel container with an absorbent such as charcoal) and transported, by road, to the CBDE for final disposal.

While the majority of the recovered munitions are of World War II origin a significant number stem from World War I and the interwar years. They include shells, mortars and grenades of various types filled with a range of toxic chemical substances. So far over 30 different types of ammunition have been identified. Typical examples of the fillings found are arsenious chloride, chloropicrin (trichloronitromethane), KSK, phosgene (carbonyl chloride), and thickened and unthickened 'sulphur mustard' (principally 2,2'-dichloroethyl sulphide). When found, the munitions are normally heavily corroded and often little, if any, indication of their probable contents remains. As such they pose both a difficult and complex waste disposal problem.

On arrival at the CBDE the munitions are carefully unpacked by the CBDE personnel wearing full protective clothing, including respiratory protection. After detailed examination each munition is tagged with a unique identification number and placed in a special well-isolated storage area to await disposal.

With the assistance of old record drawings an attempt is made to identify each munition and its contents. However, record drawings are not available for every type of munition used and in many instances, particularly during World War I, standard ammunition cases were sometimes used with a variety of fillings. The next step therefore is to X-ray the munition. This not only provides vital information on its internal structure but normally makes it possible to confirm the presence of a chemical fill. Having acquired as much information as possible about the munition the next step is to separate the chemical contents from the explosive element. Over the years a number of techniques have been developed for doing this and, based on experience and the available information, the most appropriate one is selected for each munition. The procedures can best be illustrated by taking a specific example.

Two of the most common munitions recovered are the 25-pounder and 18-pounder chemical shell. Although normally not found in the fused state these shells nevertheless do contain an explosive 'burster' charge, which is usually located in the nose of the shell. The standard procedure for dealing with these shells is to cut through the case at a point which has previously been established to be clear of explosives and then carefully to remove the contents. The cutting operation is carried out using a modified mechanical hacksaw housed in a specially designed ventilated cubicle. Owing to the presence of the explosive charge, and the fact that it is impossible to be absolutely certain that the main charge of the shell is chemical and not high explosive or white phosphorus, this operation is carried out remotely. Once the shell has been clamped to the saw-bed the operators withdraw and the cutting operation is controlled from a protected shelter using closed-circuit television. When the cut has been completed and the officer in control of the disposal operation is satisfied that it is safe to approach the shell, the contents are removed manually by operators dressed in full protective clothing. The chemical contents, which may vary from a relatively free-flowing liquid to a sticky, rubbery mass, are bulked along with the non-explosive portion of the shell into suitable containers to await final disposal.

Recovered CW material is either treated chemically or incinerated in the CBDE's licensed hazardous waste incinerator. The portion of the munition containing the explosive charge is placed in a separate container and transferred to a special burning pit, where it is destroyed using a 'destructor' incendiary. Once certified free of explosives and chemical contamination metal components, from both the pit and the incinerator, are buried in a secure landfill. The technique is safe and no major problems have been experienced with it. While the disposal rate using this method is inevitably fairly low it is ideally suited to dealing with the small number, but wide range of types, of CW munitions being found within the UK. Although the British process has been designed in such a manner that an accidental explosion would not present any risk to the staff involved, an event of this nature could result in the release of a small quantity of CW material to the atmosphere. Appropriate safety distances and decontamination procedures have therefore been established.

The CBDE's pilot-plant-scale hazardous waste incinerator was designed and constructed in the early 1980s to meet the following main requirements: (*a*) to allow the safe disposal of small quantities of hazardous chemicals and chemically contaminated wastes (in the form of solids, liquids or gases) arising from the CBDE research programme; (*b*) to enable the proofing (test firing) and disposal of pyrotechnic stores (grenades, cartridges, etc.) to be carried out under conditions which are both safe and environmentally acceptable; and (*c*) to provide a safe and effective means of destroying the contents of recovered World War I and II chemical munitions shipped to the CBDE for final disposal. It is important to note that, while the incinerator was designed to cope with the disposal of CW agents, this is not its sole purpose and it is required to and does fulfil a multi-purpose role.

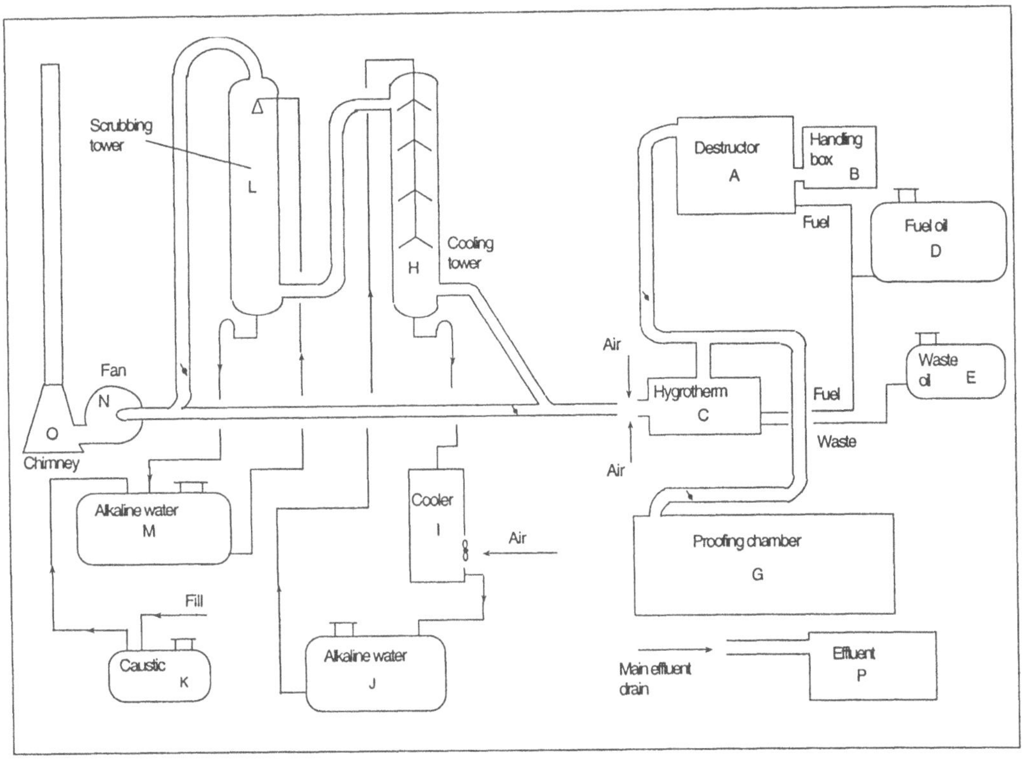

Figure 16.1. Principal components of the Chemical & Biological Defence Establishment's hazardous waste incinerator

The principal components of the plant (see figure 16.1) are a primary furnace (A), a large chamber for the functioning of pyrotechnic stores (G), a secondary furnace (C) and an exhaust gas treatment plant (H–P). A fan (N) provides an induced draught and is used to maintain the whole system, including both furnaces, at a reduced atmospheric pressure. This ensures that under all operating conditions air flows into the system and prevents the escape of potentially hazardous fumes into the working area. The furnaces are oil fired, capable of operating at temperatures up to 1200°C and provide an overall gas residence time of approximately 3 to 4 seconds. Exhaust gases from the secondary furnace pass to the gas cleaning plant where they are cooled, neutralized and scrubbed to remove acid gases and the majority of the particulates present. After passing through a demister and the induced draft fan the cleaned gases are discharged to the atmosphere via a chimney. Aqueous effluent from the proofing chamber, gas scrubbing system, demister and stack is piped to a concrete holding tank. After analysis and, if necessary, preliminary treatment the effluent is pumped to the CBDE main sewage treatment works. There, along with the rest of the CBDE aqueous and foul drainage, it is treated and subjected to conventional biological oxidation before finally being discharged.

To enable the performance of the plant to be fully monitored sensors have been fitted at critical points throughout the system. During the disposal operation a chemical agent monitor (CAM) is used to monitor the environment in the vicinity of the furnaces for 'mustard gas' vapour (the major hazard). Although no mustard gas vapour has ever been detected in the furnace area, operators working in this area routinely wear respiratory protection as an additional precaution. The exhaust gases from the plant are routinely sampled from a point just before they enter the first scrubbing tower and again from a point midway up the discharge stack. No residual chemical agent has been detected in the exhaust gases from the plant. The construction cost of the facility was approximately £0.75 million (1984 prices) and a further £1 million is planned to be spent during 1993–94 on upgrading the plant to increase its throughput and to enable it to comply with recent changes in emission standards. It should be emphasized, however, that this upgrade is primarily related to the use of the plant for the disposal of general hazard waste and not directly to its use for CW agent disposal.

Sulphur mustard which, in one form or another, is the most frequent filling found in these old munitions is also considered to be the most hazardous substance likely to be encountered. It is an extremely persistent oily liquid, a potent vesicant and is now included in the official British list of potential human carcinogens. Therefore, while the recovered material may sometimes be a complex mixture of substances, the disposal procedures have been based on the assumption that sulphur mustard is present.

There are three basic methods available for its introduction into the furnaces. The first, and preferred, option is to replace the standard effluent burner gun in the waste liquid furnace with a specially designed burner which operates on the venturi spray principle. This method enables the waste to be fed straight from its container into the burner gun and has the advantage that in the event of a system failure it will remain fail-safe. Its major limitation is that it can only cope with low or moderately low viscosity wastes, and while it can be used for the disposal of unthickened sulphur mustard it is generally unsuitable for use with the complex mixtures recovered from some of the older munitions.

The waste can also be introduced via the normal liquid waste feed system to the standard effluent burner gun. This system can readily handle liquid mixtures with kinematic viscosities up to approximately 10 $mm^2 s^{-1}$. The major disadvantage of this method is that the system operates at a pressure of approximately 6 bar, and failure of a pump gland, pipe union or control valve could lead to gross contamination of the area around the furnace.

The third approach is to dispose of the material via the solid waste furnace. Although this method can only handle a very limited throughput, this is not considered to be a problem as the rate-limiting step in the overall disposal process is the opening of the munitions. It is also the safest of the three procedures, is unaffected by variations in the viscosity of the waste and is the only method which can cope with the contaminated metal components. This is the procedure which has been adopted for the routine disposal of the high viscosity mustard-gas containing wastes.

The waste is transferred from the shell-cutting area in sealed 50-litre mild steel drums containing approximately 30 litres of waste. One or two drums of waste are loaded, by operators in full protective clothing, into the cold furnace with the induced draught fan running. The full diameter lids are then carefully removed and approximately 1 litre of diesel fuel is added to each drum. The waste mixture is ignited and allowed to burn freely. Smoke and vapour generated are drawn into the second furnace where they are reburnt at high temperature before passing on to the gas cleaning system. When the waste will no longer burn freely on its own the main burner is lit and the temperature of the solid waste furnace gradually raised to between 500°C and 600°C. The furnace is held at this temperature for one hour and then allowed to cool. The drums and metal components are then removed, checked to confirm that complete decontamination has occurred and finally buried in a secure landfill.

VI. Safety and environmental aspects

As stated above, the task of destroying these materials and their associated production and storage facilities is both difficult and hazardous. In order to keep the risk to the health of the personnel undertaking this task to a minimum the highest safety standards were required and were built into each phase of the operation. The principal criteria used when drawing up the safety procedures were as follows:

1. The hazard to health of each substance had to be identified.
2. An assessment of the risk they posed to the people involved had to be made (under both normal and abnormal operating conditions).
3. Action had to be taken to reduce any identified risk to as low a level as could reasonably be achieved.
4. Where protection from exposure was necessary this was to be achieved, wherever practicable, by primary containment and the use of engineering controls. Dependence on personal protective equipment as the primary means of protection was only acceptable if control could not reasonably be achieved by any other means.
5. Routine area monitoring was to be carried out for the substances of concern, and records were to be maintained.
6. Where appropriate, routine medical surveillance of staff was to be undertaken.
7. Staff were to be trained to understand the hazards and to cope adequately with them.
8. Emergency procedures were to be drawn up and their effectiveness regularly tested.

It is perhaps interesting to note that the UK's Control of Substances Hazardous to Health Regulations,[4] which were introduced in 1989 to govern work with all hazardous substances, were largely based on these same principles.

Whichever disposal method or methods are adopted, the aim must be to achieve appropriate high standards of safety and environmental acceptability. However, it must be remembered that these two requirements cannot be considered in isolation. With difficult and hazardous operations of this nature it is essential that external parameters, such as environmental emission standards, are set at realistic and scientifically justifiable levels. Apart from making the operation both more expensive and protracted, unjustifiably high standards may actually increase the immediate risk to the personnel undertaking the work. Equally, unnecessary delays in the destruction of these materials can only lead to a further deterioration in their condition, making the eventual disposal operation more difficult and increasing the risk of an unplanned release. It should be

[4] *Control of Substances Hazardous to Health Regulations (1988)* (Her Majesty's Stationery Office: London, 1988).

recognized that there is no perfect method for the destruction of these or any other hazardous materials.

Within the European Union (EU), although standards still vary between individual member states, the Best Available Techniques Not Entailing Excessive Cost (BATNEEC) and the Best Practical Environmental Option (BPEO) principles are the foundation on which environmental standards within the EU are being based. This is the approach which has been adopted under the UK's new Environmental Protection Act (EPA).[5] While the EPA regulations set target limits for the major pollutants, operators of plants producing these substances are required to employ the BATNEEC principle. This means that if, in the view of the Government's Pollution Inspectorates, lower limits could be achieved without entailing excessive cost they will be applied and must be met. As with similar British legislation the question of what is and is not excessive cost will, where necessary, be determined in the law courts. Operators also have to consider the impact of their processes across the complete environmental spectrum (e.g., air, water, land, etc.) and apply the BPEO principle to determine the best overall environmental option.

Under current British legislation the release of CW materials and their degradation products is not specifically proscribed. As for other similar hazardous materials the regulations require that their potential human and environmental impact be identified, the risk assessed and appropriate action taken to reduce this risk to the minimum practicable level. Again what is practicable will, if necessary, be determined in a court of law. In the present author's view the EU philosophy of BATNEEC and BPEO is a realistic and defensible approach to the setting of environmental standards for CW disposal operations.

VII. Conclusions

Since its decision in 1956 to give up its offensive chemical warfare capability the United Kingdom has developed and implemented safe and effective procedures for the disposal of its CW agents and the associated plant and storage facilities. It continues to maintain a small destruction facility and the trained personnel necessary to undertake the safe disposal of the occasional World War I and II chemical munitions still being recovered. It is the belief of the present author that the methods adopted by the UK for the destruction of its CW stocks, munitions and associated facilities were and continue to be effective, involved the minimum risk to the personnel undertaking the task and neither posed nor pose any significant threat to the environment.

[5] *Environmental Protection Act (1990)* (Her Majesty's Stationery Office: London, 1990).

17. UNSCOM's experience with chemical warfare agents and munitions in Iraq

RON G. MANLEY

I. Introduction

With its adoption of Resolution 687 (1991) of 3 April 1991 the United Nations Security Council established the terms and conditions for the formal cease-fire between Iraq and the Coalition of member states co-operating with Kuwait, thereby ending the 1991 Persian Gulf War.[1] Part C of Resolution 687 addresses Iraq's weapons of mass destruction (i.e., biological, chemical and nuclear weapons and ballistic missiles with a range greater than 150 km). It requires their identification, location and disposal and the establishment of a monitoring regime to ensure that Iraq does not reacquire these weapons. The United Nations Special Commission on Iraq (UNSCOM) was established by the UN Secretary-General in May 1991, under the Chairmanship of Ambassador Rolf Ekéus of Sweden, to oversee the implementation of the non-nuclear aspects of this part of the resolution. Its mandate is: (*a*) to carry out on-site inspections of Iraq's biological, chemical and missile capabilities; and (*b*) to take possession for destruction, rendering harmless or removal from the country of all chemical and biological weapons (BW), stocks of agents, ballistic missiles with a range greater than 150 km and all components and facilities associated with these three weapon systems. It is required to monitor and verify Iraq's compliance with its undertaking not to reacquire these weapons and also to provide assistance to the International Atomic Energy Agency (IAEA), which is tasked with dealing with the nuclear aspects of Resolution 687.

The Special Commission has 21 members, including the Executive Chairman, drawn from 21 member countries. In addition, more than 40 countries have provided personnel on temporary secondment to either work in the offices of the Special Commission in New York, Bahrain or Baghdad or to take part in inspection or destruction missions in Iraq. This chapter deals specifically with the chemical weapon aspects of UNSCOM's work.

II. Iraq's CW agents and munitions

The principal chemical agents in Iraq's arsenal were mustard gas,[2] the nerve agents tabun (GA),[3] sarin (GB),[4] and a mixture of GF[5] and GB. Bulk quantities of these agents were stored at their production site, Al Muthanna, some 80 km north-west of Baghdad. While it has been established that Iraq had carried out research on other agents, including VX,[6] the above agents are the only ones

[1] United Nations Security Council document S/RES/687 (1991), 8 April 1991.
[2] Bis(2-chloroethyl)sulphide.
[3] Ethyl-N,N-dimethylphosphoramidocyanidate.
[4] O-isopropyl methylphosphonofluoridate.
[5] O-cyclohexyl methylphosphonofluoridate.
[6] O-ethyl S-2-diisopropylaminoethyl methylphosphonothiolate.

which were declared by Iraq and found by UNSCOM in any significant quantity.

Iraq had developed a range of delivery systems for its chemical agents. These included artillery shells, aerial bombs, missile warheads and rockets. The artillery shell cases, for example, were standard 155-mm phosphorus smoke shells which were filled with mustard gas (4.4 kg) via the nose of the shell prior to the insertion of the central 'burster'[7] tube. The burster tubes were inserted using a hydraulic press thereby effectively sealing the contents into the shell. A range of both 250- and 500-gauge aerial bombs had also been developed. Some of these used imported bomb casings, designed for use with conventional fills, while others were fabricated in Iraq to Iraqi design. Both nerve agent and mustard gas fillings were used in these bombs. One bomb, the R400, was interesting in that it was a binary weapon. The bombs were initially filled and stored with the appropriate amount of the mixed alcohols (isopropanol and cyclohexanol). When required for use the second precursor chemical was added and the bomb sealed and rotated to mix the contents, which reacted to form a mixture of the nerve agents GB and GF. This technique was also used with some of the chemical warheads for the Al Hussein missile, and there were both binary and standard GB/GF-filled missile warheads in the Iraqi arsenal. The 122-mm nerve agent-filled rocket was particularly interesting in that it had clearly been specifically designed as a chemical munition. Three variants of this munition were discovered. In the most common type the agent (*c.* 6.8 kg) was held in two specially moulded high-density polythene containers which were designed to fit into the metal warhead. A central burster tube containing the explosive RDX passed through the two containers and a fuse system was fitted to the nose of the rocket. The rocket was powered by a conventional solid fuel motor.

From statements made by Iraq, and evidence found by UNSCOM, it would appear that the normal practice was to store the filled artillery shells, aerial bombs and missile warheads without explosives. The central explosive burster charge and fuse were only added when the weapons were made ready for use. The 122-mm nerve agent-filled rockets were the one exception to this practice. These were frequently stored with their solid fuel rocket motors fitted and in some circumstances they were stored complete with central explosive burster charge and fuse.

Iraq's filled chemical munitions were dispersed at airbases and storage depots throughout the country. At the time of writing the size of Iraq's CW stockpile had not been fully established. However, by early 1993 over 140 000 filled and unfilled chemical munitions, 590 tonnes of bulk agent and some 3500 tonnes of precursor chemicals had been monitored by UNSCOM inspectors, and these figures have not changed significantly since.

III. CW production facilities

Iraq's declared production facility for all its CW agents and weapons was Al Muthanna, a large site in the western desert some 80 km north-west of Baghdad between the towns Samarra and Al Fallujah. The core area of the site was

[7] A long, thin-walled steel tube, closed at one end, which holds the explosive charge used to disseminate the contents of the munition.

approximately 5 km by 5 km. It had an excellent road system and a well-developed, protected infrastructure. In addition, three nearby satellite sites (Al Fallujah 1, 2 and 3) were being developed for the production of precursor and other associated chemicals. A fifth site, Muhammadyiat, was used for storage of filled chemical munitions, empty chemical munitions, explosives for use with the munitions and associated components. Muhammadyiat was also declared by Iraq to be the location of a second facility for the filling of chemical munitions. However, owing to extensive damage at Muhammadyiat, it was not possible to establish whether or not this facility was operational or simply under development. Together these five sites were known as the Al Muthanna State Establishment.

All five sites were subjected to heavy bombing by the Coalition forces and the facilities and infrastructure at each site were severely damaged. Muhammadyiat, in particular, was almost totally destroyed by Coalition bombing. The extensive damage at these sites had major implications for the subsequent CW destruction operations.

IV. Planning the disposal of Iraq's chemical weapons

From the beginning it was apparent that the safe destruction of Iraq's chemical warfare assets would be difficult and complex. The Executive Chairman of UNSCOM therefore approached the small number of countries known to have experience in this specialized field for assistance to help plan the safe and effective destruction of Iraq's chemical weapons. Five countries—Canada, France, Russia, the UK and the USA—each agreed to provide an expert in this field, and these experts, a representative from the World Health Organization (WHO) and an expert in combustion engineering from the New Jersey Institute of Technology, USA, formed a Chemical Weapons Destruction Advisory Panel (DAP). The DAP met for the first time in the week of 24 June 1991 to assess the problem and plan the destruction operation. It subsequently met at regular intervals to review the progress of the destruction programme and to provide further technical advice as appropriate. In addition, members of the DAP have, at the time of writing, taken part in 10 missions to Iraq to oversee the start of each phase of the destruction programme and to provide technical assistance to the on-site team.

At the start of its work the DAP reviewed the available options for the safe destruction of Iraq's stockpile of chemical munitions. It was clear, given the damaged and leaking condition of many of the weapons, that action had to be taken as soon as possible. Solutions that used proven, simple and readily available technology were therefore required. The question of who would undertake the actual destruction operation also had to be addressed. In the opinion of the DAP significant experience in the destruction of these highly toxic materials was required if the operation was to be brought to a safe and effective conclusion.

The initial view was that the operation would have to be undertaken by suitably qualified personnel from one of the small number of countries with experience in this field. However, Iraq made a formal request to UNSCOM for permission to undertake the operation using its own equipment and personnel. The principal advantage of this option was that it removed the major logistical problems of transporting large amounts of equipment and personnel to Iraq and the associated delays. Detailed discussions were therefore held between the UNSCOM experts and their Iraqi counterparts to establish whether the Iraqi proposal was feasible. As a result of these and other discussions it was agreed that Iraq would construct the facilities required for the destruction of its CW stockpile and operate them under UNSCOM supervision and direction. A team of UNSCOM personnel, the Chemical Destruction Group, was established in Iraq for the life of the destruction operation to undertake this task. The team was supported, as necessary, by visits from members of the DAP.

The question of whether the destruction operation should be undertaken at a single location or at each of the individual chemical munition sites was discussed at length. The former option required the transport of chemical munitions, many of which were in poor condition, from widely dispersed sites across Iraq to a central location. The latter option required the use of mobile destruction equipment which could be moved from site to site. Although technically feasible, the use of mobile equipment is by no means straightforward, and there are the additional technical, safety and environmental problems of having to operate at several different sites. After considering all aspects of the problem it was decided to centralize the destruction operation at a single site and to transport all munitions deemed safe to be moved to this location. All the bulk stocks of chemical agent and most of the precursor chemicals were already at the CW production site at Al Muthanna. In addition, although severely damaged, much of the necessary infrastructure for the destruction operation was present there. It was therefore decided to conduct the destruction operation at Al Muthanna.

V. Safety and environmental considerations

The destruction of chemical warfare agents and munitions is always complex and hazardous. In Iraq the added risks from leaking munitions, unexploded ordnance, damaged buildings and the like made the operation especially dangerous. It was therefore necessary to ensure that safety was fully addressed at all stages of the operation. High priority was given to the establishment of a safety policy covering UNSCOM's operations in Iraq and drafting the associated safety manual. This manual has been and continues to be revised at regular intervals to ensure that it keeps pace with UNSCOM's changing operational needs in Iraq. Protection of the environment is also important and due consideration has been given to this in all aspects of the destruction operation. However, the overriding consideration has been the safety of the personnel involved in the destruction operation and that of the local civilian population.

VI. Transportation, draining and destruction of munitions

Having taken the decision to transport the chemical munitions from their various depots in Iraq to Al Muthanna, it was necessary to ensure that this was done in a safe manner. Prior to being moved each munition was checked to ensure that it was free from explosives. Leaking munitions were sealed using a variety of techniques, and then checked to confirm that they were no longer leaking. The munitions were placed in specially designed racks and packed into International Standards Organisation (ISO) transport containers which had been modified to ensure that they were both liquid- and vapour-tight. A deep layer of absorbent material was placed on the floor of each container. As an additional precaution refrigerated containers were used for the munitions containing mustard gas. This ensured that the mustard gas, which solidifies at 14°C, was maintained in a solid form throughout the transfer. A small number of munitions, judged to be in too poor condition to be moved, were set aside for on-site disposal at a later date.

Transfers were undertaken by Iraq under the direction of UNSCOM. Each shipment was accompanied by sufficient security, monitoring, decontamination and medical personnel and equipment to deal with all foreseeable incidents. Shipments were carefully monitored during the transfer process to ensure that any leaks would be promptly detected and appropriate action taken. Routes were planned, as far as possible, to avoid populated areas. On arrival at Al Muthanna the munitions were placed in storage in a remote area of the site. Using these procedures all of the munitions that could be moved were safely transferred to Al Muthanna without incident.

The next step in the process was to remove the chemical agent from the munitions and place it in bulk tanks to await disposal. This is generally regarded as the most difficult step in the destruction of chemical weapons. At Al Muthanna the operation was further hampered by the absence of a suitable, operational containment facility. As a result the draining operation had to be undertaken by Iraqi and UNSCOM personnel working in full protective clothing. This, together with the high summer temperatures in the desert and the need to develop individual draining techniques for each type of munition, resulted in a long and difficult operation.

For the aerial bombs containing mustard gas the technique used, when possible, was to carefully remove the filling plug and suck the liquid mustard gas out using a specially designed vacuum transfer system. If the filling plug could not be removed, appropriately sized holes were drilled in the casing of the bomb using a remotely operated drill. After removing the mustard gas there remained in each munition a varying quantity of viscous tar consisting primarily of polymerized material with about 7 per cent entrained mustard gas. The drained bomb cases were therefore filled with a 20 per cent solution of sodium hydroxide in isopropanol and left for several hours. An explosive charge was then placed on the bomb casing and detonated remotely. The explosion ruptured the casing and ignited the isopropanol. When the destroyed bomb casings had burnt out each was carefully checked to ensure that no trace of mustard gas remained. The technique worked well and several thousand bombs were successfully processed in this manner.

It was not possible to remove the filler plugs of the Al Hussein missile warheads. Various techniques were investigated for opening these warheads, and the technique finally used was to blow a small hole in the nose of each warhead using a 'shaped' charge. The contents were then removed using the vacuum transfer system developed for the aerial bombs. Any residual agent was neutralized by filling the warhead with alcoholic caustic soda. Once certified free from contamination the warheads were destroyed by mechanical means.

The mustard gas-filled 155-mm artillery rounds posed a particularly difficult problem as they had no filler plugs and the thick-walled cases were not easy to drill. Owing to the presence of hydrogen gas and other gaseous degradation products in many of the rounds, there was not only considerable internal pressure but also a high risk of ignition during the drilling of the case. After a number of trials a successful technique was developed for opening these rounds. It involved standing the rounds on their 'boat-tails' in specially designed racks on large concrete pads and placing a small, specially designed, explosive charge inside the burster tube of each round. The positioning of the charge was such that when it was fired the explosive force cut the burster tube just below the point where it joined the neck of the shell. The rounds were wired up and opened in batches of 200. Once opened the outside of the round was first decontaminated and then the mustard gas removed by suction in the manner developed for the aerial bombs. When the mustard gas had been removed the empty shell cases were filled with alcoholic caustic soda and, after soaking for a period of time, burnt out.

VII. Hydrolysis of the bulk nerve agent

The two established techniques for the destruction of nerve agents are direct incineration and chemical neutralization. However, neither method had previously been used for the destruction of GF or GB/GF mixtures. An added complication was that the Iraqi agent mixture also contained up to 40 per cent dichloromethane, an organic solvent. After discussion it was decided to opt for chemical neutralization as Iraq had no previous experience of building or operating the type of incinerator required to deal with the direct incineration of nerve agents. The two proven alternatives available were: (*a*) neutralization with 2-aminoethanol followed by incineration of the resulting products, or (*b*) hydrolysis with an aqueous caustic solution followed by disposal of the dried reaction product in a secure landfill. One of the disadvantages of hydrolysis with an aqueous caustic solution is the large amount of waste generated, which can be up to six times the amount of agent destroyed. Although of relatively low toxicity this waste cannot be directly dispersed into the environment owing to its high fluoride content. Discussions with Iraq established that a source of 2-aminoethanol was not readily available in Iraq but that sodium hydroxide was. It was therefore decided to evaluate neutralization with sodium hydroxide as the means of destroying Iraq's nerve agent stocks.

As this method had never been used for this particular agent mixture it was first necessary to undertake a number of pilot plant studies to establish the optimum reaction conditions and to prove that the process could achieve the required results. These studies demonstrated that, although owing to the presence of dichloromethane the reaction mixture contained two phases, with efficient mixing the neutralization of the agent proceeded rapidly to completion. Analysis of the final product showed that the residual agent concentration was less than the target figure of 1 part per million (ppm). Based on these trials a decision was taken to proceed with the construction of a full-scale plant for the destruction of Iraq's nerve agent stocks.

The majority of the chemical plants at Al Muthanna were virtually destroyed by Coalition bombing. However, one of the nerve agent plants received only light damaged and it was decided to refurbish this facility and convert it to a batch hydrolysis plant for the destruction of the nerve agent. The basic design of the plant was agreed at a meeting between UNSCOM and Iraqi experts in late November 1991 and modification work began in early 1992. After assessing the results of the pilot plant studies the final plant design was agreed during a mission to Iraq in April 1992. The plant was commissioned under the supervision of UNSCOM experts in September 1992.

The plant was housed in a fully contained area which was maintained under negative pressure throughout the operation. It was controlled from an adjacent control room which was maintained under slight positive pressure. The plant area and control room were continuously monitored for the presence of agent vapour during the operation. While traces of agent vapour were occasionally detected in the room containing the plant no agent was ever detected in the control room. Operators entering the contained area for maintenance purposes and the like did so in full impermeable protective clothing and exited via a decontamination area. The core component of the plant was a jacketed 3 cubic metre (m^3) glass-lined reactor with a working capacity of approximately 2800 litres. Using a 15 per cent solution of aqueous sodium hydroxide 300–350 litres of agent could be processed per batch. The use of higher concentrations of sodium hydroxide and correspondingly increased amounts of agent were investigated but abandoned owing to problems in transferring the reaction products to the waste holding tank. These problems arose primarily from blockages owing to the low solubility of the sodium salts formed in the neutralization process. Analytical samples were taken at regular intervals throughout the reaction period. At the point where the agent concentration fell below the detection limit of 1 ppm and the concentration of excess sodium hydroxide also remained unchanged between two consecutive samples the reaction mixture was cooled and transferred to a holding tank. After a final analytical check the hydrolysis products were pumped to a large concrete-lined lagoon where the water was allowed to evaporate naturally into the desert air. The residual dried salts were mixed with concrete and were eventually be entombed in concrete. Using this method three to four batches of nerve agent could be processed during each working day thus giving the plant a destruction capacity of 1–1.5 tonnes per day. By February 1993 the plant had successfully processed some 70 tonnes of GB and GB/GF mixture.

The plant was also used to process the tabun recovered from the 75 tonnes of impure product also held at Al Muthanna. However, in this case, as the hydrolysis product contained large amounts of sodium cyanide, it was placed in steel tanks which were in turn sealed in one of the large concrete storage bunkers at Al Muthanna. These munition storage bunkers were deemed too massive to destroy and were therefore plugged with reinforced concrete to prevent their reuse; one of them was used as a very secure long-term landfill for the tabun hydrolysis wastes.

VIII. Incineration of mustard gas

Although mustard gas can be destroyed by either chemical oxidation or neutralization the varying quality of the starting material can lead to unpredictable results and neither process has been used extensively on a large scale. However, the disposal of mustard gas by direct incineration is a well-established process which has been successfully used by a number of countries including Canada, the UK and the USA. It was therefore decided to opt for incineration as the means of destroying Iraq's stocks of mustard gas. No suitable facility existed at Al Muthanna or elsewhere in Iraq, and therefore it was necessary to design and construct a facility specifically for this purpose. The final design was agreed at a series of meetings held between UNSCOM and Iraqi experts in late November 1991 and construction began at Al Muthanna in January 1992. Construction continued throughout 1992 with UNSCOM advice and assistance. The plant was commissioned in November 1992 by Iraqi personnel working under the direction of UNSCOM experts. Iraq's achievement in constructing this plant and having it ready for commissioning in just over nine months should not be undervalued.

The plant is composed of three principal parts: the tank farm and mixing area, the furnace area and the pollution abatement system. In the tank farm the mustard gas was transferred to one of the bulk storage tanks and mixed with fuel oil and a petroleum fraction containing benzene and toluene. The purpose of this was to prevent the mustard gas from solidifying in the tanks and feed lines when the ambient temperature fell below 14°C, and to provide a uniform feed stock to the burners. Studies had shown that the mustard gas would not readily dissolve in the fuel oil and a cosolvent was necessary to produce a homogeneous solution for feeding to the burner guns. The petroleum fraction used was a by-product from Iraq's oil industry and thus readily available. The optimum mixture was found to be 12 parts mustard gas, 2 parts petroleum fraction and 1 part fuel oil.

The large, standard brick-lined furnace was constructed in a trench cut in the desert floor and, although this made it very efficient in terms of heat loss, this subsequently proved to be one of the main limiting factors governing the throughput of the plant. The furnace was brought up to temperature by operating the main burner gun and secondary burner gun on fuel oil. Once the temperature reached 1000°C the main burner was switched to the mustard gas/petroleum fraction/fuel oil mixture. The target furnace conditions were a temperature range of 1000–1200°C, an excess oxygen level of 8–14 per cent and a carbon monoxide level of 20–50 ppm. It was found that these conditions could be obtained and maintained with a throughput of 450 kg/hr of the mustard gas/fuel mixture. This gave the plant a destruction capacity of 8.6 tonnes per 24-hour operating day. The combustion efficiency of the plant was monitored by means of an emission-monitoring system which continuously measured and recorded the levels of excess oxygen, carbon monoxide and carbon dioxide in the exhaust gases.

The hot exhaust gases from the furnace are diluted by the injection of ambient air and then passed through a conventional water quench tower where the

Table 17.1. The target emission standards for the incineration plant at Al Muthanna

Emission	Stack concentration limit (mg/m^3)
SO_2 (80% removal efficiency)	500
HCl (90% removal efficiency)	100
Particulates	180
Carbon monoxide	100
Mustard gas	0.03

Source: Report of the Destruction Advisory Panel of the United Nations Special Commission on Iraq for the period 24–28 June 1991 (June 1991).

majority of the hydrogen chloride formed in the combustion process was removed. The cooled gases were then passed through a packed tower irrigated with a 5–10 per cent sodium hydroxide solution. This removed the remaining hydrogen chloride and most of the sulphur dioxide. The cleaned gases then passed through the induction fan before being finally discharged to the atmosphere via a high smokestack. The waste liquids from both the quench tower and the packed scrubbing tower were piped to a large lagoon where the bulk of the water evaporates into the desert air. At the end of the operation the residual salts are recovered and disposed of in an appropriate manner.

The Destruction Advisory Panel set the target emission standards for the plant outlined in table 17.1. These were considered values which could realistically be achieved and which were appropriate for emissions from a plant at a remote desert site. All of these emission targets were readily achieved with the exception of that set for sulphur dioxide. At the maximum throughput of the plant this limit could not be met owing to the lower than anticipated efficiency of the packed scrubbing tower. It therefore proved necessary to reduce the mustard gas feed rate to the furnace to nearer 300 kg/hr, thereby reducing the load on the packed scrubber and improving the sulphur dioxide removal efficiency.

Like all incinerators the plant suffered a number of maintenance problems. However, it continued to operate and when incineration of the mustard gas was completed in 1994 approximately 400 tonnes of mustard gas had been successfully destroyed.

IX. Disposal of 122-mm nerve agent rockets

The disposal of the 122-mm nerve agent rockets posed a particularly difficult problem. Owing to internal pressure many of the filler plugs in the plastic agent containers had been ejected and agent was therefore leaking into the metal rocket cases, which were themselves not fully sealed. In addition, a significant number of the rockets were fitted with solid fuel motors. The risks from attempting to open these rockets and drain the contents were considered to be extreme and unacceptable without the use of a purpose-built, fully contained, remotely operated plant. Constructing such a plant would have required many months and the use of technology that was not available in Iraq. During this period the rockets would have continued to release agent to the environment at an ever increasing rate. It was therefore decided that an alternative solution had to be found.

The solution finally adopted was to place the rockets, 20 at a time, on top of tanks containing diesel fuel in the bottom of a deep pit. Each rocket was wired with three small explosive charges—one to disable the solid fuel motor and one to rupture each of the two agent containers. An igniter charge was also placed in the fuel container. When preparations were complete helicopter reconnaissance was carried out to ensure that the area up to six km downwind was clear. Immediately thereafter the explosive charges on the rockets were fired and a fraction of a second later the fuel ignited. The resulting fuel–air explosion generated a high-temperature fireball and the agent released from the rockets underwent instantaneous thermal degradation to simple non-toxic molecules. During trials of this method downwind sampling arrays[8] were erected at 100-m and 200-m intervals. Although agent vapour was detected in some of these samples the levels were low (i.e., of the order of 10^{-4} mg/m^3), which is the level to which unprotected workers can be exposed for up to 8 hours per day. Burnt out rocket cases and soil samples from the pits were also checked for residual contamination. No significant contamination was found. Using this method all of the recovered 122-mm nerve agent rockets were destroyed by the spring of 1993, thus eliminating a major hazard from the site at Al Muthanna.

X. Disposal of CW agent precursors

In addition to destruction of chemical warfare agents, UNSCOM was also responsible for the destruction of the chemicals used in their production including precursors, catalysts and solvents. Large quantities of these chemicals were present at both Al Muthanna and Al Fallujah. Among these chemicals were large quantities (2000–3000 tonnes) of highly reactive inorganic halides such as phosphorus trichloride, phosphorus oxychloride, thionyl chloride, methylphosphonic difluoride and the tabun precursor N,N dimethylphosphoramidic dichloride (D4). The disposal of these chemicals was further complicated by the fact that most of their containers were highly corroded and, in many cases, leaking. It was thus necessary to adopt a disposal method which minimized the need to handle or transport these containers, primarily 205-litre drums. The chosen solution was to construct a series of large lagoons close to each of the storage sites. Each lagoon was filled with a mixture of crushed limestone and water. Wherever practical the contents of each container were piped to the bottom of the lagoon where they reacted with the limestone–water mixture to form a non-toxic, relatively insoluble calcium salt. When it was deemed too dangerous to handle a container manually it was picked up by a crane and lowered into the lagoon and, if not already leaking, the container was ruptured with a mechanical prod. Once the operation had been completed the lagoons were allowed to evaporate to dryness under the desert sun. Eventually the dried salts will be permanently sealed in concrete. The stocks of combustible organic precursors such as thiodiglycol, isopropanol and the like were destroyed by incineration once destruction of the mustard gas stocks had been completed.

[8] A series of air samplers arranged in the form of a vertical two-dimensional grid.

XI. Destruction of production facilities

Destruction of production facilities was the major remaining task in UNSCOM's chemical weapon destruction programme. With the exception of the old sarin production facility, used for the destruction of the nerve agent stocks, and two pilot plants all former CW production facilities, the laboratories at Al Muthanna and the satellite precursor production sites were severely damaged during the Persian Gulf War and were no longer able to be operated. However, in order to fulfil its mandate UNSCOM is required to ensure that all process equipment which is considered critical to the production of chemical agents is also destroyed. Given the condition of the plants and laboratories this was a difficult and hazardous task. First, the critical components had to be identified and a decision taken as to whether they could be recovered and destroyed or whether they had to be destroyed *in situ*. In the latter case the preferred method of component destruction was explosive demolition using appropriately placed, shaped cutting charges. Where components could be recovered they were first decontaminated and then destroyed either by mechanical deformation, cutting into sections or, in the case of some tanks and reactors, by filling with concrete. Having completed its work, the hydrolysis plant was first decontaminated and then destroyed to prevent its possible reconversion to a CW production plant.

Iraq was permitted to retain some dual-purpose items of chemical plant which could not be clearly identified with its CW programme. However, each item was permanently marked and will be subject to regular ongoing monitoring by UNSCOM.

XII. Completion of CW destruction operations

The final phase of the CW destruction operation was completed in May 1994. At that time UNSCOM conducted a final sweep of the entire Al Muthanna site to identify and dispose of any item that might have escaped earlier notice. All areas of the site where destruction operations had been undertaken were cleaned up, checked to ensure that they were free from contamination and, to the extent possible, restored to their original condition. Two inspections were then conducted in May and June 1994: first, to undertake an environmental survey of the site, and second, to document the status of the site prior to its reversion to Iraqi control. The full list of chemicals and chemical munitions destroyed has been published elsewhere.[9] On 16 June 1994 the Chemical Destruction Group was disbanded, thus marking the successful conclusion of one of UNSCOM's largest, most hazardous and most important tasks under Resolution 687.

XIII. Future compliance monitoring

UNSCOM's CW-related operations will now be concentrated on verifying that Iraq continues to comply with its obligations not to use, retain, possess, develop

[9] Trevan, T., 'UNSCOM: activities in 1993', *SIPRI Yearbook 1994* (Oxford University Press: Oxford, 1994), chapter 19, pp. 750–51.

or otherwise acquire chemical weapons. The proposed means of carrying out this undertaking are set out in a plan[10] approved by the Security Council in Resolution 715.[11] Under this plan Iraq is obliged to provide, at regular intervals, full, correct and timely information on activities, sites, facilities and imports, both military and civilian, that might be used for purposes forbidden under Resolution 687. The plan approved under Resolution 715 also confers on UNSCOM the right to carry out inspections at any time and at any site within Iraq without hindrance. Iraq's acknowledgement of Resolution 715 was essential if UNSCOM was to fulfil its obligations under Resolution 687 to verify that Iraq does not once again become a CW possessor. On 26 November 1993 Iraq acknowledged its acceptance of Resolution 715, and UNSCOM has now begun this phase of its operations. Notwithstanding the many problems that remain UNSCOM's operations in Iraq have demonstrated that chemical weapons can be monitored, transported and destroyed in a safe, environmentally acceptable and timely manner.

[10] Contained in United Nations Security Council document S/22871/Rev.1, 2 Oct. 1991.
[11] United Nations Security Council document S/RES/715(1991), 11 Oct. 1991.

Part VI. Sea dumping of chemical weapons

18. Legal problems related to old chemical munitions dumped in the Baltic Sea

HANS-JOACHIM HEINTZE

I. Introduction

Old, dumped chemical munitions pose a significant pollution problem in the Baltic Sea area, and one which was long neglected owing to political reasons.[1] In recent years public concern about chemical munitions dumped after World War II has increased, especially in Scandinavia.[2] The incidents that have occurred merely suggest the potential for disaster. For instance, it has become known that 'surprisingly high levels of arsenic in the Baltic Sea led recently to the discovery that 7000 tons of arsenic had been dumped almost 40 years ago in concrete containers, reportedly enough to kill the population of the world three times over if properly administrated'.[3]

Marine pollution has been the focus of political and legal interest in recent years. International agreements and customary international law regulating the marine environment have developed on the basis of the common law related to relations between neighbours,[4] and some lawyers view these agreements as laws regulating marine pollution.[5] These laws are international. There are others that apply specifically to the Baltic Sea area, and they are among the most strict such laws. The legislation which governs the Baltic Sea area is an important example of European cooperation with regard to the problem of old, dumped chemical munitions. This cooperation should be seen in relation to the norms established by international law—the 1974 Helsinki Convention on the Protection of the Marine Environment of the Baltic Sea Area[6] and the 1972 Oslo Convention for the Prevention of Marine Pollution by Dumping from Ships and Aircraft[7]—and the continuing process of the development of a human right to a clean, balanced and protected environment under international law.[8]

[1] Boczek, B. A., 'The Baltic Sea: a study in maritime regionalism', *German Yearbook of International Law*, vol. 23 (Duncker & Humblot: Berlin, 1980), p. 196 ff.

[2] Laurin, F., 'Scandinavia's underwater time bomb', *Bulletin of the Atomic Scientists*, vol. 47, no. 2 (1991), p. 47. See also Stock, T., 'Chemical and biological weapons: developments and proliferation', *SIPRI Yearbook 1993: World Armaments and Disarmament* (Oxford University Press: Oxford, 1993), chapter 7, p. 283.

[3] Schacter-Serwer, A., 'Marine pollution: potential for catastrophe', *American Journal of International Law*, vol. 65 (1971), p. 84.

[4] Bothe, M., 'International law and the protection of the environment', eds M. Bothe, T. Kurzidem and C. Schmidt, *Amazonia and Siberia* (Graham & Trotman/Martinus Nijhoff: London, Dordrecht and Boston, 1993), p. 238.

[5] Timagenis, G. J., *International Control of Marine Pollution*, vol. 1 (Oceana Publications: Dobbs Ferry, N.Y./Sitjhoff & Noordhoff International: Alphen aan den Rijn, 1980), p. iii.

[6] 'Convention on the Protection of the Maritime Environment of the Baltic Sea Area of 22 Mar. 1974', *International Legal Materials*, vol. 544 (1974).

[7] *Bundesgesetzblatt 1977*, II, p. 165.

[8] Symonides, J., 'The human right to a clean, balanced and protected environment', *International Journal of Legal Information*, vol. 20 (1992), p. 24.

II. International law and old, dumped chemical munitions

Unlike the situation in other areas, the marine environment is governed more by treaty law than by customary international law.[9] However, states are obliged under general customary international law to cooperate on the high seas.[10] The 1982 Law of the Sea Convention,[11] which provides a general framework of state obligations for the protection and preservation of the maritime environment, reflects only the applicable customary international laws.[12] Legally the high seas are regarded as an open area for the common use of all states. States share common responsibility for those parts of the sea which lie outside their territorial waters. One of the consequences of this is that all states are obliged to respect the environment regardless of the legal system in a particular state. This rule applies to all aspects of the environment.[13] It would appear contrary 'to the very principle of freedom of the sea to permit actions which could interfere with the undisturbed and safe use of the sea by the community of nations'.[14] However, the obligations of states to protect the marine environment are more fiction than fact because the standards which should be applied by states are not specified.

None the less, the international community seems to have recognized its obligation to prevent significant harm to the high seas.[15] This was stressed at the 1992 United Nations Conference on Environment and Development (UNCED) in Rio de Janeiro, the so-called Earth Summit. Chapter 17 of Agenda 21, the action programme adopted by the Rio Conference, calls for the protection of coastal and marine areas, including exclusive economic zones (EEZ).[16] It is thus necessary, as far as is possible, to provide concerned individuals, groups and

[9] Salomon, J. A., 'Marine environment, protection and preservation', ed. R. Bernhardt, *Encyclopedia of Public International Law*, instalment 11 (Elsevier Science Publishers: Amsterdam, New York and Oxford, 1989), p. 200.

[10] Article 197 of the 1982 Law of the Sea Convention states: 'States shall co-operate on a global basis and, as appropriate, on a regional basis, directly or through competent international organizations, in formulating and elaborating international rules, standards and recommended practices and procedures consistent with this Convention, for the protection and preservation of the marine environment, taking into account characteristic regional features'. United Nations document A/CONF.62/122.

[11] United Nations document A/CONF.62/122.

[12] For the legal consequences of the Law of the Sea Convention, see Wolfrum, R., 'The legal order for the seas and oceans', eds. M. H. Nordquist and J. N. Moore, *Entry into Force of the Law of the Sea Convention* (Martinus Nijhoff: The Hague, Boston, London, 1995), p. 164: 'In establishing the relevant national laws and regulations the States are not totally free. Such legislation must not be less effective than global rules, regulations and procedures, hence such regulations describe an international minimum standard. Or at least internationally agreed rules and standards have to be taken into account.'

[13] See Antoine, P., 'International humanitarian law and the protection of the environment in time of armed conflict', *International Review of the Red Cross*, no. 291 (Nov./Dec. 1992), p. 519.

[14] Legault, J., 'The freedom of the high seas: a license to pollute?', *University of Toronto Law Journal*, vol. 21 (1971), p. 217 f.

[15] Smith, B. D., *State Responsibility and the Marine Environment: The Rules of Decision* (Clarendon Press: Oxford, 1988), p. 90.

[16] United Nations document A/CONF.151/4, part 2, 1 May 1992. The draft was adopted by consensus. In general, this aspect of the Earth Summit was rather disappointing: 'An assessment of the results achieved at the Conference on Environment and Development concerning the protection of the marine environment gives no indication as to the development of new environmental policies the implementation of which would seem to be more appropriate or more promising. Hence, the Rio Conference does not initiate an embarkation for new shores as was the case with the Third UN Conference on the Law of the Sea', quoted in Wolfrum, R., 'The protection of the marine environment after the Rio Conference: progress or stalemate?', eds U. Beyerlin, M. Bothe, R. Hofmann and E.-U. Petersmann, *Festschrift für Rudolf Bernhardt* [Essays in honour of Rudolf Bernhardt] (Springer: Berlin, 1995), p. 1017.

organizations with access to relevant information.[17] The basis of the Helsinki Convention obliges parties to cooperate.

Even if it is accepted[18] that the dumping of old chemical munitions in the Baltic Sea after World War II was not prohibited by international law at that time, there is none the less an obligation for states to avoid, minimize and repair transnational damage.[19] The United Nations International Law Commission (ILC)[20] adopted principles of law, applicable to transnational environmental damage, under the heading 'International liability for injurious consequences arising out of acts not prohibited by international law'.[21] Article 8 requires the states of origin to take appropriate measures to prevent or, when necessary, minimize the risk of transnational harm. In this respect international law is governed by the principle of causality or strict liability. International liability is the key concept as regards transnational environmental injury.[22] This was also recognized by the 1972 United Nations Stockholm Conference on the Human Environment.[23] Under the principle of international liability a state which inflicts damage is obliged to make reparation for appreciable harm caused and to provide information about the risk of the damage and relevant technical data to other states that could be affected. In part these provisions of the ILC represent international customary law; the extent to which this is so is debated by international lawyers.[24] The old chemical munitions that were dumped in the Baltic Sea are of particular relevance in this regard, since the most important ILC principle is the obligation of states to cooperate in order to minimize harm.

III. International law and dumping at sea

The sea has traditionally been used for the disposal of waste, a practice which has led to problems, as various incidents have illustrated. Public interest was aroused in 1970 when the US Army sank a ship loaded with nerve gas rockets in the Atlantic Ocean: 'It clearly demonstrated at the same time, however, the frightening inadequacy of both municipal and international law to provide even minimum safeguards for the protection of the marine environment and to cope effectively with the growing number of reported dumping incidents'.[25] The failure of national and international law to adequately address dumping at sea is

[17] Johnson, S. P. (ed.), *The Earth Summit: The United Nations Conference on Environment and Development (UNCED)* (Graham & Trotman/Martinus Nijhoff: London, Dordrecht and Boston, 1992), p. 309.

[18] Bothe, M., 'Völkerrechtliche Situation bei militärischen Altlasten' [Military waste and international law], eds K. J. Thomé-Kozmiensky *et al.*, *Management zur Sanierung von Rüstungsaltlasten* [Managing the clean-up of old armament wastes] (EF Verlag für Energie und Umwelttechnik: Berlin, 1992), p. 142.

[19] 'Fourth report on international liability for injurious consequences arising out of acts not prohibited by international law', United Nations document A/CN.4/373, para. 4.

[20] The ILC is a body of the United Nations that consists of experts from various legal systems. It is an organ of the General Assembly.

[21] United Nations document A/CN.4/373.

[22] Harndt, R., *Völkerrechtliche Haftung für die schädlichen Folgen nicht verbotenen Verhaltens, Schadensprävention und Wiedergutmachung* [International liability for injurious consequences arising out of acts not prohibited by international law] (Duncker & Humblot: Berlin, 1993), p. 324 ff.

[23] Sohn, L. B., 'The Stockholm Declaration on the Human Environment', *Harvard International Law Journal*, 1973, p. 423.

[24] Schachter, O., *International Law in Theory and Practice* (Martinus Nijhoff: Dordrecht, Boston and London, 1991), pp. 364–65.

[25] Timagenis (note 5), p. 110.

not surprising since the quantity of waste dumped at sea is less than 10 per cent of all ocean pollution.[26]

The provisions of traditional international law, in particular those of the 1958 Convention on the Territorial Sea and Contiguous Zone,[27] are inadequate with regard to the dumping of waste. The 1958 Convention on the High Seas[28] focuses on the freedom of the high seas and permits states to dispose of wastes. States are therefore not restricted from dumping chemical waste into the sea, which creates an immanent future risk. This has caused serious concern in countries which are largely dependent on fishing, such as the Scandinavian states. In 1971 the Government of Norway presented a draft convention for the prevention of the dumping of toxic substances into the sea from ships.[29] General public international law regards dumping at sea as a mixed form of pollution. The pollutants are transported from land by ship or aircraft for disposal at sea, and relevant legislation must thus consider both aspects (i.e., national legislation and international law).

The Oslo Convention for the Prevention of Marine Pollution by Dumping from Ships and Aircraft[30] is a regional instrument open to the states of the region specified in the North–East Atlantic Fisheries Convention.[31] The Oslo Convention strengthens the traditional law on waste disposal. Traditional law allowed dumping on the high seas, but the Oslo Convention amends it by adding prohibitions and requirements: 'The Contracting Parties pledge themselves to take all possible steps to prevent the pollution of the sea by substances that are liable to harm living resources and marine life, to damage amenities or to interfere with other legitimate uses of the sea'.[32]

This forms the basis of an obligation for contracting parties to harmonize their policies to prevent the pollution of the sea by dumping. Of importance for the Baltic Sea is Article 3 which states: 'The Contracting Parties agree to apply the measures which they adopt in such a way as to prevent the diversion of dumping of harmful substances into seas outside the area to which this Convention applies'.[33]

The Oslo Convention is limited with regard to the area covered and the type of pollution, but it covers a broad range of substances and materials that are listed in annexes to the convention and which are categorized in black, grey and white lists, based on their potential threat to the environment. A Commission, established under Article 16 of the Convention, supervises implementation, reviews the dumping permits that have been issued and evaluates dumping activities that have taken place.

[26] Leitzell, A., 'The ocean dumping convention: a hopeful beginning', *San Diego Law Review*, vol. 10 (1973), p. 502.

[27] *United Nations Treaty Series*, vol. 516, p. 205.

[28] *United Nations Treaty Series*, vol. 450, p. 11.

[29] Heintschel von Heinegg, W., 'Internationales Öffentliches Umweltrecht' [International public environmental law], ed. K. Ipsen, *Völkerrecht* [*International law*], 3rd edn (C. H. Beck: Munich, 1990), p. 817.

[30] *Bundesgesetzblatt 1977*, II, p. 1492.

[31] *Bundesgesetzblatt 1976*, II, p. 1.

[32] *Bundesgesetzblatt 1977*, II, p. 1492.

[33] *Bundesgesetzblatt 1977*, II, p. 1493.

IV. The Helsinki Convention

The 1974 Helsinki Convention, which establishes the Baltic Sea regime, is the most comprehensive regional environmental agreement. It is the only one 'that deals with all major sources of pollution'.[34] The conclusion of the Convention is a consequence of the exceptional ecological characteristics of the Baltic Sea that make it particularly vulnerable to pollution.

The accumulation of toxic substances in living and non-living components in an ecosystem is generally considered to be of special international concern. The Preamble of the Helsinki Convention notes therefore 'with deep concern the increasing pollution of the Baltic Sea Area, originating from many sources such as discharges through rivers, estuaries, outfalls, and pipelines, dumping and normal operations of vessels'.[35] The old, dumped chemical munitions are undoubtedly among the most dangerous pollutants in the Baltic Sea. This section analyses the obligations assumed by states under the Convention and the manner in which these obligations are implemented.

The Helsinki Convention covers all of the Baltic Sea, except for the internal waters of the parties,[36] and all pollutants. The approach to the sources of pollution is rather general, but it is complemented by numerous annexes that detail the measures to be taken for individual pollutants. It is significant that the annexes also contain lists of hazardous and noxious substances. According to Article 3 of the Convention, States Parties are obliged to take appropriate measures, both individually or jointly with others, to abate and prevent pollution and to protect and enhance the maritime environment.

Implementation of the Helsinki Convention

The success of the Helsinki Convention and its importance is largely the result of the Helsinki Commission (HELCOM).[37] The Commission is an implementing organ and has been active in research, monitoring and fostering regional cooperation. Sections (*b*) and (*c*) of Article 13 give HELCOM the tasks of reviewing the content of the Helsinki Convention and recommending amendments. HELCOM has the authority to adopt amendments to the annexes and substance lists. It then communicates such amendments to the States Parties to the Convention; HELCOM considers the amendments accepted unless a State Party submits a written objection.

The ministers for the environment of the Baltic states provided important support[38] for the work of HELCOM and the implementation of the Helsinki Convention by adopting the Declaration on the Protection of the Marine Environment of the Baltic Sea in 1988.[39] However, it is difficult to assess the legal

[34] Copithorne, M. D., 'Circumpolar environmental management and regulation', E. McWhinney *et al.*, *From Coexistence to Cooperation* (Martinus Nijhoff: Dordrecht, Boston and London, 1991), p. 183.

[35] See note 6. It has been in force for the Federal Republic of Germany since 3 May 1980. *Bundesgesetzblatt 1980*, II, p. 1449.

[36] Note 6, Article 4[3].

[37] Tulokas, M., 'The Baltic Sea and pollution', *Scandinavian Studies in Law*, vol. 25 (1981), p. 205 ff.

[38] Fitzmaurice, M., *International Legal Problems of the Environmental Protection of the Baltic States* (Graham & Trotman/Martinus Nijhoff: London, Dordrecht and Boston, 1992), p. 142.

[39] Fitzmaurice (note 38), annex III, p. 265.

effects of such a declaration, and international lawyers disagree. The Baltic Sea Declaration is generally viewed as a so-called 'soft law'.[40]

Although it has limited regulatory power, HELCOM has been surprisingly successful.[41] Its work is carried out by three committees: the Scientific–Technological Committee (STC), the Maritime Committee and the Combating Committee. The numerous HELCOM working groups are important in the context of the current discussion, in particular the Ad Hoc Working Group on Dumped Chemical Munition of the Baltic Marine Environment Protection Commission of the Helsinki Commission (HELCOM CHEMU).

HELCOM and old, dumped chemical munitions

The first HELCOM CHEMU meeting took place in St Petersburg in April 1993.[42] The aim of the meeting was to enable HELCOM to appropriately plan a strategy for addressing the issue of dumped chemical munitions. HELCOM CHEMU defines a chemical munition as follows: '[A] chemical munition has the same design and is used in the same weapons systems as [a] conventional munition. Instead of explosives they are filled with chemical warfare agents used for their direct toxic effect for people. A blasting or pyrotechnical charge is put in to spread out the agent. [A] chemical munition consists of warheads for missiles, artillery grenades and rockets, aircraft bombs, spray tanks, land mines and hand grenades'.[43] Chemical munitions 'can be stored either empty or filled with chemical warfare agents. A part of the chemical agents dumped in the Baltic were stored in drums and containers'.[44]

HELCOM CHEMU began by addressing chemical munitions, afterwards focusing on other dumped toxic munitions. As with other Helsinki Convention working groups the approach of HELCOM CHEMU is to collect reports from states in order to acquire national information on the relevant topic. The history of old chemical munitions that were dumped in the Baltic Sea is such that little is known about the location of the munitions or the state of the marine environment where the dumping occurred. Important data have also been concealed making international cooperation essential in order to deal with the problem. From the perspective of public international law it is encouraging that many States Parties have submitted reports and compilations of national information on dumped chemical munitions to HELCOM CHEMU.[45] Obviously, this is only the first step in obtaining reliable data on the state of the dumped munitions. The reports, however, vary in quality.

[40] Heusel, W., *'Weiches' Völkerrecht, Eine vergleichende Untersuchung typischer Erscheinungsformen* ['Soft' law: a comparative analysis of typical categories] (Nomos Verlagsgesellschaft: Baden-Baden, 1991), p. 290 ff.

[41] Brunnée, J. A., 'The jigsaw puzzle of international environmental protection: international approaches to atmospheric pollution and the Baltic Sea area', *International Journal of Legal Information*, vol. 20 (1992), p. 6.

[42] It was attended by Denmark, Finland, Germany, Lithuania, Poland, Russia and Sweden. Observers from Coalition Clean Baltic, Greenpeace International, Latvia, Norway, the UK and the USA took part.

[43] Baltic Marine Environment Protection Commission, Helsinki Commission, Ad Hoc Working Group on Dumped Chemical Munition (HELCOM CHEMU), Report of the 1st Meeting, St Petersburg, Russia, 19–21 Apr. 1993, HELCOM CHEMU 1/8, p. 2. The following discussion is based on this report.

[44] HELCOM CHEMU (note 43) para. 2.2.

[45] HELCOM CHEMU (note 43), item 3.

The successor states of the former USSR were unable to submit reliable information. The delegation from Russia, for example, provided no information on the location of munitions dumped after 1947, the current status of these munitions or their impact on the marine environment of the Baltic Sea.[46] Latvia presented publicly available information such as naval charts, press information and correspondence between Latvian and Russian authorities.[47] Information of a more concrete nature was provided by Poland which stated that there are 16 dumping sites in the Polish exclusive economic zone.[48]

The information given by the victorious Allied Powers in World War II was rather disappointing. The British delegation reported that following World War II the Allied Control Council in Berlin had discussions on the disposal of war *matériel.* Allied Control Council documents mention a time-limit for destruction of the chemical munitions, but there is no information on 'any requirement to identify the locations or methods used for disposal' and 'records of locations may never have existed'.[49] The United States stated that US archives contain information only on chemical munitions that were dumped outside the Baltic Sea area. At the first HELCOM CHEMU meeting Russia was unable to provide information on its activities in the Allied Control Council relating to the dumping of chemical munitions in the Baltic Sea.

Russia submitted a report to the second HELCOM CHEMU meeting which was held in Vilnius, Lithuania, in September 1993.[50] According to the information from Russian archives, approximately 12 000 tonnes of chemical agents were dumped in the area covered by the Helsinki Convention. This corresponds with the amount of dumped chemical *weapons* (35 000–45 000 tonnes), of which these agents are a component, mentioned in other Russian reports. The Russian report covered only dumping that was carried out prior to 1947. The Russian delegation stated that more information on dumping could become available if and when a decision were taken to disclose documentation that is currently secret. Limiting the Russian report to dumping prior to 1947 was clearly inadequate.

Other delegations expressed regret about the limited period of time covered by the Russian report.[51] Nevertheless, the Russian report did provide new information both about the amount and type of chemical warfare agents dumped in the Baltic Sea and the potential risk to the environment posed by the corrosion of the chemical weapons. The meeting requested that Russia submit additional information on dumping activities after 1947 as soon as possible.[52]

The HELCOM CHEMU meetings have exhibited a significant level of cooperation. At the second HELCOM CHEMU meeting, on the basis of national reports, Denmark presented a draft report on chemical munitions that have been dumped in the sea in the area covered by the Helsinki Convention. The report and subsequent discussion pointed to the need for further research.

[46] HELCOM CHEMU (note 43), para. 6.2.
[47] HELCOM CHEMU (note 43), para. 3.3.
[48] HELCOM CHEMU (note 43), annex 7.
[49] HELCOM CHEMU (note 43), p. 4.
[50] HELCOM CHEMU 2/2/1 and 2/2/1/Rev.1.
[51] HELCOM CHEMU 2/8, para. 2.3.
[52] Stock, T. and De Geer, A., 'Chemical weapon developments', *SIPRI Yearbook 1994* (Oxford University Press: Oxford, 1994), chapter 9, p. 340.

The third HELCOM CHEMU meeting was held in January 1994 in Copenhagen, Denmark. At the meeting the Russian delegation stated that no official information was available about the dumping of chemical weapons in the Baltic Sea after 1947. Experts have expressed scepticism about this since there are allegations of dumping by the USSR after 1947.[53]

The HELCOM CHEMU meetings illustrate the importance of non-governmental organizations (NGOs) participating in such meetings. This parallels developments in the implementation of human rights obligations under international conventions. At the first April 1993 HELCOM CHEMU meeting Greenpeace provided information on a 1990 pilot study on chemical munitions which was conducted in the vicinity of the Danish island Bornholm and noted that Russian authorities still considered information on chemical munitions secret.[54] It is important from the viewpoint of international law that Greenpeace has formal observer status in HELCOM and the associated right to participate in its work. Cooperation with Greenpeace facilitates an 'unusual' readiness for joint activities.[55]

In March 1994 a third meeting of the Helsinki Commission at the ministerial level was held to discuss HELCOM CHEMU proposals. Additional studies were commissioned to gather information from States Parties concerning undocumented dumping and to recommend actions to be taken by the States Parties to the Helsinki Convention.

V. Conclusions

The HELCOM CHEMU meetings have emphasized cooperation, and this has allowed identification of key questions related to the dumping of munitions in the Baltic Sea. These include: assessing the state of the corrosion of dumped chemical munitions, developing guidelines for fishermen and the authorities for dealing with such munitions, and establishing plans for handling incidents involving munitions. The activities of HELCOM CHEMU also reflect the growing cooperation among states around the Baltic Sea. Further study is needed, however, to make recommendations for future action. In this way it seems possible that binding obligations on the States Parties to the Helsinki Convention can be created.

[53] Stock, T. and De Geer, A., 'Chemical and biological weapons: developments and destruction', *SIPRI Yearbook 1995: Armaments, Disarmament and International Security* (Oxford University Press: Oxford, 1995), chapter 10, p. 353.

[54] HELCOM CHEMU (note 43), para. 3.19.

[55] Fitzmaurice (note 38), p. 176.

19. The Baltic and North Sea dumping of chemical weapons: still a threat?

FREDRIK LAURIN

I. Introduction

In May 1945 when Germany surrendered and after almost six years the war in Europe ended, most of the nations involved in the war had stockpiles of hundreds of thousands of tonnes of a weapon that was never used—chemical munitions. There had been little enthusiasm for chemical weapons among the nations which participated in the war, and the primary reason for not deploying CW was probably the special nature of these weapons. In addition, wind and other weather conditions are important factors in chemical warfare, as is the ability of an opponent to strike back with the same means.

It seems clear that at the end of 1939 all of the parties to the conflict suspected that the other nations involved were prepared to use chemical weapons. However, this was not the case, and it was not until 1941 or 1942 that adequate stockpiles of CW existed for anything more than last resort actions.[1]

Whatever the reason for not deploying the CW grenades and bombs, their existence left postwar Europe with an enormous problem. Chemical weapons had been produced by both the Allies and the Axis powers in an escalating process in preparation for an all-out chemical war that never began. If used in combat the poison in these weapons would have been sufficient to kill approximately 1–10 per cent of the population.

A total of approximately 302 857 tonnes of chemical warfare munitions were found in Germany by the Allies.[2] Small amounts were burnt, buried, sold as scrap or shipped to the UK and the USA, but the rest were dumped at sea. The dumping operations required many trips to the dumping sites and several years to complete. Convoys with as many as 60 ships left various ports such as Kiel, Emden and Flensburg, Germany.[3] The majority of the munitions captured by the Allies were dumped in the Norwegian Trench, a 700-metre deep trench 25 nautical miles south of Arendal, Norway. Munitions were also dumped in the Skagerrak, 25 nautical miles west of the lighthouse on Måseskär on the Swedish west coast (see table 19.1 and figure 19.1). Several ships were scuttled in the North Sea, and one dumping site was located close to the island of Helgoland. One ship was sunk in the Mediterranean, and at least three vessels were sunk in the Bay of Biscay. Additionally, the UK formed its own convoys of ships loaded with some of its stockpile (approximately 120 000 tonnes), which it scuttled in the Atlantic Ocean.[4]

[1] Baltic Marine Environment Protection Commission, Helsinki Commission, UK, Report on sea dumping of chemical weapons by the United Kingdom in the Skagerrak waters post World War II, HELCOM CHEMU 2/2/5, 28 Sep. 1993.

[2] *Armed Forces Chemical Journal*, vol. 3, no. 4 (1949).

[3] Åkerlund, B., 'Gasfartygen i Skagerack' [Gas ships in the Skagerrak], Kungsbacka, Sweden, 1990. See also chapter 20 in this volume.

[4] Knightley, P., 'Dumps of death', *Sunday Times Magazine* (London), 5 Apr. 1992; and Baltic Marine Environment Protection Commission (note 1).

Table 19.1. Known dumping sites in and near Scandinavia[a]

Location	Position	Amount	Comments
Confirmed dumping			
Norwegian Trench	58.50N 9.40E	150 000 tonnes (t)	As many as 38 ships scuttled
Måseskär, Sweden	58.10N 10.45E	20 000 -50 000 t	At least 9 ships scuttled
Bornholm/Christiansø, Denmark			
Main dumping area	Circle with radius of 3 nautical miles, centred on 55.21.0N 15.37.02E	30 000–50 000 t	Most CW dumped over the side from barges
Danish ship *Hano*	55.15N 15.45E	Unknown	Scuttled at position given
Minesweepers	North-east of Bornholm	Unknown	4–5 ships scuttled
Lille Bælt (Little Belt), Denmark	54.48.22N 10.12.22E	69 000 shells filled with tabun, and additional 5 000 t of CW munitions, tabun and phosgene	Sunk in 2 ships, which were later raised; munitions then sunk in Bay of Biscay
North-eastern coast of Germany	Various positions	Unknown	Dumping from vessels *en route* to the Bornholm site
South-east of Swedish island of Gotland and west of Liepaja, Latvia	55.50N 18.30E[c] and 56.40N 20.00E[c]	2 000–5 000 t[b]	Total for both sites
Unconfirmed dumping			
Gulf of Finland			
Coast of Estonia			

[a] The dumping sites at Bornholm, Måseskär and the Norwegian Trench are the largest. Approximately 200 000 tonnes of CW munitions, mainly mustard gas, were dumped there.

[b] The 5000 tonnes amount may be a mistranslation since the position given is closer to Wolgast, Germany.

[c] Positions are approximate.

Sources: *Chemische Kampfstoffmunition in der südlichen und westlichen Ostsee: Bestandsaufnahme, Bewertung und Empfehlungen*, Bericht der Bund/Länder-Arbeitsgruppe Chemische Kampfstoffe in der Ostsee [Chemical munitions in the southern and western Baltic Sea: compilation, assessment and recommendations, Report by the Federal/Länder Government Working Group 'Chemical Munitions in the Baltic Sea'] (Federal Maritime and Hydrographic Agency: Hamburg, May 1993); Sjöfartsverket, *Rapport om kartläggning av förkomsten av dumpade kemiska stridsmedel på den svenska delen av kontinentalsockeln* (Sjöfartsverket: Norrköping, Sweden, 1992) and National Maritime Administration, *Summary of a Report on Investigation of the Existence of Dumped Chemical Weapons on the Swedish Part of the Continental Shelf* (National Maritime Administration: Norrköping, Sweden, 1992); Tørnes, J. A. *et al.*, *Investigation of Ships Carrying Chemical Ammunition Sunk in Norwegian Waters after World War II*, FFI/Report-89/6007 (Forsvarets Forskningsinstitutt (FFI) [Norwegian Defence Research Establishment]: Kjeller, Norway, 1989); *Redegjørelse vedrørende de miljø-. sundheds- og sikkerhetsmessige forhold i forbindelse med dumpet giftgasammunition i farvandene omkring Danmark* [An account of the environmental, health and security situation in connection with the dumped poison gas ammunition in the waters surrounding Denmark] (Miljøstyrelsen: Copenhagen, 7 May 1985); Åkerlund, B., 'Gasfartygen i Skagerack' [Gas ships in the Skagerrak],

Kungsbacka, Sweden, 1990; US Arms Control and Disarmament Agency (ACDA), *Special Study on the Sea Disposal of Chemical Munitions* (ACDA: Washington, DC, 19 Jan. 1993); and Baltic Marine Environment Protection Commission, Helsinki Commission, UK, Report on sea dumping of chemical weapons by the United Kingdom in the Skagerrak waters post World War II, HELCOM CHEMU 2/2/5, 28 Sep. 1993.

While these operations were conducted by the Western powers, the USSR mounted its own operation in what was to become the German Democratic Republic. Trainloads of CW were dispatched from storage sites and manufacturing plants mainly to the Baltic ports of Peenemünde and Wolgast. There the chemical weapons were loaded on barges or, in some cases, hulks. The USSR chose an area north-east of Bornholm, Denmark, as its main dumping ground, and the Gotland Basin, south-east of the Swedish island of Gotland, was also used as a Soviet dumping site. The barges were towed to sea and the munitions were dumped over the side, sometimes *en route* to the dumping site.

After being generally forgotten for almost 40 years these dumping sites have received attention and aroused concern in the past 10 years. The international community focused its attention on the issue of sea-dumped CW munitions first in the 1950s, which resulted in the 1958 Convention on the High Seas.[5] In 1972 the USA prohibited the dumping of chemical weapons,[6] but the issue was never explicitly addressed until the substantive negotiations on a Chemical Weapons Convention began in the 1980s. When the CWC was signed in January 1993[7] it contained a provision that specifically prohibits dumping CW munitions at sea as a way to get rid of stockpiles of chemical weapons. It had been difficult to negotiate this provision, however, and a compromise was achieved which excluded CW munitions that had been dumped at sea prior to 1985.

Public pressure has forced some governments to conduct investigations, and surveys (some more thorough than others) have been done of the dumping sites. Although samples of the sediment at these sites have indicated the presence of chemical agents in the seabed, the conclusion has always been that the sites are best left undisturbed. None the less, fishermen have been severely hurt when bombs, grenades or lumps of mustard gas have been brought on deck in their trawls. All surveys have indicated that the chemical agents are still present at the dumping sites.[8]

[5] *United Nations Treaty Series*, vol. 450, p. 11. See also chapter 18 in this volume.

[6] US Arms Control and Disarmament Agency (ACDA), *Special Study on the Sea Disposal of Chemical Munitions* (ACDA: Washington, DC, 19 Jan. 1993)

[7] The text of the 'Convention on the Prohibition of the Development, Production, Stockpiling and Use of Chemical Weapons and on their Destruction' is reproduced in *SIPRI Yearbook 1993: World Armaments and Disarmament* (Oxford University Press: Oxford, 1993), appendix 14A, pp. 735–56.

[8] Sjöfartsverket, *Rapport om kartläggning av förkomsten av dumpade kemiska stridsmedel på den svenska delen av kontinentalsockeln* (Sjöfartsverket: Norrköping, Sweden, 1992) and National Maritime Administration, *Summary of a Report on Investigation of the Existence of Dumped Chemical Weapons on the Swedish Part of the Continental Shelf* (National Maritime Administration: Norrköping, Sweden, 1992); *Chemische Kampfstoffmunition in der südlichen und westlichen Ostsee: Bestandsaufnahme, Bewertung und Empfehlungen*, Bericht der Bund/Länder-Arbeitsgruppe Chemische Kampfstoffe in der Ostsee [Chemical munitions in the southern and western Baltic Sea: compilation, assessment and recommendations, Report by the Federal/Länder Government Working Group 'Chemical Munitions in the Baltic Sea'] (Federal Maritime and Hydrographic Agency: Hamburg, May 1993); *Redegjørelse vedrørende de miljø-. sundheds- og sikkerhetsmessige forhold i forbindelse med dumpet giftgasammunition i farvandene omkring Danmark* [An account of the environmental, health and security situation in connection with the dumped poison gas ammunition in the waters surrounding Denmark] (Miljøstyrelsen: Copenhagen, 7 May 1985); and Tørnes, J. A. *et al.*, *Investigation of Ships Carrying Chemical Ammunition Sunk in Norwegian*

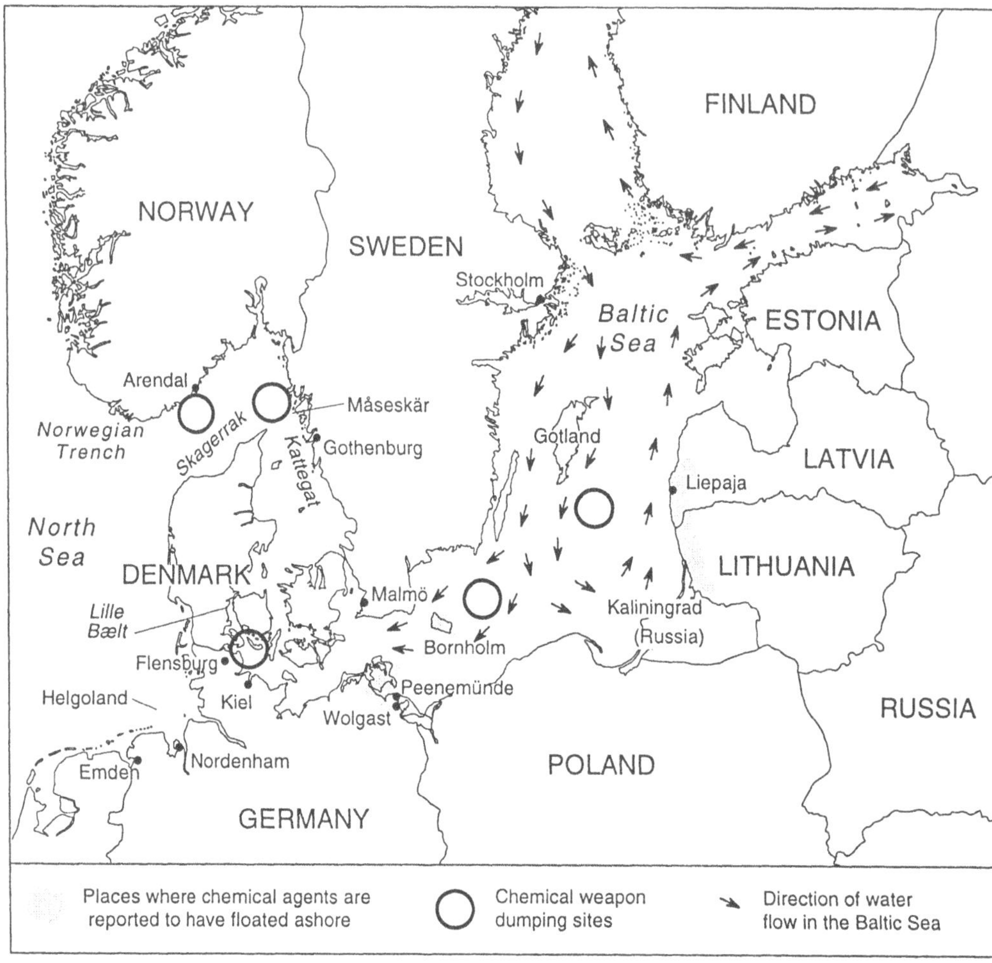

Figure 19.1. Sites of chemical munition dumping after World War II

Source: Based on maps by Stieg Larsson, Tidningarnas Telegrambyrå (TT), the Swedish news agency.

The majority of the CW munitions which were dumped were in artillery shells with steel casings that were several centimetres thick. Some agents, such as the nerve gas tabun, will have dissolved in the sea water if they have leaked out of their original casings, but most of the agents are probably still inside the shells or their original casings. A thorough study has not yet been made of the effect that these thousands of tonnes of chemical agents will have on the environment when they eventually enter the water and the sediment of the Skagerrak, the Kattegat and the Baltic Sea.

II. Historical overview

The dumping of chemical munitions began at the end of World War II. Aware that they were fighting a losing battle, the German command decided to try to move (and in some cases destroy) part of the German CW stockpile, in particular tabun, which Germany believed was still a secret weapon. Much of the stock

Waters after World War II, FFI/Report-89/6007 (Forsvarets Forskningsinstitutt (FFI) [Norwegian Defence Research Establishment]: Kjeller, Norway, 1989).

was moved from areas about to be captured by the Allies as they advanced. Several secret transports took place, and some of the tabun (69 000 shells with an additional 5000 tonnes of chemical munitions, phosgene and tabun) was dumped in the southern entrance to Lille Bælt (Little Belt), Denmark, and off the north-eastern coast of Germany in the North Sea in an operation during the last days of the war.[9]

There were also efforts to remove markings from other types of CW grenades in order to make them more inconspicuous. Many of these operations were never documented because of the chaotic situation in Germany in the final year of the war.[10]

The Potsdam Agreement

Dumped chemical weapons are a problem today for people living on the shores of the Baltic Sea, the Kattegat, the Skagerrak and the North Sea. Chemical weapons were dumped in these areas in operations undertaken by the forces which were victorious in World War II.

The Potsdam Agreement of 2 August 1945[11] outlined in general terms the way in which Germany was to be demilitarized and the manner in which its weapons and munitions were to be destroyed or to become the property of the Allied forces.[12] The Allied Control Commission and its various sub-committees negotiated the terms for these actions, but the process was so slow that on 17 August 1945, the Combined Chiefs of Staff (British and US), issued a general order to their zone commanders to initiate their own measures to destroy German war material.[13]

In November 1945 the Allied Control Commission's Standing Committee on War Material submitted its 13th report, which contained specific recommendations that CW munitions should be dumped at sea as 'dumping was the only practicable method of disposal'.[14] On 3 January 1946 an order was issued by Headquarters Allied Sea Command to the commanding generals of the various theatres stating that they now had the authority to dispose of CW munitions by dumping them at sea or by demolition.[15]

There were numerous bodies responsible for the operations that took place, including the Allied Dumping Commission, the Disposal Group, the OMGUS (Office Military Government United States) and the Continental Dumping Commission. The Continental Dumping Commission was the British organization which was responsible for organizing British sea dumping operations.[16] It is difficult to determine whether any one body in the Allied forces had overall responsibility for the various dumping operations.

[9] *Redegjørelse vedrørende de miljø-. sundheds- og sikkerhetsmessige forhold i forbindelse med dumpet giftgasammunition i farvandene omkring Danmark* (note 8); and *Chemische Kampfstoffmunition in der südlichen und westlichen Ostsee: Bestandsaufnahme, Bewertung und Empfehlungen* (note 8).

[10] Sjöfartsverket (note 8).

[11] von Oppen, B. R. (ed.), *Documents on Germany under Occupation 1945–1954* (Oxford University Press: London, 1955), pp. 40–50; and *The Tehran, Yalta and Potsdam Conferences: Documents* (Progress Publishers: Moscow, 1969), pp. 320–33.

[12] Baltic Marine Environment Protection Commission (note 1).

[13] Baltic Marine Environment Protection Commission (note 1).

[14] Baltic Marine Environment Protection Commission (note 1).

[15] US Arms Control and Disarmament Agency (note 6).

[16] Baltic Marine Environment Protection Commission (note 1).

In the 26 April 1946 declassified document 'Allied Control Council Directive no. 28, Reports on Disposal of German War Material in Germany', it was decided that the Allied Dumping Commission should report every six months on the progress of the demilitarization and disposal of the chemical weapons.[17] Some sources state that these reports were submitted until the end of 1947, but other sources claim that the information never reached the Allied Control Commission.[18]

If any body had an overview of the situation it was probably the Berlin-based Allied Control Council, but it appears that there was no organization which knew what was being done when and where with the captured German stockpile of CW munitions.[19] Apparently, each occupation authority was responsible for disposing of the stocks found in its zone, and, consequently, each formed its own organization. This is reflected in the abundance of names of different organizations and abbreviations for them which exist in the material on this topic.

One other organization is of relevance: the Tripartite (UK, USA and USSR) Naval Commission, which was established in the secret protocols of the Potsdam Conference. Its task was to organize the disposal of German naval vessels, many of which were used in the dumping operations.[20]

Chemical munitions captured by the Allies

As mentioned above, approximately 300 000 tonnes of CW munitions were found in occupied Germany (94 000 tonnes in the US zone, 123 000 tonnes in the British zone, 9000 tonnes in the French zone and 63 000 tonnes in the Soviet zone[21]), and most of these munitions were disposed of by 1948. (By the end of 1947, 280 000 tonnes of CW munitions had been found, of which 269 390 tonnes had been destroyed.[22]) In 1945–48 most of these large stockpiles of chemical weapons were dumped at sea. As far as can be concluded, the methods employed did not contradict the treaties or international law in force at the time.[23]

According to a May 1993 report by the German Maritime and Hydrographic Agency the British and US occupation authorities sank one ship in the European polar zone and 42 ships in the Skagerrak (a total of 130 000 tonnes of CW munitions).[24] France dumped 1500 tonnes at unspecified sites.[25] The USSR dumped chemical weapons in the Baltic Sea from barges which had departed from Wolgast and Peenemünde in northern Germany.[26]

[17] Knightley (note 4).

[18] *Redegjørelse vedrørende de miljø-. sundheds- og sikkerhetsmessige forhold i forbindelse med dumpet giftgasammunition i farvandene omkring Danmark* (note 8).

[19] *Redegjørelse vedrørende de miljø-. sundheds- og sikkerhetsmessige forhold i forbindelse med dumpet giftgasammunition i farvandene omkring Danmark* (note 8).

[20] Baltic Marine Environment Protection Commission (note 1).

[21] *Chemische Kampfstoffmunition in der südlichen und westlichen Ostsee: Bestandsaufnahme, Bewertung und Empfehlungen* (note 8). These figures are rounded.

[22] Sjöfartsverket (note 8).

[23] Sjöfartsverket (note 8).

[24] *Chemische Kampfstoffmunition in der südlichen und westlichen Ostsee: Bestandsaufnahme, Bewertung und Empfehlungen* (note 8).

[25] Sjöfartsverket (note 8).

[26] *Redegjørelse vedrørende de miljø-. sundheds- og sikkerhetsmessige forhold i forbindelse med dumpet giftgasammunition i farvandene omkring Danmark* (note 8); and *Chemische Kampfstoffmunition in der südlichen und westlichen Ostsee: Bestandsaufnahme, Bewertung und Empfehlungen* (note 8).

The accuracy of data from the immediate postwar period was affected by the circumstances under which the disposal operations were conducted, and it is necessary to allow for mistakes, lack of data and inaccurate reports. However, if approximately 300 000 tonnes of CW munitions were disposed of after World War II, the amount of chemical agents would be 10–25 per cent of the total weight of the munitions. It can therefore be concluded that 30 000–75 000 tonnes of chemical agents, mostly mustard gas in various forms, were dumped at sea.[27]

Operations by the Western Allies

The preferred method of disposal for the Western Allies was to load the weapons on damaged or unfinished ships and then to sink them in deep water. Bombed and burned hulks were collected from all over northern Europe and patched up for their last journey, at the end of tugboat cables, to two main dumping grounds—the Norwegian Trench and a site in the Skagerrak 25 nautical miles off Måseskär, Sweden. As mentioned above, the UK conducted extensive operations in the Atlantic Ocean to dispose of its stockpile, as did the USA off its Atlantic coast.[28]

Baltic coastal states informed

It has often been assumed that the countries closest to the dumping areas did not know about these activities, but they were in fact aware of them and did not protest. In June 1948 the US Naval Authority addressed the following request to the US embassy in Stockholm: 'Request you obtain clearance respective governments for scuttling approximately eight thousand tons ammunition in area enclosed by lines through following points: A, B, C, D.[29] Ammunitions will be loaded in hulks which will be sunk by gunfire. Munitions and containers will not float. Program will commence late July or early August'.[30]

Despite protests in the Swedish Parliament by the Swedish Fishermen's Union the Swedish authorities did not oppose the US request.[31] Seven cases of poisoning caused by mustard gas bombs, which had washed ashore from the munitions dumped by the USSR in the Baltic Sea, had already occurred on the Swedish south coast in the vicinity of Kalmar. This did not appear to concern the Swedish authorities. The Norwegian authorities, however, requested information from Sweden about these incidents in 1948 so that they could give the USA 'some advice' as to where Norway thought the scuttling of ships should take place. The site preferred by Norway was in the Arctic Ocean.[32] The Norwegian proposal was not accepted, but Norway did not object to the operations.

[27] Interview with CW expert Johan Santesson, at the Swedish National Defence Research Establishment, FOA 4, Umeå, Sweden, Sep. 1993.

[28] Knightley (note 4); and *Armed Forces Chemical Journal* (note 2).

[29] The coordinates given indicate an area on land. However, if the location '9 degrees east of Greenwich', which is mentioned in the request, should have been '10 degrees east of Greenwich', the area indicated would be a thin stretch of water in the mouth of the Oslo fjord.

[30] Swedish Ministry for Foreign Affairs memo concerning questioned dumping of poison gas in the Skagerrak, July 1948.

[31] Swedish Ministry for Foreign Affairs, Telegram to the Swedish Ministry for Foreign Affairs from the Swedish Fishermen's Union, 21 June 1946.

[32] Swedish Ministry for Foreign Affairs memo concerning information request from Norwegian Foreign Office, 10 July 1948.

The Norwegian Chief of Staff is cited in a Swedish Ministry of Foreign Affairs report in 1948 as saying : 'With reference to the issue of dumping of ammunition we have suggested to the Defence Department that we answer—Yes'.[33] A later telegram from the US Department of State confirmed the start of dumping operations: 'Loading begins nine July and departure date for scuttling convoys set at twenty July'.[34]

Scuttling ships loaded with CW munitions

The independent Swedish researcher Björn Åkerlund, together with other amateur European naval researchers, concluded from extensive investigations in various archives that 65 merchant ships and 15 naval ships loaded with chemical munitions were sunk intentionally by the UK and the USA after World War II.[35] Åkerlund reported that ships were scuttled as late as 1956, but most ships were sunk in the period from the end of the war to September 1948.[36]

Åkerlund found the names of 46 ships which were loaded with chemical munitions and sunk in the Skagerrak in 15 scuttling operations/convoys between May 1945 and September 1948; the total loading capacity of the ships was approximately 190 000 tonnes. The average age of the ships was 25 years for the merchant ships and 15 years for the naval vessels. Some of the ships were uncompleted hulks taken directly from the builder's yard. According to Åkerlund approximately 20 ships were sunk in the Norwegian Trench and at least 9 ships were scuttled west of Måseskär. If these ships were loaded to 90 per cent of capacity, as witnesses claim, the amount dumped near Måseskär could have been as much as 55 000 tonnes. British authorities claim, however, that the ships in question were part of scuttling operations by the Tripartite Naval Commission, and that they did not contain CW munitions.[37]

In an interview by the author of this chapter with Einar Høvding, a retired manager of a Norwegian salvage company which bought and dismantled a number of German ships after the war, Høvding stated that the scuttling operations were directed by a US general named Gotthard, working out of the Allied Dumping Commission offices in Hamburg.[38] The men doing the dangerous and sometimes deadly work were Germans under the command of British and US officers.

A Norwegian newspaper account from July 1946 quotes an English commander who led one of the first convoys to the Norwegian Trench: 'We are not used to daylight at night, but out there we really turned night into day. We opened all the bottom vents and to get things going we used our cannons'.[39]

Not all of the ships reached their destination. Høvding noted that 'Certainly they had orders to go to this or that position, but if you want the truth they sank them all over the place: If the weather was bad or if they were low on fuel, they

[33] Swedish Ministry for Foreign Affairs memo concerning questioned dumping of poison gas in the Skagerrak (note 30).

[34] Swedish Ministry for Foreign Affairs memo concerning information request from Norwegian Foreign Office (note 32).

[35] Åkerlund (note 3).

[36] Åkerlund (note 3).

[37] Baltic Marine Environment Protection Commission (note 1).

[38] Interview by the author with former salvage operator Einar Høvding, Oslo, Sep. 1989.

[39] *Fædreneslandet*, July 1947 (in Norwegian).

just sank them on the way'.[40] Høvding added that because of a severe shortage of ships, those available were loaded to close to maximum capacity.

Karl Gänsch, a German, signed on as a deck hand on the *Monte Pascoal*, a war-damaged German passenger ship, on 26 November 1946, at Emden. He described his experiences in an interview with *Ostfriesen Zeitung*:

The ship was loaded from barges with some 3000 tons of chemical warfare ammo. Grenades, drums, air-bombs went to the holds all marked with the rings for CW-ammo or just plain Gift [poison] signs on them. When we got to Kiel, charges were mounted on the inside of the hull by American engineers. Concrete was put into the holds and the cargo doors welded together. Then we were towed by tugs from Kiel to Skagen at the northern tip of Denmark, and on the twenty-first of December we were moved over to the tugs and a navy ship. Half an hour later the charges went off and she went down slowly. It was about 20 nautical miles off Skagen.[41]

Harald Francke, a retired shipyard worker now living in the Swedish village Smögen, 150 km north of Gothenburg, was 16 years old and living with relatives in Hamburg at the end of the war.

In September 1945 my uncle got a job for me and a friend working on the docks in Kiel; there we became part of a large crew which was loading a whole dock full of ships with grenades from railroad cars. We were being guarded by Polish soldiers under British command. After a few weeks on the docks I got a job as a stoker on the *Jantje Fritzen*, one of the ships that we had been loading. The pay was food for the day and a few cigarettes. Together with about a dozen other ships she was anchored off Kiel waiting for better weather when one day yellow smoke started pouring from one of the holds. We were all brought to the hospital. Some of the crew died from their injuries, but I got out after a couple of weeks. My eyes were burned, and I couldn't swallow for some time, but it all went away with time. The convoy left Kiel without *Jantje Fritzen*, which was scuttled on a later occasion in Skagerrak.[42]

US disposal activities

The chemical weapon munitions captured in the US zone of occupation in Germany were transferred to five former German ammunition depots. From these depots operations were undertaken to dispose of the munitions by the following methods: burning, detonating or dumping on land (32 500 tonnes), scuttling at sea (31 000–39 000 tonnes) and shipping to Italy and the UK (27 000 tonnes). In addition, some munitions were sold as scrap, and much of the CW stock was shipped to the USA (12 000 tonnes). The total amount of German CW munitions handled by US forces was therefore 110 000 tonnes, according to the available US information.[43]

Operation Davey Jones Locker

The US scuttling operation was code-named 'Operation Davey Jones Locker' and was conducted in Scandinavia from June 1946 through August 1948.

[40] Interview by the author with former salvage operator Einar Høvding (note 38).
[41] The report of an interview with Gänsch in Emden, Germany, is in *Ostfriesen Zeitung*, 8 Feb. 1990 (in German).
[42] Interview by the author with Harald Francke, Sep. 1989.
[43] US Arms Control and Disarmament Agency (note 6).

Eleven ships were scuttled in this operation: 9 in the Norwegian Trench and 2 in the North Sea. The German cruiser *Leipzig* was one of the latter two ships. All of the ships which were scuttled in the US operation sailed from the Midgard Docks in Nordenham, Germany.[44]

British disposal activities

There were approximately 100 000–130 000 tonnes of German CW munitions in the British zone of occupation in Germany. According to a British Ministry of Defence spokesman, 'These were dumped in the Skagerrak'. The spokesman also confirmed that an additional 120 000 tonnes of British chemical weapons were dumped in the Atlantic Ocean.[45]

British and US archives have remained largely closed to the public until recently. Secrecy in the UK has made investigation difficult, while some US material has been destroyed because of lack of storage space. Both the UK and the USA submitted reports to the September 1993 meeting of the Ad Hoc Working Group on Dumped Chemical Munition of the Baltic Marine Environment Protection Commission of the Helsinki Commission (HELCOM CHEMU) in Vilnius, Lithuania. The British report stated that at least 127 000 tonnes of German CW munitions were disposed of by dumping at sea.[46] The report listed 32 ships as 'CW vessels' and gave their names and additional data about them.

Monthly Intelligence Reports

There is little information about British dumping activities which is publicly available. However, two reports in a spring 1947 *Monthly Intelligence Review* from the British admiralty can be cited:

> After the Allied occupation of Germany immense quantities of 'Yellow Cross' poison-gas shells and other highly poisonous materials for chemical warfare were found. Of these about 200 000 tons were in the British zone of occupation, either in factories in the Ruhr or in dumps scattered about on the Luneburger Heide. It was decided to get rid of these accumulations by loading them into unwanted ships and sinking them in water of such depth that their slow decomposition would not adversely affect surface fisheries.
>
> An area in the deep rift in the Skagerrak, with depths ranging from 300 to 440 fathoms, was selected (543–783 metres). German prisoners were turned on to load the poison-gas shells and other materials in the elderly German ships *Jantje Fritzen*, 6582 tons,[47] *Taglia*, 2682 tons, *Edith Howaldt*, 2067 tons, *Sesostris* 2013 tons[48] and *Taurus*, 1158 tons which had been built respectively in 1920, 1906, 1903, 1918 and 1895.
>
> At the end of November these five ships assembled outside Kiel, where they had to ride out a severe gale for five days. They were then sailed in convoy under the command of HMS *Cotillion* (Lieutenant Commander H. J. Howard, R.N.V.R.). Four of the shell-ships proceeded under their own steam manned by German crews, but the cargo

[44] US Arms Control and Disarmament Agency (note 6).

[45] Knightley (note 4).

[46] Baltic Marine Environment Protection Commission (note 1).

[47] This was the ship on which Harald Franke had served as a crew member.

[48] The *Sesostris* was located and positively identified at 58.19.44.9N, 9.39.43.6E at 666 metres by the Norwegian Government survey in 1989. Tørnes *et al.* (note 8) and chapter 20 in this volume.

of the largest, the *Jantje Fritzen*, had begun to leak during the gale, and the engine room had been flooded with poison gas. She therefore had to be towed to the position appointed for the whole convoy to be sunk by means of scuttling-charges of which the strength had to be so nicely calculated as to produce explosions sufficient to sink the ships but not affect the dangerous cargo.

A scuttling-party under Lieutenant-Commander Staines, O.B.E, R.N.V.R., fixed the charges and the German crews were taken off and brought home in the motor vessel *Oberhausen*, 1261 tons, which at one time had been used as a hospital ship. Of the ships scuttled in this operation, the *Sesostris*, which disappeared only 90 seconds after the scuttling-charge had gone off, had been a passenger steamer, and it is believed that one of those landed from her last voyage as such to Cuxhaven was 'Lord Haw-Haw' [William Joyce] with his wife and bicycle.

Pictures in the report show the *Jantje Fritzen*, a steamer with a midship superstructure, being blown up by internal charges.[49] The following appears in a later *Monthly Intelligence Report*:

. . . it appears from recent complaints made by the coastal authorities in Sweden, that cases of explosives which had been dumped in the Skagerrak, in the hope that they would sink harmlessly to the bottom of the 400 fathom 'deep', have . . . refused to drown and are being washed up on the beaches of Göteborg and Bohuslän. Most of the cases which have made this unwelcome reappearance contain burning-charges for 15 cm (5.9 in) shells, or tear gas. It seems that as a result of immersion these substances became unexpectedly buoyant, and that other means than drowning will have to be devised for getting rid of them.[50]

Soviet disposal activities

The Soviet Military Administration in Germany (SMAD) assembled the Soviet share of the German CW munitions at a depot in the Baltic port of Wolgast. The Tripartite Naval Commission put eight German merchant ships at the disposal of the SMAD, and German soldiers under Soviet command were given the task of fulfilling the Soviet undertakings in the Potsdam Agreement.[51]

A report prepared by the Danish Government concluded from Soviet and German information that 35 000–50 000 tonnes of chemical weapons were dumped north-east of Bornholm and south-east of Gotland.[52] The munitions had usually been dumped over the sides of the vessels, but four or five minesweepers and several non-military hulks were loaded with munitions and sunk in the area north-east of Bornholm (the approach preferred by the Western Allies).[53]

Denmark first became aware that the USSR was dumping CW munitions off Bornholm when a ship docked in Nexö on Bornholm asked for medical assistance following an accident with CW munitions. Danish diplomatic efforts to

[49] British Admiralty, *Monthly Intelligence Report for March 1946*, 10 Apr. 1946 (Naval Intelligence Division, Naval staff, British Admiralty: London, 1946).

[50] British Admiralty, *Monthly Intelligence Report for April 1946* (Naval Intelligence Division, Naval staff, British Admiralty: London, 1946).

[51] *Redegjørelse vedrørende de miljø-. sundheds- og sikkerhetsmessige forhold i forbindelse med dumpet giftgasammunition i farvandene omkring Danmark* (note 8).

[52] *Redegjørelse vedrørende de miljø-. sundheds- og sikkerhetsmessige forhold i forbindelse med dumpet giftgasammunition i farvandene omkring Danmark* (note 8).

[53] *Chemische Kampfstoffmunition in der südlichen und westlichen Ostsee: Bestandsaufnahme, Bewertung und Empfehlungen* (note 8); and *Redegjørelse vedrørende de miljø-. sundheds- og sikkerhetsmessige forhold i forbindelse med dumpet giftgasammunition i farvandene omkring Danmark* (note 8).

stop the dumping in this area of rich fishing grounds were unsuccessful. Soviet operations continued until 30 December 1947, when the USSR declared that they had ceased; the GDR continued to dump an additional several hundred tonnes of CW munitions in the area up to 1965.[54]

According to a German Hydrographic Institute report, which includes information that supposedly came from Russian authorities, the dumped munitions (all of which were filled with mustard gas) included approximately 400 000 grenades, 70 000 aerial bombs (250 kg) and 1000 drums (1.5 tonnes).[55] The report notes that the USSR also dumped munitions filled with adamsite, Clark, cyclon (cyanide) and phosgene in the Baltic. Tabun may also have been dumped in the Bornholm area since bombs of this type have been found in and around the ports where the loading of the ships occurred.[56] Several eyewitnesses tell of crews starting to dump munitions long before reaching the designated dumping site. It must therefore be assumed that, since several trips were made, significant amounts of dangerous chemicals are on the bottom along a route from Peenemünde to Christiansø north-east of Bornholm. Very little is known about the composition of the approximately 2000 tonnes of chemical munitions dumped off Gotland.

Soviet postwar dumping

The German report mentioned above estimates the Soviet share of the German CW stockpile at 63 000 tonnes. If 35 000–50 000 tonnes of the stockpile were dumped north-east of Bornholm and approximately 2000 tonnes were sunk south-east of Gotland, approximately 10 000–30 000 tonnes remain unaccounted for. Soviet secrecy on the subject was total, and the new Russian leadership has failed to disclose more information.

The Soviet Union produced its own chemical weapons during and after World War II,[57] and there are indications that these weapons were also dumped in the Baltic Sea and that these operations continued well into the 1980s. The St Petersburg newspaper *Tjas Pik* cited a top secret document which reveals that Soviet dumping took place from 1946 until 1978.[58] These operations were unrelated to the dumping operations that were conducted by the Allies after World War II. The document states that the Soviet Baltic Sea Fleet dumped 356 872 tonnes of chemical warfare agents, including: 408 565, 75-mm and 150-mm grenades filled with mustard gas; 71 469, 250-kg aerial bombs filled with mustard gas; 17 543 aerial bombs filled with adamsite and arsine; 1564, 1500-kg containers of mustard gas; 10 420, 100-mm chemical mines; 7295 barrels of CW grenades; 7860 barrels of cyclon; and 189 000 kg of cyanide salt.

Two dumping sites are mentioned in the document: one north of Bornholm, and one 70 km from Liepaja, Latvia. (The exact coordinates are said to be given

[54] *Redegjørelse vedrørende de miljø-. sundheds- og sikkerhetsmessige forhold i forbindelse med dumpet giftgasammunition i farvandene omkring Danmark* (note 8).

[55] *Chemische Kampfstoffmunition in der südlichen und westlichen Ostsee: Bestandsaufnahme, Bewertung und Empfehlungen* (note 8).

[56] *Chemische Kampfstoffmunition in der südlichen und westlichen Ostsee: Bestandsaufnahme, Bewertung und Empfehlungen* (note 8).

[57] Stock, T., 'Chemical and biological weapons: developments and proliferation', *SIPRI Yearbook 1993* (note 7), chapter 7, p. 277

[58] *Tjas Pik* [Journal of the local journalist association in St Petersburg], 10 Feb. 1993, p. 5 (in Russian).

on the Navy battle fleet charts which were approved for use on 12 July 1985.) The validity of both the document and the data in it has been questioned. Several St Petersburg groups have attempted to obtain Western aid to organize search operations; these groups advocate removal of the dumped munitions.

Tjas Pik also cited Dr Lev Fedorov, a Russian chemist who was arrested on charges of revealing state secrets when he and a colleague disclosed that Russia was developing a new nerve gas that would be 5–8 times more toxic than the chemical agent VX.[59] Fedorov was quoted as saying that in admiralty circles 12 dumping sites are mentioned, 4 of which are located in the Baltic. Only 2 dumping sites have been disclosed thus far.

A St Petersburg water inspector is reported by *Tjas Pik* to have said that there are three dumping sites in the Gulf of Finland. The sites are north-north-east of Tolbuchin lighthouse, close to the island of Nerva, 17 km from Finnish territorial waters and at a depth of 46–58 metres.

In an article in the Russian newspaper *Izvestia*, Konstantin Terskov, an officer of the Soviet fleet, stated that he personally took part in the dumping of 45 000–50 000 tonnes of chemical aircraft bombs,[60] presumably of German origin. The majority were dumped close to Christiansø, near Bornholm.[61]

Soviet Air Force Major-General Boris Surikov conducted a search of the archives of the Soviet military forces after the collapse of the Soviet Union. His findings are incorporated in a memo:

> In the summer of 1945, the Soviet military forces in Germany received instructions from the Allied Dumping Committee to sink all German munitions found in their occupation zone. A division of Soviet troops, specialized in chemical weapons, did an enormous amount of work in the period of 1945–1947 to find and concentrate chemical munitions to the port of Wolgast in Northern Germany. About 35 000 tons of chemical munitions were collected, this was about 14 per cent of the substances found in the zone.[62]

If Surikov's information is correct, the Soviet Army had 250 000 tonnes of CW munitions at its disposal at the end of World War II. All of these chemical weapons cannot have been German, since the total German output is estimated to have been approximately 300 000 tonnes.[63] Surikov's research in the Soviet archives indicated that 5000 tonnes of chemical munitions were sunk west of Liepaja, Latvia, in May–December 1947, either over the side or in eight scuttled ships, and that 30 000 tonnes were sunk in the northern Baltic.[64]

III. Conclusions

There are five confirmed chemical munition dumping sites in the Baltic and the Skagerrak in addition to several other suspected sites. Public concern has led to surveys of the area, which have varied in their degree of thoroughness, by the

[59] Stock (note 57), pp. 266–67; see also chapter 9 in this volume.

[60] Non-chemical bombs were also dumped.

[61] Terskov, K., *Izvestia*, 26 Jan. 1991 (in Russian).

[62] 'Memo by Major-General Boris Surikov (ret.), Russian Air Force', trans. M. Nielsen, Greenpeace Russia.

[63] *Armed Forces Chemical Journal* (note 2).

[64] Surikov (note 62). The 5000 tonnes amount may be a mistranslation since the position given is closer to Wolgast, Germany, in the southern Baltic.

Danish, German, Norwegian and Swedish governments. There has been no other known damage caused by the dumps than the injuries, some serious, suffered by a few fishermen who have come in contact with leaking grenades and bombs.

The responsibility for the dumping sites lies with the governments of the states in whose waters the dumping occurred and with those governments which dumped CW munitions. In the future, the dumps may present significant or possibly lethal threats to people or the environment. There is profound unwillingness to take steps to prevent the dumps from becoming a greater problem than they are today.

Those mainly to blame are the governments of Russia (as the successor to the Soviet Union), the UK and the USA, which were responsible for the scuttling and dumping operations. Much information has been withheld by these nations, and Russia and the UK continue to withhold vital archive material from the public and even from government investigators from the Baltic countries that have conducted surveys of the area.

HELCOM CHEMU

The Ad Hoc Working Group on Dumped Chemical Munition of the Baltic Marine Environment Protection Commission of the Helsinki Commission was formed in 1993 to gather information that is available about dumped chemical weapons and chemical warfare materials from the states concerned: Denmark, Estonia, Finland, Germany, Latvia, Lithuania, Poland, Russia and Sweden. The first HELCOM CHEMU meeting took place in St Petersburg, Russia, in April 1993. The second meeting was held in Vilnius, Lithuania, in September 1993. In 1994 meetings were held in January, June, September and December.

All member states are supposed to supply reports on the information which they possess. It was hoped that Russia would use HELCOM CHEMU as a means to disclose the activities of the former Soviet Union regarding the dumping at sea of CW munitions, but the Russian delegation has submitted little data. The UK and the USA have submitted reports on their respective and combined operations in Scandinavian waters. It is particularly noteworthy that for the first time the UK presented significant information about its postwar CW operations.

The conclusions of the January 1994 HELCOM CHEMU meeting were submitted to a March 1994 ministerial meeting of the Helsinki Commission for further action.[65] The ministerial meeting decided to prolong the mandate of the group. The June and September 1994 meetings focused on the chemistry of CW agents, their effect on the environment, corrosion of the dumped munitions and guidelines both for fisherman who come in contact with chemical agents and the authorities of their countries.[66] The December meeting focused on information about dumping activities.

[65] Stock, T. and De Geer, A., 'Chemical weapon developments', *SIPRI Yearbook 1994* (Oxford University Press: Oxford, 1994), chapter 9, p. 340.

[66] Stock, T. and De Geer, A., 'Chemical and biological weapons: developments and destruction', *SIPRI Yearbook 1995: Armaments, Disarmament and International Security* (Oxford University Press: Oxford, 1995), chapter 10, pp. 353–54.

Much remains to be done, however, before agreement is reached on a plan to chart the sites or on the drafting of common legislation which clarifies who is responsible for injuries to fishermen or damage to the environment.

Environmental concerns

None of the governments which are affected by the problem of dumped CW munitions have conducted a thorough environmental study of the effect of these chemical agents (of which there may exist as much as 75 000 tonnes) on the marine environment. The Swedish survey mentioned above found traces of mustard gas in sediment from the seabed near the wrecks off Måseskär[67] and concluded that further investigation was needed. However, no formal proposal for action was made to the government.

Contrary to what most governmental reports claim, little is actually known about the effect of these dumps on marine and coastal life in the affected areas. Among the issues in need of investigation are the effects that water pressure at the depths where the wrecks lie has had on the solubility and hydrolysis of the chemical agents. More geologic and biosphere data about the dumping sites is also needed.

The present author took part as an observer in three surveys of known and suspected dumping areas near Måseskär and off the coast of Estonia. The conclusion which can be drawn from these costly and time-consuming operations is that substantial work remains to be done to chart the exact positions of the scuttled ships in order to clarify which areas are hazardous. This is the only way in which further injuries to fishermen can be avoided. Trawlers from the Scandinavian countries as well as from Germany, Poland, the Baltic states and Russia all fish in the areas where CW munitions were dumped and in adjacent areas. Trawlers from as far away as the Faeroe Islands also fish these waters, and the fishermen who man them have little knowledge of the problem. This is also the case for fishermen from the eastern side of the Baltic who were not informed of the problem before the dissolution of the USSR.

Legislation

Coordinated legislation is needed to regulate fishing near the dumped chemical weapons. In 1968–85, 150 cases were reported involving Danish fishermen who had come in contact with CW munitions. In the vast majority of these cases, the agent was mustard gas.[68] Numerous additional cases have occurred since then at a rate of approximately 12 per year.

Currently, only Danish fishermen are covered by a programme which is designed to meet this problem. Each Danish fishing boat is issued adequate first-aid equipment, and Danish fishermen receive full compensation from the Danish Government for lost catch and time lost in port for decontamination. This programme is designed to lessen injuries to fishermen and to prevent them from dumping suspicious items and possibly contaminated fish back into the

[67] Sjöfartsverket (note 8).

[68] *Redegjørelse vedrørende de miljø-. sundheds- og sikkerhetsmessige forhold i forbindelse med dumpet giftgasammunition i farvandene omkring Danmark* (note 8).

sea in areas far from the known and charted dumping sites. Fishermen from other Baltic nations often dump dangerous materials back into the sea because of the cost and bother of reporting their finds to the coast guard.

Swedish fishermen, for example, have only been given a leaflet about the danger and are left to their own resources if they wish to supply their vessels with adequate CW first-aid equipment. In addition, they must bear the cost of decontaminating their vessels and equipment.

In the Skagerrak the situation is worse. The area near the dumping site at Måseskär is a popular fishing ground, and there is extensive fishing every day near the wrecks which are suspected to contain chemical weapons. Some fishermen have plotted trawl patterns as close as a few metres from the wrecks. Presumably, no major accident has occurred because the hulks have not yet broken up. However, according to calculations of the amount of oxygen in the water, the salinity of the sea and the thickness and quality of the steel used in the hulls of these ships, they are likely to begin to break up within a few years.[69]

Salvage: raising the munitions

The US Army has been given the task of destroying the US stockpile of chemical weapons by the year 2004. The cost of this operation was initially estimated at $7.9 billion—an indication of the funding needed for any CW destruction operation.[70] The cost to destroy one kg of CW agent is estimated at 500–1000 Swedish crowns (approximately $65–130). If 10 000 tonnes of dumped CW agents were raised from the seabed the cost of incineration would be approximately $650–1300 million.[71] Numerous technical problems would need to be solved to bring the munitions to the surface, and the cost of safely locating and raising them would be great. Governmental funds might more profitably be used to: (*a*) conduct research on the effect of these chemicals and the products released by their decay on the marine environment; (*b*) chart the dumping sites; and (*c*) establish a programme of compensation and education about the danger posed by these weapons for fishermen and other affected individuals.

Additional sources

Swedish Ministry for Foreign Affairs, 'PM rörande ifrågasatt sänkning av giftgaser m m i Skagerack' [Memo concerning disputed dumping of poisonous gases in the Skagerrak], 1948.

Telex from Aivars Lembergs, Chairman of the Ventspils City Council, Latvia, to the author, 12 Mar. 1992.

Interview by the author with Gerhard Seifert, retired poison gas expert of the Schwerin Police Department of the former GDR, Schwerin, Mar. 1992.

Letter from the Swedish Fishermen's Union to the Swedish Ministry for Foreign Affairs, 21 June 1946.

Swedish Ministry for Foreign Affairs, 'Note of telephone message from The Foreign Service of the United States of America to the Swedish Foreign Office', 4 July 1948.

SIPRI, *The Problem of Chemical and Biological Warfare*, vol. 1, *The Rise of CB Weapons* (Almqvist & Wiksell: Stockholm, 1971), pp. 207, 302–309.

Donnelly, P. J., 'Environmental fate of chemical agents in deep seawater', *Literature Review*, 3 Aug. 1990.

[69] Åkerlund (note 3).
[70] US Arms Control and Disarmament Agency (note 6).
[71] Sjöfartsverket (note 8).

20. Investigation of the ships filled with chemical munitions which were sunk off the Norwegian coast after World War II

FRODE FONNUM

I. Introduction

The Potsdam Agreement, which was signed by the UK, the USA and the USSR on 2 August 1945, following the defeat of Germany in World War II, stated that Germany should be demilitarized and that all war materials should either be removed or destroyed.[1] Implementation of the agreement led to the dumping of chemical weapons which took place after the war.

Chemical weapons were found in all four military occupation zones. The Western allies, at least the UK and the USA, cooperated as regards the dumping at sea of the chemical munitions. The USSR dumped most of the weapons recovered in its zone near the Danish island of Bornholm and the Swedish island of Gotland, and the Western powers dumped the majority of the CWs which had been found in their zones in the Skagerrak, particularly in the Norwegian Trench and outside Måseskär, on the Swedish west coast. The dumping of these chemical munitions was not secret, but following the turmoil at the end of World War II it was forgotten for many years.

In Norway there was public discussion of the fate of the dumped munitions in 1970 and again in 1984. In Sweden fishermen in the vicinity of Måseskär were well aware of the dumping, but apparently Swedish authorities did not know about these activities until 1990. However, the dumping outside Bornholm and Gotland was well known to both Danish and Swedish authorities. Detailed information about CW dumping was difficult to obtain from the Allied countries until recently.

II. Information from British sources

In November 1945 the British marine attaché notified the Danish Marine Ministry that the UK planned to dump large quantities of chemical weapons in international water, 15–20 nautical miles south of the Norwegian coast between 8.45E and 10.40E.[2] The depth in that area is more than 450 metres. The Norwegian Directorate of Fishing objected to the size of the area but agreed that dumping could be carried out in a smaller area 21–24 nautical miles south of the Norwegian coast between 9.27E and 9.40E. The depth there is more than 650 metres. The official message read:

[1] See von Oppen, B. R. (ed.), *Documents on Germany under Occupation 1945–1954* (Oxford University Press: London, 1955), pp. 40–50; and *The Tehran, Yalta and Potsdam Conferences: Documents* (Progress Publishers: Moscow, 1969), pp. 320–33.

[2] *Redegjørelse vedrørende de miljø-. sundheds- og sikkerhetsmessige forhold i forbindelse med dumpet giftgasammunition i farvandene omkring Danmark* [An account of the environmental, health and security situation in connection with the dumped poison gas ammunition in the waters surrounding Denmark] (Miljøstyrelsen: Copenhagen, 1985).

From: Searail 7
To: Chairman Continental Ammunition Dumping Committee

Following for quartermaster general Norwegian Army begins ref No J nr 3758/45 JR/SK the Director of Fisheries maintains that the ammunition ought to be taken out and dumped outside the continental platform either in the Atlantic Ocean or north of the North Sea alternatively in Skagerrak within an area enclosed by lines joining the following positions at 58 DEGS 14 MINS N 09 DEGS 27 MINS E B 58 DEGS 16 MINS N 09 DEGS 27 MINS E C 58 DEGS 19 MINS N 09 DEGS 40 MINS E D 58 DEGS 17 MINS N 09 DEGS 40 MINS E[3] should this area be chosen the area must be marked with buoys beforehand the ministry cannot agree to the dumping in any other place in Norwegian waters than those mentioned by the Director of Fisheries.[4]

The Store Nordiske Telegrafselskap (the Norwegian telephone company) also objected since the main telephone cable between Norway and Hirtshals, Denmark, passed through the so-called 'larger area'.[5] Norwegian authorities stipulated that all CW munitions dumped by the UK should be loaded on ships that would then be scuttled.

Information about the number of ships which were scuttled in Norwegian waters is insufficient, and the available sources provide contradictory data. In 1946 the British Navy stated that five ships containing mustard gas and phosgene had been scuttled in the Skagerrak at 58°15' and 9°30'.[6] In July 1984 a letter from the Hydrographic Department of the Ministry of Defence (MOD) stated that five possible 'gas ships' had been sunk in the so-called 'small area' and that one ship had been sunk south of Lista, Norway:

Hydrographic Department
Ministry of Defence
17 July 1984
Our ref H4015/83

(a) The German Light Cruiser *LEIPZIG* (6710 dwt) was reported to have been scuttled in approx. position 57.53N, 06.13E on 16 December 1946 with about 1000 tons of poisonous gas (presumably in canisters) onboard.

(b) Five ships, with chemical warfare cargoes, were scuttled in February 1946 as follows:

SS *PATAGONIA*	in 58.15N	9.35E	
SS *DUBORG*	in 58.14.5N	9.31E	These positions fall within or
SS *LOUIS SCHROEDER*	in 58.15N	9.27E	near the charted Dumping Ground
SS *BALKAN*	in 58.16N	9.26E	(explosives) on BA chart 2289.
SS *TAURUS*	in 58.16N	9.30E	

(c) A further convoy of vessels, number and names unspecified, were scuttled in May 1946 in approx. position 58.08N, 10.48E. Cargoes unknown but presumed to be of a dangerous nature.[7]

[3] This area is referred to as the 'small area' in this chapter.
[4] Teleprint from Searail 7 to Chairman Continental Ammunition Dumping Committee, 20 Aug. 1945.
[5] *Redegjørelse vedrørende de miljø-. sundheds- og sikkerhetsmessige forhold i forbindelse med dumpet giftgasammunition i farvandene omkring Danmark* (note 2).
[6] *Redegjørelse vedrørende de miljø-. sundheds- og sikkerhetsmessige forhold i forbindelse med dumpet giftgasammunition i farvandene omkring Danmark* (note 2).
[7] Hydrographic Department, Ministry of Defence, ref. H4015/83, 17 July 1984.

Another report stated that 12–15 ships filled with mustard gas and grenades had been scuttled in 1946–47, but these ships were said to have been sunk near the German island of Helgoland or in the Baltic Sea.[8]

At a meeting of the Ad Hoc Working Group on Dumped Chemical Munition of the Baltic Marine Environment Protection Commission of the Helsinki Commission (HELCOM CHEMU) it was reported that in 1945–48 the UK had scuttled 34 ships containing a total of 151 425 tonnes of gas munitions and conventional munitions 25 nautical miles south-east of Arendal, Norway.[9] The ratio between conventional and gas munitions was not known. The ships carried cargo from both the British and US occupation zones. They were loaded in Emden, Flensburg, Kiel and Lübeck, Germany, and sailed in convoys. According to other reports the convoys were met by Norwegian pilots. In addition, six concrete vessels and some wrecks were scuttled, but little information about these activities was given to the Norwegian authorities.

III. Norwegian information

On 2 July 1946 the newspaper *Fædrelandsvennen* published an interview with Commander Joachim Flood of the US Navy.[10] He had led a convoy of three German ships filled with 3800 tonnes of the nerve agent tabun and mustard gas to a dumping site in the Skagerrak. The interview contained the first detailed information from a participant in the dumping operation. Flood explained that the ships were sunk by opening their bottom valves; water then flooded in covering the 100 000 containers which were filled with gas. In addition, US ships fired on the German ships to make them sink faster. Flood noted that tabun was very poisonous and that it killed in three minutes.

In 1970 the newspaper *Aftenposten* published a list of ships which had been scuttled in the Skagerrak and the North Sea (see table 20.1).[11] The list was compiled by Einar Høvding, the owner of a ship salvage company, and is of interest since, in an arrangement with the Norwegian authorities, Høvding had assumed ownership of all of the ships from World War II which had been scuttled off the Norwegian coast. Since the ships sailed in a convoy, it would be expected that all ships scuttled on the same day would have been sunk in the same area. For example, ships that were scuttled on 4 October 1945 were probably all scuttled at the same place, although both the North Sea and the Skagerrak were indicated in Høvding's list.

According to an article in *Hamburger Abendblatt*, the *Jantje Fritzen*, *Theda Fritzen*, *Edith Howaldt* (2067 gross tonnes), *Tagila*, *Taurus* and *Sesostris* were

[8] *Redegjørelse vedrørende de miljø-. sundheds- og sikkerhetsmessige forhold i forbindelse med dumpet giftgasammunition i farvandene omkring Danmark* (note 2).

[9] Baltic Marine Environment Protection Commission, Helsinki Commission, UK, Report on sea dumping of chemical weapons by the United Kingdom in the Skagerrak waters post World War II, HELCOM CHEMU 2/2/5, 28 Sep. 1993. Appendix 1 of the report is a table entitled 'Vessels sunk in Skagerrak as part of UK CW disposals'. It lists the ships and the dates they were sunk (including those in table 20.1 in this chapter). According to the report the ships were transported in 10 convoys. The positions of 15 of the ships in convoys 1, 2, 3 and 6 correspond to the area examined (see figure 20.1).

[10] '48 000 tonn giftgass skal senkes i Skagerakk' [48 000 tons poisonous gas scuttled in Skagerrak], *Fædrelandsvennen*, 2 July 1946.

[11] 'Skipene med gass en velkjent sak' [Gas ships well known], *Aftenposten*, 18 Aug. 1970.

Table 20.1. List of ships scuttled in the sea provided by Einar Høvding

Ship	Gross tonnage	Dead-weight	Length overall	Beam	Draught	Date scuttled	Position
Eider	3 288	4 420	109.4	15.3	6.1	12 Apr. 1945[a]	North Sea
Duburg	2 675	4 950	120.0	16.6	9.9	4 Oct. 1945	North Sea
Luise Schrøder	1 327	. .	74.2	11.0	4.6	4 Oct. 1945	North Sea
Patagonia	5 898	8 890	137.0	17.4	7.6	4 Oct. 1945	Skagerrak
Pillau	1 308	2 169	73.8	11.3	4.3	4 Oct. 1945	Skagerrak
Balkan	2 269	3 440	85.2	12.2	5.8	4 Oct. 1945	Skagerrak
Triton	1 620	2 055	73.6	11.1	4.6	4 Oct. 1945	West of Jylland
Emmy Friederich	5 066	9 500	133.7	17.4	7.7	17 Oct. 1945	Skagerrak
Olga Siemers	3 347	5 050	101.4	14.2	6.3	17 Oct. 1945	Skagerrak
Taurus	1 158	1 980	73.0	10.4	5.0	17 Nov. 1945	North Sea
Jantje Fritzen	6 594	11 000	125.7	16.9	10.5	17 Nov. 1945	North Sea
Theda Fritzen	2 882	5 000	100.6	14.5	6.1	17 Nov. 1945	Skagerrak
Drau	5 241	4 560	122.0	15.9	8.6	1945	Skagerrak
Oderstrom	1 511	. .	. .	. .	. .	1945	Skagerrak
Sesostris	2 013	2 920	111.3	15.3	7.3	1945	North Sea
Tagila	2 683	4 425	93.2	13.6	6.5	1945	North Sea
Trude Schunemann	1 260	1 980	69.8	10.9	4.6	1956	North Sea
Karl Leonardt	6 042	7 000	126.3	16.9	8.1	6 Mar. 1946[b]	Skagerrak
Hugo Ollendorf	1 876	2 700	85.0	11.7	5.4	16 Mar. 1946	Skagerrak
Lotte	2 116	2 800	90.2	12.4	5.5	26 Mar. 1946	North Sea
Røhn	1 778	3 050	79.5	12.5	5.3	8 Sep. 1946	South-west of Arendal
Ludwigshafen	1 923	3 000	87.9	13.5	4.8	12 Oct. 1946	North Sea
Schwalbenland	8 188	11 300	142.7	18.4	10.0	31 Dec. 1946	Skagerrak
Fechenheim	8 116	11 150	142.9	17.8	10.0	1946	South-west of Arendal
Gertrud Fritzen	2 999	5 100	98.8	14.2	6.1	1946	North Sea
Dora Ollendorff	2 730	4 450	95.5	13.2	6.4	1946	Bay of Biscay
Dessau	5 933	9 632	120.0	16.6	9.6	17 May 1947	North Sea
Harm Fritzen	4 818	8 100	119.0	16.3	8.0	Mar. 1948	Bay of Biscay
Approximate total	**100 000**	**150 000**					

[a] The correct date is probably 15 Oct. 1946.

[b] A British report on the convoy (HELCOM CHEMU 2/2/5, 28 Sep. 1993) lists the date as 16 Mar. 1946.

Source: Based on a list published in *Aftenposten*, 18 Aug. 1970, and later corrected by Einar Høvding, owner of the ship salvage company which has the right of ownership for all ships from World War II that were scuttled off the Norwegian coast.

all scuttled in the Norwegian Trench on 17 November 1945.[12] The article was based on a statement by Friedrich Passehl, who was captain on board the *Taurus* on its final expedition.

[12] 'Gasing Geleitzüge voll Gasgranaten versenkt' [Convoy of gas ships full of gas grenades scuttled], *Hamburger Abendblatt*, 19 Aug. 1970; and 'Tyske marineoffiserer bekrefter dumpingen utenfor Arendal: 200 000 tonn krigsgass i "Den norske graven"' [German naval officer confirms dumping outside Arendal: 200 000 tonnes war gas in 'the Norwegian Trench'], *Agderposten*, 24 Apr. 1984.

Table 20.2. Positions of ships sunk in Operation 'Davey Jones Locker', 1946

Name and type of ship	Tonnes	Date scuttled	Depth in metres	Latitude and longitude
Sperrbrecher (German minebreaker)	1349	1 July	650	58.14.0N 9.15.0E
T-65 (German flak ship)	1526	1 July	650	58.17.9N 9.37.1E
U-J 305 (German trawler)	671	2 July	650	58.16.4N 9.29.0E
Alco Banner (Hog islander)	2765	14 July	650	58.18.7N 9.36.5E
James Otis (US Liberty ship)	3653	30 Aug.	650	58.16.0N 9.32.0E

Source: US Arms Control and Disarmament Agency (ACDA), *Special Study on the Sea Disposal of Chemical Munitions* (ACDA: Washington, DC, 19 Jan. 1993), p. 18.

The 11 April 1984 *Agderposten* listed several other ships that were not mentioned by Høvding but which were reported by laymen interested in marine history.[13] Those vessels were: *Helgoland* (2947 tonnes) sunk in 1948; *Z29* and *Z37*, two German destroyers, scuttled on 16 December 1946 and 26 March 1946, respectively; *Sperrbrecher 2* (also known as the *Karl Leonardt*) sunk on 16 March 1946; *Sperrbrecher 27* (also called *H. C. Horn*) scuttled on 26 May 1946; and *Sperrbrecher 36* (or *Eider*) sunk on 15 October 1946. (Sperrbrechers were civilian ships which Germany modified in World War II so that they could safely enter harbours which might have been laid with mines.)

One possible explanation for the large number of ships which are reported as scuttled in the Skagerrak is that the lists include ships which were scuttled both in the Norwegian Trench and outside Måseskär, Sweden.

IV. US reports[14]

Approximately 102 500–110 000 tonnes of chemical weapons were captured in the US occupation zone in Germany. These munitions were disposed of as follows: 12 000 tonnes were shipped to the USA; 8000 tonnes were shipped to the UK; 10 500 tonnes were shipped to Italy; 8500 tonnes were sold to the German corporation STEG for destruction; 31 000–39 000 tonnes were scuttled at sea; and 32 500 tonnes were destroyed on site.

The US dumping operation was called 'Davey Jones Locker'. Five ships were scuttled in 1946 (see table 20.2), and additional ships were scuttled later (see table 20.3). (The *Sperrbrecher*, *T-65* and *U-J 305* are obviously the ships

[13] '48 000 tonn giftgass på bunnen utenfor Arendal: Tyskerne turde ikke selv bruke gassen' [48 000 tons poisonous gas on the sea bottom outside Arendal: the Germans did not dare to use the gas], *Agderposten*, 11 Apr. 1984.

[14] US Arms Control and Disarmament Agency (ACDA), *Special Study on the Sea Disposal of Chemical Munitions* (ACDA: Washington, DC, 19 Jan. 1993).

Table 20.3. Positions of ships sunk in Operation 'Davey Jones Locker', June 1947–Aug. 1948

Name and type of ship	Tonnes	Date scuttled	Depth in metres	Latitude and longitude
James Sewell (US Liberty ship)	4000	6 June 1947	725	58.15.2N 9.30.6E
James Harrod (US Liberty ship)	3000	20 June 1947	665	58.10.0N 9.33.0E
George Hawley (US Liberty ship)	1000	30 June 1947	685	58.18.5N 9.38.0E
Nesbitt (US Liberty ship)	6000	18 July 1947	580	58.18.5N 9.30.0E
Philip Heiniken (German freighter)	2000	24 July 1948	1035	62.57.0N 1.32.0E
Marcy (German freighter)	2500	24 Aug. 1948	1180	62.59.0N 1.23.0E

Source: US Arms Control and Disarmament Agency (ACDA), *Special Study on the Sea Disposal of Chemical Munitions* (ACDA: Washington, DC, 19 Jan. 1993), p. 20.

which are described in the *Fædrelandsvennen article.*[15]) The ships were scuttled by opening their bottom valves, and shells were fired at the *T-65* and *U-J 305* at the waterline to breach the ships where no toxic agents were stored. The *Alco Banner* took 45 minutes to sink and was sent to the bottom by shell fire. The chemical agents loaded on these ships were mustard gas, chloracetophenone, phosgene and other unknown chemical agents of which at least part was tabun, according to Norwegian newspapers.[16]

V. The Norwegian investigation

In 1989 the Division for Environmental Toxicology of the Norwegian Defence Research Establishment (NDRETOX) was tasked by the Norwegian Ministry of Defence to investigate the wrecks of the ships which were scuttled 25 nautical miles off Arendal, Norway.[17] The NDRETOX contracted ships from Geoconsult A/S, and the investigation was carried out in two steps: (*a*) locating the wrecks and their positions using a sidescan sonar; and (*b*) investigating the status of the ships and their cargo by means of a remote-operated vehicle (ROV) with a television camera and a Nansen water-sample collector.

The so-called 'small area' mentioned above was 4 x 14 km, but the area to be investigated by the NDRETOX was increased to 8 x 16 km. This was done in order to allow for errors in navigation when the ships were scuttled and also so that any vessels which had been sunk just outside the small area would not escape detection.

[15] '48 000 tonn giftgass skal senkes i Skagerakk' (note 10).

[16] '48 000 tonn giftgass skal senkes i Skagerakk' (note 10); and 'Skipene med gass en velkjent sak' (note 11).

[17] Tørnes, J. A. *et al.*, *Investigation of Ships Carrying Chemical Ammunition Sunk in Norwegian Waters after World War II*, FFI/Report-89/6007 (Forsvarets Forskningsinstitutt (FFI) [Norwegian Defence Research Establishment]: Kjeller, Norway, 1989).

Table 20.4. Positions of wrecks identified by sidescan sonar in the Norwegian Trench[a]

Wreck number	North coordinate	East coordinate	Latitude	Longitude	Comments on scan interpretation
1	6 454 855	531 481	58.13.56.5	9.32.09.7	*c.* 100 metres (m) long, *Duborg*?
2	6 457 016	533 767	58.15.05.8	9.34.31.0	> 70 m, *Patagonia*?
3	6 457 465	542 405	58.15.20.9	9.43.21.1	80–100 m
4	6 458 316	541 262	58.15.45.5	9.42.11.5	
5	6 458 543	539 147	58.15.53.6	9.40.01.9	110 m
6	6 459 015	540 229	58.16.08.5	9.41.08.6	2 parts, main part 80–100 m
7	6 458 695	530 562	58.16.00.9	9.31.15.2	70–80 m, *Taurus*?
8	6 461 904	539 768	58.17.42.0	9.40.42.0	*c.* 100 m
9	6 462 242	540 611	58.17.52.7	9.41.34.0	*c.* 100 m
10	6 462 219	537 875	58.17.52.8	9.38.46.0	70 m
11	6 461 428	532 386	58.17.28.8	9.33.08.5	130 m
12	6 461 416	524 620	58.17.30.2	9.25.11.7	
13	6 463 906	539 002	58.18.47.0	9.39.56.2	120 m
14	6 463 437	540 138	58.18.31.5	9.41.05.7	100 m
15	6 465 694	538 780	58.19.44.9	9.39.43.6	

[a] Positions in universal transverse mercator (UTM) zone 32 coordinates and geographical positions. Wrecks 3, 4, 5 and 6 were at 550–650 m; the others were localized at 650–700 m.

Source: Tørnes, J. A. *et al.*, *Investigation of Ships Carrying Chemical Ammunition Sunk in Norwegian Waters after World War II*, FFI/Report-89/6007 (Forsvarets Forskningsinstitutt (FFI) [Norwegian Defence Research Establishment]: Kjeller, Norway, 1989).

A sidescan sonar (type EG & G model 990), was towed at 3.4 knots 50 metres above the seabed. An area of the seabed 500 metres to each side of the sonar was examined; the area to be searched was divided by 10 search lines each 800 metres long. Several wrecks were therefore observed twice. A Trisponder navigation chain was established on shore and used together with Decca (a radio navigation system). This gave an accuracy of ±10 metres for the surface vessel and 10±20 metres for the sonar and the wrecks. In some cases, however, the distance to the Trisponder stations on shore was too large, which decreased accuracy. The wrecks were investigated repeatedly, and the recorded positions should be sufficiently accurate for practical use. The positions of the wrecks are given in table 20.4 and in figure 20.1.

Examination of five wrecks with the ROV[18]

Five of the shipwrecks (numbers 6, 13, 5, 14, and 11 in table 20.4) were selected for closer examination by the ROV using the following criteria:

1. Wrecks which appeared to have broken in several pieces were selected since the cargo would be readily available for examination.
2. Larger wrecks were regarded as more important than smaller wrecks.
3. Wrecks that gave particularly distinct echoes on sidescan sonar were selected since the probability of identifying the scuttled ships was considered high.

[18] The author of this chapter was responsible for planning the investigation.

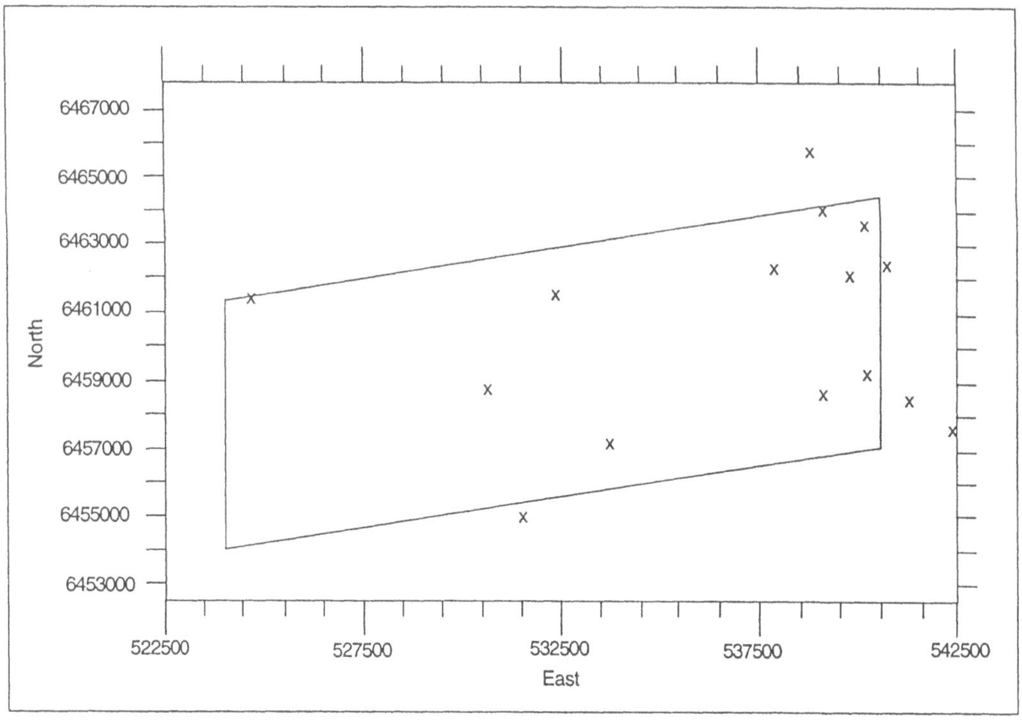

Figure 20.1. Positions of the shipwrecks as determined by the use of sidescan sonar, a Trisponder navigation chain and Decca

Source: Tørnes, J. A. *et al.*, *Investigation of Ships Carrying Chemical Ammunition Sunk in Norwegian Waters after World War II*, FFI/Report-89/6007 (Forsvarets Forskningsinstitutt (FFI) [Norwegian Defence Research Establishment]: Kjeller, Norway, 1989).

Examination of the wrecks was not as easy as had been expected because there was turbulence near the seabed, and small floating particles limited visibility. In addition, the absence of light made inspection of the cargo difficult.

Wreck number 6 was found at a depth of 585 metres. It consisted of several parts and may have been one of the ships which was sunk by having shells fired at it. Several bombs were discovered on the seabed, some by themselves and some in cases. The diameter of the bombs was 30–40 cm, and each weighed 200–250 kg. Some of the bombs were corroded, and their contents had probably leaked out. Water samples were collected near the bombs.

Wreck number 13 was located at 671 metres. It was intact and little corrosion had occurred. No weapons were detected outside the wreck. It was not possible to make observations inside the cargo hold because it was too dark in the hold. However, water samples were collected near the cargo hold.

Wreck number 5 was found at 594 metres and appeared intact. It was possible to examine the cargo room of this wreck, and bombs of the same size as those found in wreck number 6 (200–250 kg) were observed. One bomb was corroded. A metal bar from the ship was taken to the surface and was also found to be heavily corroded. Water samples were collected in the cargo hold.

Wreck number 14 was found at 666 metres. It was intact, but the superstructure had been removed. The ship was labelled *Sesostris*, a name which is found in several reports.[19] This gave credence to the German report that the *Taurus*,

[19] 'Gasing Geleitzüge voll Gasgranaten versenkt' (note 12); and 'Tyske marineoffiserer bekrefter dumpingen utenfor Arendal: 200 000 tonn krigsgass i "Den norske graven"' (note 12).

Jantje Fritzen, *Theda Fritzen*, *Tagila* and the *Edith Howaldt* had been scuttled in this area. The cargo room was open, but it was impossible to see the cargo. Turbulence caused the water near the wreck to be filled with small particles and hindered further study of the ship.

Wreck number 11 was located at 702 metres. It was intact but heavily corroded. There was a hole in the superstructure which could have been caused by artillery fire. A bomb, larger than those found previously, was observed on the deck. It was not possible to investigate the cargo hold, but water samples were collected close to the bombs and above the cargo hold.

Sampling and chemical analysis

Ten samples were collected near the wrecks, and another two samples were collected far from them to serve as a control. The samples were taken with a Nansen water-sample collector mounted on the ROV, and the water was immediately tested with a water analysis kit. The detection limits were 10 μg/l for nerve gas and 5 mg/l for mustard gas.[20] No chemical agents were detected.

Following these initial tests, samples were frozen and transported to the NDRETOX laboratory where they were concentrated on a C_{18} column and analysed with gas chromatography–mass spectroscopy (GCMS) using selected ion monitoring. The detection limits during the analysis were 200 ng/l for mustard gas, 10 μg/l for thiodiglycol (the hydrolysis product of mustard gas) and 2 μg/l for tabun. None of the agents were detected in the samples. (For comparison, in wartime the concentrations allowed in drinking water are 20 μg/l for tabun and 2 mg/l for mustard gas.)

VI. Analysis of the data

Four questions are raised by the information presented above:

1. What is known about the ships scuttled in the Skagerrak?
2. What chemical warfare agents were dumped off the coast of Norway?
3. What risk do these agents constitute?
4. What precautions should be taken in the future?

Number of scuttled ships

The investigation launched by the NDRETOX in the Norwegian Trench located 15 wrecks close to or in the area used by the Allied European Advisory Commission to dump gas ships. The most conservative estimate of the number of ships scuttled in the Norwegian Trench is 8 US ships (the 5 ships in table 20.2 are the ones most often named) and 5–9 British ships, a total of 13–17 ships. This number is apparently accurate since small vessels (e.g., trawlers) and vessels that were in poor shape, and which broke up after being fired on, may have escaped detection. Some ships may also have been scuttled outside the designated area. As Høvding stated, the ships were generally in poor condition, and many could have been lost *en route* to the dumping site because of bad weather.

[20] Karlsen, P. J. and Johnsen, B. A., *Testing av Vannanalysesett-Giftstoffer i Vann–M272* [Testing of water analysis kit], report no. FFI/NOTAT-86/6004 (FFI: Kjeller, Norway, 1986).

There have been estimates that as many as 34 British and 9 US vessels were scuttled in the Skagerrak. Only 15 wrecks have been found there, which could indicate that the vessels were sunk far outside the designated area or that they never reached their destination. Alternatively, these large numbers could be accounted for if they include the vessels sunk outside the Swedish and Norwegian coasts, and perhaps also those ships sunk outside Helgoland. Approximately 16 vessels have been located outside Måseskär, Sweden,[21] and an unknown number may have been sunk near Helgoland.

There is also the disturbing information that the *Leipzig*, filled with 1000 tonnes of chemical munitions, was scuttled in a different area, namely, outside Lista, Norway. This raises the question of whether other vessels were scuttled in areas which remain unknown. Other information indicates that the *Leipzig* did not contain chemical munitions,[22] and an attempt to localize it and other ships in the area mentioned failed to find the *Leipzig* at the position where it was assumed to have sunk.[23] (The two ships which the USA sank in 1948, the *Philip Heiniken* and the *Marcy*, were scuttled in the Atlantic at 1000 metres.)

Types of chemical agents

The 1946 report in the Norwegian newspaper *Fædrelandsvennen* clearly stated that tabun was present in the cargo of the three vessels—the *Sperrbrecher*, the *T-65* and the *U-J 305*—which were scuttled on 1–2 July 1946.[24] It does not seem surprising that this deadly poison was dumped at the greatest depth. However, the generally held view is that most of the tabun was taken by the USSR and either transported to the Soviet Union or dumped in the Baltic Sea. The total amount of tabun in Germany in 1945 was apparently 12 000 tonnes.

Since large quantities of mustard gas, normal or thickened, were found in Germany, it can be assumed that some of it was dumped in the Norwegian Trench. It has also been reported that chloracetophenone, a tear gas, and phosgene were dumped there. There is no information about any of the chemical warfare gases which contain arsenic, such as Clark I, Clark II and adamsite. This does not, however, exclude the possibility that they, too, were dumped in the Norwegian Trench.

Current and future risks

The chemical analysis of water samples, even those collected with the ROV close to the cargo holds of the wrecks, showed no traces of chemical agents such as tabun, mustard gas or its hydrolysis product, thiodiglycol, or arsenic. In many cases the wrecks were intact and in good condition, although some of the bombs discovered outside the wrecks were severely corroded. Corrosion is usually retarded at great depths because of the decreased temperature and the low oxygen content of the water. A general survey carried out by the Flødevigen

[21] 'Här ligger tusentals ton farlig stridsgas begravd [Here thousands of tons of dangerous war gas lie buried], *Sundsvalls tidning*, 28 Oct. 1990.

[22] '48 000 tonn giftgass på bunnen utenfor Arendal: Tyskerne turde ikke selv bruke gassen' (note 13).

[23] Private communication with the author on investigation by Norwegian oil directorate, Sep. 1993.

[24] '48 000 tonn giftgass skal senkes i Skagerakk' (note 10).

biological station showed that, on average, there were six ml of oxygen per litre of sea water at a temperature of 5°C at this depth (seabed).[25]

Bombs are usually constructed with double walls of thin metal which are 2.5 mm or 4.0 mm thick. By the end of World War II the standards for military material had dropped, and some shells may therefore have had only one wall. Photographs of the 200–250 kg bombs on the seabed show that some are corroded, which may have resulted in the leaking out of their contents.

Tabun

The nerve agent tabun is a liquid whose density is close to that of water (1.077 grams/cm^3 at 20°C) and which is soluble in water (12 per cent at 20°C). It readily mixes with and is diluted by sea water. In water it decomposes with a half-life of 12–24 hours into cyanide and non-toxic components. In sea water it can be toxic to marine animals at a concentration of 100 mg/cubic metre (m^3). Tabun is more poisonous than cyanide; this means that when the agent decomposes it is degraded to a non-toxic concentration.

To illustrate this, the example can be taken of a bomb containing 100 kg of tabun which corrodes and from which tabun leaks out through a small hole over a 24-hour period, which would be a rapid rate of leakage for a small hole. If the sea stream was 10 cm/second and the tabun flowed out in a 5°-wide stream, after 24 hours the tabun would be diluted over 1 km and would cover an area of 50 metres x 50 metres at the widest point (i.e., at the point farthest from the bomb).[26] The concentration of agent 1 km from the bomb would be 50 mg/m^3, just reaching a non-toxic level. Parts of the stream would contain a toxic concentration of tabun which could affect animals that live on the seabed, although fish and other fast-moving species would be able to move away from the toxic section.

Technical mustard

Technical mustard[27] has a melting point of 5–10°C and is much heavier than water (1.27 grams/cm^3), in which it is barely soluble (0.3 grams/litre). The small amount which is soluble in water decomposes rapidly to non-toxic compounds. On the bottom of the sea mustard becomes waxlike and is only slowly dissolved. Some of the German mustard gas from World War II was thickened with polymers which form a solid crust protecting the mustard from further degradation. It can therefore be expected that mustard-filled munitions will remain in an active state on the seabed for a long time. Marine animals which are contaminated by mustard will presumably develop skin lesions similar to those developed by humans, but the fish can still be consumed safely. Experiments were carried out in which eels and flounders were placed in beakers with 10 ppm (10 grams/m^3) of mustard.[28] The eels died, but the flounders survived.

[25] Dahl, E., Danielssen, D. S. and Hognestad, P. T., Fiskeridirektorates havforskningsinstitutt, Statens biologiske stasjon Flødevigen, *Flødevigen meldinger* [Flødevigen announcements], no. 5 (1987).

[26] van Weering, T. C. E., Berger, G. W. and Kalf, J., *Netherlands Journal of Sea Research*, no. 21 (1987), pp. 177–89.

[27] Technical mustard is the name used for the product obtained in large-scale production. It is not pure, but this has few or no consequences for its military use.

[28] Bundesforschungsanstalt für Fischerei, Hamburg, *Veröffentlichungen des Instituts Für Küsten- und Binnenfischerei* [Publications of the Institute of Coastal and Domestic Fishing], no. 60 (1976).

There have been reports that sea currents could move the mustard gas to another location, but most experts no longer find this credible.[29]

Phosgene and arsenicals

The chemical agent phosgene dates from World War I. It is a gas which rapidly decomposes in water to carbon monoxide and hydrochloric acid. Adamsite and other arsenic-containing agents decompose to arsenic trioxide and arsenic-containing organic compounds. Some of these are poisonous to marine animals at a concentration of 0.2 mg/m^3. Clark I and II, which are phenylarsenicals, may also accumulate in some marine organisms and thus create environmental problems. It is not known what quantity of these agents is present in the sea, and the current level of arsenic is also unknown.

Precautions and future action

It would be hazardous to try to recover the ships or munitions today. The wrecks could break up, releasing all their contents at once and creating a major disaster. If the munitions are left undisturbed in the scuttled ships the chemicals will be released slowly, probably over several years. The environmental risk in such a case is considered low for phosgene and tabun, mainly because of their high solubility and instability in water. Although tabun is very poisonous and lethal to marine animals, it does not, as shown above, affect a large area for a long time. Arsenic-containing agents, on the other hand, will remain stable, and particularly the phenylarsenic compounds may accumulate. The hazards presented by these compounds depend on the amount that was dumped.

Technical mustard is not very soluble, and it will remain dangerous for decades. It is impossible to evaluate the effect this may have on the environment over such a long period. Further investigation is needed, and regular surveillance and sampling should be carried out to detect any major release from the scuttled ships. However, the munitions will not corrode at the same rate, and a sudden disaster is therefore unlikely. Experts continue to debate whether it will be necessary to destroy the agents in the future.

Underwater technology is rapidly improving, and it may be less expensive to deal with the problem in the future. It is already possible to encase gas pipes and oil pipes in the North Sea, and the scuttled ships and their toxic contents might also be similarly covered with a mixture that would degrade the agents. This would inhibit both corrosion and leakage into the sea but would cover the wrecks and prevent surveillance of them. Although some of the munitions are corroded and their contents therefore have leaked out, the general impression formed by the NDRETOX investigation was that the ships and their cargo are intact and that several years remain before a decision need be made about their future. Fishing in the area is prohibited to prevent accidents involving fishermen, not because their catch would be contaminated by toxic agents and unfit for consumption.

[29] Poland, National report on war gases and ammunition dumped in the Polish Economic Zone of the Baltic Sea, Paper presented at Baltic Marine Environment Protection Commission, Helsinki Commission, Ad Hoc Working Group on Dumped Chemical Munition (HELCOM CHEMU), Vilnius, Lithuania, HELCOM CHEMU 2/2/4, 17 Sep. 1993.

Part VII. Legal issues

21. The Chemical Weapons Convention and old and abandoned chemical weapons

THOMAS STOCK

I. Introduction

There is general agreement that the destruction of chemical weapons owned or possessed by a State Party to the Chemical Weapons Convention after 1 January 1946 will be a major focus of the Convention in the first 10 years after its entry into force. Destruction must be completed within 10 years. (In exceptional cases a 5-year extension may be granted.) In Europe only Russia, the inheritor of the former Soviet CW stockpile, faces the challenge of destroying chemical warfare agents, of which it possesses at least 40 000 tonnes.

The CWC allows variation in the application of verification and destruction measures to the chemical weapons which a state owns or possesses based on the date of their production. There is a time schedule for the destruction of any chemical weapons which a state has owned or possessed (essentially, those produced by it) since 1 January 1946, and such activities are subject to full and systematic verification.[1] Chemical weapons produced before 1 January 1946 may be subject to a less stringent verification and destruction regime.

Chemical weapons produced before 1925 and those produced between 1925 and 1946 and that are no longer usable as CW are classified as 'old chemical weapons' (OCW). Chemical weapons which were left on the territory of a state without its consent after 1 January 1925 are classified as 'abandoned chemical weapons' (ACW). There are four categories of OCW: chemical weapons abandoned on the territory of another state, chemical weapons dumped at sea, buried chemical weapons and chemical weapons which have been recovered.

Although OCW and ACW might pose less of a risk to the CWC than do chemical weapons produced after 1 January 1946, there are special technical and legal questions associated with their destruction. Chemical weapons which are old and no longer usable for military purposes (e.g., those recovered from former battlefields or weapons accidentally netted by fishermen) continue to pose a risk to people and the environment. A party to the CWC could unwittingly declare that it has no old chemical weapons, when in fact it does. Such weapons could have been stockpiled in World War I or World War II or abandoned under unknown circumstances. Finally, although the shell of a CW munition might be unusable, the chemical fill could still be active. Agreement has not been reached about how to define the risk such weapons pose to the CWC or the type of destruction regime which ought to be applied to them.

[1] This is addressed in paragraphs 2 and 3 of Article I, paragraph 1 (a) (i) and (v) of Article III and paragraph 6 of Article IV, Conference on Disarmament document CD/1170, 26 Aug. 1992, appendix, pp. 8, 13, 16; and in Part IV (A), Destruction of Chemical Weapons and its verification pursuant to Article IV, of the Verification Annex, Conference on Disarmament document CD/1170, 26 Aug. 1992, appendix, annex 2, pp. 88–104.

II. The CWC provisions on old and abandoned chemical weapons

The problem of old and abandoned chemical weapons has long been a controversial issue. In the negotiation of the CWC there were various views about how the declaration and destruction requirements for these weapons should differ from those for CW produced after 1 January 1946. (The requirements are outlined in Articles III and IV and Part IV (A) of the Verification Annex of the CWC.)

Responsibility for the destruction of abandoned chemical weapons was a primary concern during the CWC negotiations at the Conference on Disarmament in Geneva. Some members of the Group of 21, an informal organization of neutral and non-aligned states that no longer exists, expressed apprehension that chemical weapons which had been abandoned or lost by former colonial powers might be discovered on their territories. Some states argued that the original possessor should be responsible for destroying the weapons and that the country on whose territory the abandoned chemical weapons were found should not be forced to pay for their destruction. Other countries questioned the extent to which the state that had abandoned the weapons should be held responsible for them. The possible lack of technical know-how and resources to destroy the abandoned chemical weapons was also discussed.

The discussion was influenced by a dispute between China and Japan over Japan's abandonment of CW in China. In February 1992 China submitted a document to the CD detailing its long-standing dispute with Japan about Japanese chemical weapons abandoned on Chinese territory.[2] The document listed 18 dumping sites. Approximately 2 million chemical munitions and 120 tonnes of bulk CW agents have been discovered, including mustard gas, mustard gas and lewisite mixtures, diphenylcyanoarsine, hydrogen cyanide, phosgene and chloroacetophenone. Japan responded that it was making efforts to resolve the issue,[3] and the Japanese Government announced just before the January 1993 CWC signing conference that a team from Japan would be sent to China to make an inventory of the abandoned weapons.[4] Joint Chinese–Japanese fact-finding missions visited ACW sites in China in 1995, and Japan also officially admitted that it had left chemical weapons in China. Chinese newspapers estimated the cost of recovery and destruction at more than $1 billion.[5]

There was debate about the definition in the CWC of 'jurisdiction and control', and the problem of old and abandoned chemical weapons is of relevance in that context. Questions were raised as to whether it would be possible to clearly define responsibility for ACW or to determine the degree of financial and technical support which the abandoning state should provide to the state on whose territory the weapons had been abandoned. Chemical weapons which were dumped at sea, particularly in the Baltic Sea and the North Sea, are another aspect of the problem. Should the countries which dumped CW at sea after World War II (e.g., Russia , the UK and the USA) be forced to retrieve

[2] Conference on Disarmament document CD/1127, 18 Feb. 1992.

[3] Conference on Disarmament document CD/PV. 614, 27 Feb. 1992, pp. 18–19.

[4] 'Tokyo to assess chemical weapons left in PRC', in Foreign Broadcast Information Service, *Daily Report–East Asia* (*FBIS-EAS*), FBIS-EAS-93-006, 11 Jan. 1993, pp. 7–8.

[5] Beal, T., 'China wants disposal of Japanese bombs', *UPI via Nexis*, 17 Aug. 1995.

these weapons, and to what extent is Germany responsible for the dumped chemical weapons which once belonged to it?

The CWC addresses this problem by: *(a)* defining old and abandoned chemical weapons, *(b)* outlining verification and destruction procedures for them, *(c)* establishing guidelines for declaration and provisions to deal with old and abandoned CW which are discovered after entry into force of the CWC, and *(d)* assigning legal responsibility for old and abandoned chemical weapons, including a cooperative mechanism for sharing destruction costs and verification.

The CWC framework for dealing with OCW and ACW

The final version of the CWC is a compromise designed to satisfy the concerns of parties to it. It does not treat old and abandoned chemical weapons in the same manner as chemical weapons produced after 1 January 1946. In addition, the CWC recognizes that a party to it could possess OCW or ACW and not necessarily be a possessor of chemical weapons. It also differentiates between the destruction responsibilities for old and abandoned chemical weapons.

The CWC provides a general framework for dealing with old and abandoned chemical weapons and procedures for dealing with subsequent discoveries of them. Those provisions are contained in: Article I, General Obligations; Article II, Definitions and Criteria; Article III, Declarations; Article IV, Chemical Weapons; and Part IV(B) of the Annex on Implementation and Verification. These obligations are outlined in table 21.1.

Paragraph 2 of Article I defines the general obligation of a State Party which: *(a)* possesses old or abandoned chemical weapons on its territory; *(b)* has abandoned chemical weapons on the territory of another state without the consent of that state at any time since 1 January 1925; or *(c)* has owned or possessed CW after 1 January 1946 as follows: 'Each State Party undertakes to destroy chemical weapons it owns or possesses, or that are located in any place under its jurisdiction or control, in accordance with the provisions of this Convention'. Paragraph 3 completes the definition: 'Each State Party undertakes to destroy all chemical weapons it abandoned on the territory of another State Party, in accordance with the provisions of this Convention'.[6]

Paragraphs 5 and 6 of Article II contain definitions that are of relevance for parties which possess old or abandoned chemical weapons. Paragraph 5 defines OCW as: '(a) Chemical weapons which were produced before 1925; or *(b)* Chemical weapons produced in the period between 1925 and 1946 that have deteriorated to such an extent that they can no longer be used as chemical weapons'. Paragraph 6 defines ACW as 'Chemical weapons, including old chemical weapons, abandoned by a State after 1 January 1925 on the territory of another State without the consent of the latter'.[7]

[6] All chemical weapons produced and/or abandoned since 1 Jan. 1946 must be treated as 'normal' chemical weapons, declared according to the declaration provisions of Articles III and IV, and destroyed according to provisions in Article IV. The requirements for verification of these declaration and destruction obligations are in Part IV (A) of the Verification Annex.

[7] The date 1925 was chosen because the Geneva Protocol (Protocol for the prohibition of the use in war of asphyxiating, poisonous or other gases, and of bacteriological methods of warfare) was signed then. After 1946 all chemical weapons are treated as normal and are covered by other parts of the CWC.

Table 21.1. Obligations and definitions related to old and abandoned chemical weapons for parties to the Chemical Weapons Convention

Article of the CWC	Obligation or definition
Article I, General Obligations	General obligation to destroy chemical weapons (paragraph 2); obligation to destroy chemical weapons abandoned on the territory of another party (paragraph 3)
Article II, Definitions and Criteria	Old chemical weapons are those produced before 1925 or between 1925 and 1946 and determined to be 'unusable'; abandoned chemical weapons are chemical weapons, including old chemical weapons, which have been abandoned by a state after 1 January 1925 on the territory of another party without its consent
Article III, Declarations	After entry into force of the CWC for the party the following information must be submitted within 30 days: declaration of whether it has on its territory old or abandoned chemical weapons or has abandoned chemical weapons on the territory of other states, with all available, relevant information in accordance with the Verification Annex of the CWC
Article IV, Chemical Weapons	General obligation to destroy chemical weapons, with the exception that 'The provisions of this Article and the relevant provisions of Part IV of the Verification Annex shall not, at the discretion of a State Party, apply to chemical weapons buried on its territory before 1 January 1977 and which remain buried, or which had been dumped at sea before 1 January 1985'.

Source: Conference on Disarmament document CD/1170, 26 Aug. 1992.

The question of 'usability' was not addressed at length in the CWC and therefore the Preparatory Commission (PrepCom) for the Organisation for the Prohibition of Chemical Weapons assumed responsibility for its further elaboration. It is not clear how the definition of abandoned chemical weapons will affect states which are not signatories to the CWC.

Paragraph 1 (b) (i) of Article III of the CWC requires that parties provide the following data on old chemical weapons: 'Declare whether it has on its territory old chemical weapons and provide all available information in accordance with Part IV (B), paragraph 3, of the Verification Annex'. With respect to abandoned chemical weapons, under paragraphs 1 (b) (ii) and (iii), a party must: 'Declare whether there are abandoned chemical weapons on its territory and provide all available, information in accordance with Part IV (B), paragraph 8, of the Verification Annex' and 'Declare whether it has abandoned chemical weapons on the territory of other States and provide all available information in accordance with Part IV (B), paragraph 10, of the Verification Annex'.

If a party to the CWC has abandoned chemical weapons on its territory that were left there by another state the Technical Secretariat should receive two declarations of ACW. However, this may not occur since the state which abandoned the weapons may not be a party to the CWC. All initial declaration requirements must be completed not later than 30 days after the CWC enters into force for the State Party.

III. Chemical weapons buried or dumped at sea

The CWC negotiators faced the problem of reconciling the wish of some states not to be held responsible for past dumping and burial activities, some of which occurred more than 50 years ago, with provisions of the CWC related to such activities. The solution was a cut-off date and the following text in Articles III and IV of the CWC: 'The provisions of this Article and the relevant provisions of Part IV of the Verification Annex shall not, at the discretion of a State Party, apply to chemical weapons buried on its territory before 1 January 1977 and which remain buried, or which had been dumped at sea before 1 January 1985'.

Consequently, chemical weapons which were buried on the territory of a state must be declared only if they were buried after 1 January 1977, and CW that were dumped at sea must be declared only if the dumping took place after 1 January 1985. The destruction requirements of Article IV do not apply to chemical weapons which were buried or dumped before these dates. More importantly, parties to the CWC which dumped chemical weapons at sea before 1 January 1985 or which buried them before 1 January 1977 do not need to retrieve and destroy the weapons.[8]

IV. Provisions of the CWC related to former CW production facilities

Under Article III each party to the CWC is obliged to declare 'whether it has or has had any chemical weapons production facility under its ownership or possession, or that is or has been located in any place under its jurisdiction or control at any time since 1 January 1946'.[9] The procedures for closing, destroying or converting such facilities are outlined in Article V and Part V of the Verification Annex. There are no declaration requirements for production sites which operated before 1 January 1946. The CWC does not address environmental damage or long-term contamination of soil and water resulting from the operation of CW production sites or damage caused by improper disassembly operations before 1 January 1946.[10] Such contamination must be dealt with by national legislation on toxic waste disposal.[11] The CWC does specify how CW production sites built after 1 January 1946 are to be closed.[12]

[8] The 1977 date is related to a CW burial programme which was conducted by Austria in 1949–50. Old German World War II CW munitions were collected, sorted and 'temporarily' buried. Between 1974 and 1976 more than 28 000 grenades and rocket projectiles were placed underground for 'long-term' storage. The programme is described in Old chemical weapons: description of a long-term storage facility under safe conditions, Conference on Disarmament document CD/CW/WP. 397, 5 May 1992.

[9] Paragraph 1 (c) (i) of Article III, Conference on Disarmament document CD/1170 (note 1), pp. 13–15.

[10] If these cases were covered by the CWC damage from former World War II CW production sites on the territory of another state (e.g., the Third Reich CW production site at Dyhernfurth, Poland) would have legal implications.

[11] Stock, T., 'Chemische Kampfstoffaltlasten unter der zukünftigen Chemiewaffenkonvention: Probleme der Vernichtung und Kontrolle' [Old chemical armament wastes under the future Chemical Weapons Convention: problems of destruction and control], Spyra, W. *et al.*, *Verfahren zur Sanierung von Rüstungsaltlasten* [Clean-up methods for old armament wastes] (EF Verlag für Energie und Umwelttechnik: Berlin, 1992), pp. 131–149; Stock, T., 'Internationale Mechanismen zur Beseitigung von militärisch-chemischen Rüstungsaltlasten unter der zukünftigen Chemiewaffenkonvention' [International mechanism for inspection of old chemical armament wastes under the future Chemical Weapons Convention], Thomé-Kozmiensky, K. J. *et al.*, *Management zur Sanierung von Rüstungsaltlasten* [Managing the clean-up of old armament wastes] (EF Verlag für Energie und Umwelttechnik: Berlin, 1992), pp. 91–104.

[12] Conference on Disarmament document CD/1170 (note 1), appendix, pp. 19–21.

V. Provisions for declaration, identification, verification and destruction of old and abandoned chemical weapons

Old chemical weapons

Part IV (B) of the Verification Annex to the CWC provides general destruction guidelines for both old and abandoned chemical weapons.[13] The time schedules for declarations distinguish between old chemical weapons which are known to exist at the time the CWC enters into force and OCW that are discovered later. The following conditions apply to old chemical weapons:

1. If the old chemical weapons were produced before 1925 a party to the CWC is obliged to submit all available relevant information 'including, to the extent possible, the location, type, quantity and the present condition'[14] to the Technical Secretariat not later than 30 days after the CWC enters into force for that state.
2. Each party to the CWC must submit a declaration to the Technical Secretariat within 30 days after entry into force of the CWC about CW produced between 1925 and 1946 that the party determines to be 'unusable' (and thus categorized as OCW). The declaration must acknowledge the presence of the OCW and provide all available information about them.[15]
3. If old chemical weapons are discovered by a party to the CWC after its entry into force the party is obliged to submit the required information about them to the Technical Secretariat not later than 180 days after the discovery.[16]

The Technical Secretariat is required to conduct an initial inspection to verify that the data which it has received are correct and to confirm that the weapons meet the definition of old chemical weapons. If the weapons are confirmed to have been produced before 1925 they are treated as toxic waste. The state which possesses the OCW is required to inform the OPCW of the steps it will take to destroy or dispose of the toxic waste in accordance with its national legislation.[17] However, if it is ascertained that the weapons were produced between 1925 and 1946 and that they have deteriorated so that they can no longer be used as chemical weapons[18] they are to be destroyed in accordance with the guidelines in Part IV(A) of the Verification Annex.

The time-limits for destruction and the level of verification of destruction to be applied can be modified at the request of a State Party if the Executive Council consents.[19] It has still to be determined how to tailor the level of verification necessary to the threat that the old chemical weapons pose to the CWC.

[13] Conference on Disarmament document CD/1170 (note 1), pp. 105–107.

[14] Paragraph 3, Part IV (B) of the Verification Annex, Conference on Disarmament document CD/1170 (note 1), appendix, annex 2, p. 105.

[15] Conference on Disarmament document CD/1170 (note 1), pp. 88–90.

[16] Paragraph 9, Part IV (B) of the Verification Annex, Conference on Disarmament document CD/1170 (note 1), appendix, annex 2, p. 106.

[17] Paragraph 6, Part IV (B) of the Verification Annex, Conference on Disarmament document CD/1170 (note 1), appendix, annex 2, p. 105.

[18] The Preparatory Commission was required to develop: 'Guidelines to determine the usage of chemical weapons produced between 1925 and 1946, in accordance with Part IV (B), paragraph 5, of the Verification Annex', Conference on Disarmament document CD/1170 (note 1), appendix, annex 2, p. 105.

[19] Paragraph 17, Part IV (B) of the Verification Annex, Conference on Disarmament document CD/1170 (note 1), appendix, annex 2, p. 107.

The destruction of OCW produced before 1925

A state which is a party to the CWC must destroy or dispose of any old chemical weapons produced before 1925 which it possesses. The State Party is required to inform the Technical Secretariat by submitting: (*a*) an annual report describing the implementation of plans for destruction of OCW,[20] (*b*) information about any destruction facilities, and (*c*) the steps being taken to destroy or dispose of chemical weapons classified as 'toxic waste'.[21]

The destruction of OCW produced between 1925 and 1946

Paragraph 7 of Part IV (B) of the CWC Verification Annex states 'a State Party shall destroy old chemical weapons that have been confirmed by the Technical Secretariat as meeting the definition in Article II, paragraph 5 (b),[22] in accordance with Article IV and Part IV (A) of this Annex'. A party to the CWC may request that the Executive Council allow it to change the time-limit for destruction or the order of destruction. Part IV (A) of the Verification Annex addresses the destruction and verification of chemical weapons produced after 1 January 1946. The same procedures apply to chemical weapons produced between 1925 and 1946 which have been determined to be usable.

The situation will differ for each country which possesses OCW produced between 1925 and 1946. Some countries will not have a CW destruction facility, and they could instead use destruction facilities for highly toxic wastes. The verification regime would then be applied to such facilities.

Old chemical weapons may be discovered after the initial declaration by a party to the CWC. If a decision is taken that it is safest to destroy the chemical weapons where they have been found, a report must be submitted in accordance with paragraph 4 of Part IV (B) of the Verification Annex of the CWC in the state's biannual declaration.[23]

Abandoned chemical weapons

The CWC addresses the problem of abandoned chemical weapons by distinguishing between the requirements which must be met by a State Party on whose territory chemical weapons have been abandoned (the territorial State

[20] Paragraph 36, Part IV (A) of the Verification Annex, Conference on Disarmament document CD/1170 (note 1), appendix, annex 2, p. 98.

[21] Paragraph 6, Part IV (B) of the Verification Annex. Information on national legislation on the destruction and location of such facilities may also be provided. Conference on Disarmament document CD/1170 (note 1), appendix, annex 2, p. 105.

[22] 'Old chemical weapons' are: 'Chemical weapons produced in the period between 1925 and 1946 that have deteriorated to such an extent that they can no longer be used as chemical weapons'. Paragraph 5 (b), Article II, Conference on Disarmament document CD/1170 (note 1), appendix, p. 10.

[23] 'A State Party which discovers old chemical weapons after this Convention enters into force for it shall submit to the Technical Secretariat the information specified in paragraph 3 not later than 180 days after the discovery of the old chemical weapons'. Paragraph 3 , Part IV (B) states: 'A State Party which has on its territory old chemical weapons as defined in Article II, paragraph 5 (a), shall, not later than 30 days after this Convention enters into force for it, submit to the Technical Secretariat all available relevant information, including, to the extent possible, the location, type, quantity and the present condition of these old chemical weapons'. Conference on Disarmament document CD/1170 (note 1), appendix, annex 2, p. 105.

Party) and one which has abandoned CW on the territory of another state without the consent of the latter (the abandoning State Party).

The territorial State Party is obliged to 'submit to the Technical Secretariat all available relevant information' including 'to the extent possible, the location, type, quantity and the present condition of the abandoned chemical weapons as well as information on the abandonment' not later than 30 days after entry into force of the CWC.[24] If abandoned chemical weapons are discovered after the CWC enters into force for a party to it, this information must be submitted to the Technical Secretariat not later than 180 days after their discovery. The abandoning State Party is obliged to 'provide all necessary financial, technical, expert, facility as well as other resources'[25] and to submit 'all available, relevant information concerning the abandoned chemical weapons' within 30 days after the CWC enters into force for it to the Technical Secretariat.[26]

The Technical Secretariat must conduct an initial inspection to verify the information, determine the level of verification required and identify the abandoning state, if necessary. The CWC gives the territorial State Party the right to request that the abandoning State Party enter into consultation with it for the purpose of destroying the ACW in a cooperative effort. Such consultation should begin not later than 30 days after the request has been made. A mutually agreed plan for destruction must be submitted to the Technical Secretariat not later than 180 days after consultation starts. The abandoning State Party is obliged to provide 'all necessary financial, technical, expert, facility as well as other resources' to destroy the abandoned chemical weapons.[27] The territorial State Party may request that the OPCW and other parties to the CWC provide assistance in the destruction of the ACW.[28] (It may not be possible to identify the abandoning state or the state may not be a party to the CWC.)

The provisions for destruction of chemical weapons produced after 1 January 1946 also apply to abandoned chemical weapons. If the ACW were produced between 1925 and 1946 and are unusable the Executive Council may modify or, in exceptional cases, suspend the application of the provisions for destruction on request of the territorial State Party. If the ACW cannot be defined as old chemical weapons produced between 1925 and 1946 and as unusable, the territorial State Party may ask the Executive Council to modify the provisions governing the time-limit and the order of destruction of ACW. The territorial and the abandoning States Parties may conclude agreements or arrangements for destruction of the ACW between themselves. The Executive Council has the authority to decide whether such agreements complement the CWC's provisions for destruction of ACW.

[24] Paragraph 8, Part IV (B) of the Verification Annex, Conference on Disarmament document CD/1170 (note 1), appendix, annex 2, p. 106.

[25] Paragraph 15, Part IV (B) of the Verification Annex, Conference on Disarmament document CD/1170 (note 1), appendix, annex 2, p. 107.

[26] Paragraph 10, Part IV (B) of the Verification Annex, Conference on Disarmament document CD/1170 (note 1), appendix, annex 2, p. 106.

[27] Paragraph 15, Verification Annex IV (B), Conference on Disarmament document CD/1170 (note 1), appendix, annex 2, p. 107.

[28] Paragraph 16, Verification Annex IV (B), Conference on Disarmament document CD/1170 (note 1), appendix, annex 2, p. 107.

The destruction of abandoned chemical weapons

Destruction of abandoned chemical weapons must be carried out in accordance with the provisions of Part IV (B) of the Verification Annex. The territorial State Party and the abandoning State Party must plan the destruction of the ACW and take into account the risks posed to the CWC. The Executive Council must approve the plan and agree to any proposed suspension of the provisions for verification. Flexibility about the level of verification that is required may be possible. Destruction is to be carried out in accordance with the principles and methods for destruction specified under Article IV[29] and Part IV (A) of the Verification Annex.[30] The Executive Council must approve requests for extension of destruction deadlines. Annual reports from both the territorial and abandoning States Parties on the progress of destruction must be submitted to the Technical Secretariat.

VI. The function of the Preparatory Commission

The Text on the Establishment of the Preparatory Commission, which was adopted by the international conference that met in Paris in January 1993 to sign the CWC, gave the PrepCom the task of developing 'guidelines to determine the usability of chemical weapons produced between 1925 and 1946, in accordance with Part IV (B), paragraph 5, of the Verification Annex'. The PrepCom in turn established an Expert Group on Old and Abandoned Chemical Weapons, which first met in November 1993.[31]

The Expert Group focused on: *(a)* developing declaration formats for old chemical weapons produced before 1925, for OCW produced between 1925 and 1946 and for abandoned chemical weapons; *(b)* identifying criteria to determine the usability of chemical weapons produced between 1925 and 1946; *(c)* elaborating the verification regime for OCW produced between 1925 and 1946; *(d)* outlining inspection-equipment requirements and *(e)* addressing the health and safety aspects of inspections involving OCW and ACW.

In developing criteria and guidelines for determining usability a distinction was made between the body of a munition and its chemical fill. The Expert Group had to agree on usability guidelines that were based on the definition of

[29] 'Each State Party, during transportation, sampling, storage and destruction of chemical weapons, shall assign the highest priority to ensuring the safety of people and to protecting the environment. Each State Party shall transport, sample, store and destroy chemical weapons in accordance with its national standards for safety and emissions'. Paragraph 10, Article IV, Conference on Disarmament document CD/1170 (note 1), appendix, p. 17.

[30] Paragraphs 12–14 of Section C, 'Principles and methods for destruction of chemical weapons', of Part IV (A) of the Verification Annex state: Paragraph 12: '"Destruction of chemical weapons" means a process by which chemicals are converted in an essentially irreversible way to a form unsuitable for production of chemical weapons, and which in an irreversible manner renders munitions and other devices unusable as such'. Paragraph 13: 'Each State Party shall determine how it shall destroy chemical weapons, except that the following processes may not be used: dumping in any body of water, land burial or open-pit burning. It shall destroy chemical weapons only at specifically designated and appropriately designed and equipped facilities'. Paragraph 14: 'Each State Party shall ensure that its chemical weapons destruction facilities are constructed and operated in a manner to ensure the destruction of the chemical weapons; and that the destruction process can be verified under the provisions of this Convention'. Conference on Disarmament document CD/1170 (note 1), appendix, annex 2, p. 92.

[31] Expert Group on Old and Abandoned Chemical Weapons, First report, PrepCom document PC-V/B/WP.14, 26 Nov. 1993.

chemical weapons in Article II[32] and which took into account the nature of the toxic chemicals, precursors, munitions, submunitions and devices.

In 1993 the Expert Group requested that the Executive Secretary of the Provisional Technical Secretariat assess the implications of two possible interpretations of the envisaged verification regime for OCW produced between 1925 and 1946.[33] The first interpretation would require the full application of destruction and verification measures as outlined under Part IV (A) of the Verification Annex for CW. These provisions apply to CW produced after 1 January 1946 and may include the application of systematic verification and continuous, on-site monitoring by inspectors. The second interpretation would mean application of a more flexible verification regime that would be dependent on risk assessment and might not apply the time requirements for initial inspection in all instances.

Either interpretation could have major budgetary impact by increasing the number of inspections and, consequently, inspectors. In addition, it is essential to ascertain the number of cases which might be covered under the regime for old chemical weapons. There is still inadequate data on the amount of OCW which could be declared in accordance with Article III after the CWC enters into force. The Executive Secretary therefore asked the signatory states to provide information on old chemical weapons.[34]

The evaluation by the Executive Secretary estimated that an additional $14 million would need to be added to the OPCW budget for the first year after entry into force if the first approach (application of both the destruction obligations and the verification provisions as outlined under Part IV (A) of the Verification Annex) were fully applied.[35] Additional inspectors would be needed to conduct initial inspections, negotiate facility agreements and carry out continuous, on-site monitoring of destruction of the old chemical weapons.

The Expert Group on Old and Abandoned Chemical Weapons developed declaration formats for both the territorial and the abandoning State Party and verification procedures for ACW based largely on Article II and Part IV (B) of the Verification Annex. The Expert Group suggested that if the abandoned chemical weapons met the definition of old chemical weapons produced between 1925 and 1946 (i. e., not able to be used) the level of verification of their destruction should be correlated to the risk they pose to the CWC. Those that pose the greatest risk would be subject to more verification, which would result in higher costs. The Expert Group continued to work on developing usability guidelines for OCW produced between 1925 and 1946 and the regimes for verification of both OCW and ACW, but consensus was not

[32] Chemical weapons include the following, together or separately: '(a) Toxic chemicals and their precursors, except where intended for purposes not prohibited under this Convention, as long as the types and quantities are consistent with such purposes; (b) Munitions and devices, specifically designed to cause death or other harm through the toxic properties of those toxic chemicals specified in subparagraph (a), which would be released as a result of the employment of such munitions and devices; (c) Any equipment specifically designed for use directly in connection with the employment of munitions and devices specified in subparagraph (b)', Conference on Disarmament document CD/1170 (note 1), appendix, p. 9.

[33] Expert group on old and abandoned chemical weapons: first report, PrepCom document PC-V/B/WP.14, 26 Nov. 1993.

[34] Note by the Executive Secretary: request for information on old chemical weapons, PrepCom document PC-V/B/6, 7 Dec. 1993.

[35] Note by the Executive Secretary, Verification regime for old chemical weapons produced between 1925 and 1946, PrepCom document PC-V/B/9, 11 Dec. 1993.

reached. The limited progress of the Expert Group was mainly attributable to political, not technical, differences of opinion.

Guidelines for 'usability'

The Expert Group had to determine usability based on the definitions of chemical weapons and old chemical weapons in Article II of the CWC. Chemical weapons are defined as:

(a) Toxic chemicals and their precursors, except where intended for purposes not prohibited under this Convention, as long as the types and quantities are consistent with such purposes;
(b) Munitions and devices, specifically designed to cause death or harm through the toxic properties of those toxic chemicals specified in subparagraph (a), which would be released as a result of the employment of such munitions and devices; [and]
(c) Any equipment specifically designed for use directly in connection with the employment of munitions and devices specified in subparagraph (b).

An item is a chemical weapons if it falls under at least one of these three subparagraphs.

The definition of chemical weapons provoked debate in the Expert Group on how to assess the level of deterioration of the chemical fills and the bodies of the munitions and how to determine whether a weapon is usable if only the fill and not the body of the munition is functional, or vice versa. How to ascertain whether or not equipment is functional has not been adequately explored, but if the usability of a component is unknown or cannot be demonstrated, it should be assumed that the component can be used.[36] The Expert Group therefore suggested that assessment of the usability of chemical weapons produced between 1925 and 1946 should be based on the following assumptions: *(a)* that old chemical weapons are chemical weapons produced between 1925 and 1946, as determined by the Technical Secretariat on the basis of a usability assessment; *(b)* that usability is established by assessing the components of a chemical weapon; *(c)* that the configuration and composition of a stockpile of old chemical weapons determine its usability; and *(d)* that assessment of usability is independent of such factors as location, quantity, and the like.[37]

The assessment of the usability of the CW produced between 1925 and 1946 will serve to decide what inspection regime to apply. The Expert Group developed guidelines for the initial inspection of a site (e.g., OCW present, location, degree of deterioration, period of production and usability). The Technical Secretariat will also evaluate the level of risk. The assessment of deterioration should evaluate the properties of the toxic chemicals in the OCW, their degree of deterioration and the extent of deterioration of the munitions and equipment.

In confirming that the OCW were produced between 1925 and 1946, the following must be done: *(a)* checking for recognizable labels, markings, design features, and the like; *(b)* reviewing production information and evidence provided by witnesses, as available; *(c)* cross-checking with data in the Technical

[36] However, the question of whether to provide examples of chemical weapons in the Declaration Handbook has been raised.

[37] Expert Group on Old and Abandoned Chemical Weapons: second report, PrepCom document PC-VI/B/WP.5, 4 Feb. 1994.

Secretariat's OCW database; and *(d)* assessing circumstantial evidence such as historical information (e.g., data on former testing activities or incidents of past CW use) and types of toxic chemicals present in the OCW.

The final usability assessment should be based on the declaration made by the state which possesses the chemical weapons and must be done by the Technical Secretariat. Final confirmation of the declaration is the responsibility of the OPCW, which assesses the total amount and types of OCW present, their operational characteristics, ease of extraction of the toxic agent, utility of any such extracts as CW and the hazards involved in the extraction process.

The Technical Secretariat assesses risk and decides the frequency, duration and intensity of any subsequent inspections in order to ensure that the old chemical weapons are destroyed in accordance with the destruction plan, that none of them have been removed except for destruction and that no process has been conducted or activities undertaken that would modify their status (e.g., draining the agent, storing it in bulk containers, and the like). If usable chemical weapons are present they are subject to full and systematic verification. Chemical weapons that were produced before 1 January 1946 and which have been determined to be unusable will vary in the degree of risk they pose to the CWC, and it has been suggested that a risk assessment be made of unusable CW that were produced between 1925 and 1946. The level of verification would vary with the level of risk, but these weapons would not be subject to full and systematic verification by the Technical Secretariat. All questions about the assessment of usability which remained open at the time of entry into force of the CWC have been transferred to the first Conference of States Parties for final consideration.

VII. The role of the individual State Party

A state which is either a territorial State Party or an abandoning State Party has obligations and rights related to the identification, declaration and destruction of old and abandoned chemical weapons. The active co-operation and participation of the parties to the CWC is essential to the effectiveness of the OPCW. Article VII of the CWC calls on each party to undertake national implementation measures 'to adopt the necessary measures to implement its obligation under this Convention'. The party also shall 'not permit in any place under its control any activity prohibited to a State Party under this Convention'.

Old chemical weapons produced before 1925 are to be treated as toxic waste, and a party to the CWC is obliged to provide information to the Technical Secretariat about their destruction and disposal. Paragraph 5 of Article VII contains the requirement that: 'Each State Party shall inform the Organization of the legislative and administrative measures taken to implement this Convention'. Each party must also 'assign the highest priority to ensuring the safety of people and to protecting the environment'.[38] In other words, the destruction or possible disposal of the toxic wastes (i.e., the chemical weapons produced before 1925) must be carried out in a manner which minimizes risk to the environment and to people. The Expert Group on Old and Abandoned Chemical Weapons recommended that States Parties should present information to the Technical

[38] Paragraph 10, Article IV, Conference on Disarmament document CD/1170 (note 1), appendix, p. 17.

Secretariat about the methods used for the destruction and disposal of OCW produced before 1925 and about relevant national legislation.[39]

Article VII also obliges each State Party to 'designate or establish a National Authority to serve as the national focal point for effective liaison with the Organization and other States Parties' and 'to afford the appropriate form of legal assistance'. The National Authorities will play an important role in resolving possible disputes related to abandoned chemical weapons.

A party to the CWC that possesses old or abandoned CW must accomplish the following: (*a*) submit the initial declarations (which must be completed 30 days after entry into force) based on the declaration formats; (*b*) prepare for the initial inspection and escort the inspectors to the site where old or abandoned chemical weapons are located; (*c*) destroy or dispose of the declared OCW if they meet the definition of toxic waste and inform the Technical Secretariat of this; (*d*) destroy old chemical weapons if they meet the definition for OCW (i.e., produced between 1925 and 1946) using an approved destruction plan which includes systematic verification by OPCW inspectors (according to Verification Annex, Part IV (A), paragraphs 41–43); (*e*) negotiate the destruction plan with the abandoning state if abandoned chemical weapons are present; (*f*) request assistance from the OPCW or other States Parties if the abandoning state cannot be identified; and (*g*) submit further declarations (180 days after discovery) if additional old or abandoned chemical weapons are discovered.

Some states have compiled lists of possible sites on their territory where old chemical weapons might be present or expected to be found. In the USA, for example, the US Army Chemical Materiel Destruction Agency was created in 1992 to address such issues. It has two programme managers, one for Chemical Demilitarization and one for Non-Stockpile Chemical Materiel (NSCM), and the latter is responsible for administering the Non-Stockpile Chemical Destruction Programme. In November 1993 the Army submitted to Congress its final survey and analysis report on the NSCM, as called for in the Defense Authorization Act for fiscal year 1993. The survey indicated that there may be 215 sites in 33 states where munitions have been buried.[40] The costs of destroying all the NSCM is estimated at $17.7 billion, and it will take approximately 40 years to complete the destruction.

VIII. Destruction technology

The discovery, dismantling and safe destruction of chemical weapons has been a significant problem since these weapons were first used in World War I. Until the end of the 1960s chemical agents and munitions were destroyed by methods now considered environmentally unsafe. These methods included open-pit burning, burial and dumping at sea, all of which are no longer in use owing to environmental and safety legislation. The CWC also prohibits such measures. Destruction poses various challenges including: (*a*) safe disassembly of

[39] Expert group on old and abandoned chemical weapons: first report, PrepCom document PC-V/B/WP.14, 26 Nov. 1993, p. 15.

[40] US Army Chemical Materiel Destruction Agency, *Non-Stockpile Chemical Materiel Program: Survey and Analysis Report* (US Army Chemical Materiel Destruction Agency: Aberdeen Proving Ground, Edgewood, Md., Nov. 1993).

Table 21.2. Technologies for the destruction of chemical agents

Type of technology	Method
Thermal detoxification	Pyrolysis, incineration at high temperature, plasma detoxification, 'glazing'
Chemical detoxification	Hydrolysis, catalytic decomposition, chemical oxidation, specific decontamination by absorption
Physical detoxification	Decomposition by radiation, absorption or detonation
Biological detoxification	Enzyme-based or metabolic detoxification
Dumping techniques	Stabilization in cement or organic material, underground disposal

Source: Compiled by the author.

explosives and propellants, (*b*) disposal of disassembled explosive components and propellants, (*c*) access to the chemical agent, and (*d*) disposal of the toxic agent, the bodies of the munitions and the by-product wastes generated by destruction.

The destruction technologies which can be used for old chemical weapons are well-known and already in use. (Table 21.2 summarizes some relevant destruction technologies.) High standards of health and environmental safety must be maintained when chemical agents and munitions are destroyed, and protecting the people involved must be given top priority. The capability to destroy chemical weapons, in general, and old chemical weapons, in particular, is limited.

Old chemical weapons are often found in small quantities and in various types of munitions. Traditional destruction techniques (e.g., incineration and chemical detoxification) have advantages, but so-called alternative destruction technologies[41] may provide valuable new approaches to OCW destruction.

Mobile destruction units combined with technology for on-site dismantling may be one way to deal with small quantities of OCW that are discovered at a number of locations. Chemical agent detoxification processes (neutralization or hydrolysis) would be optimal in such cases, although electrochemical oxidation technologies also have some advantages.[42]

IX. Conclusions

The problems related to the discovery of old and abandoned chemical weapons and their subsequent destruction will not be major in the first years after the CWC enters into force on 29 April 1997. Instead, the destruction of chemical weapons produced and stockpiled after 1 January 1946 will be the top priority

[41] Committee on Alternative Chemical Demilitarization Technologies, Board on Army Science and Technology, Commission on Engineering and Technical Systems, *Alternative Technologies for the Destruction of Chemical Agents & Munitions* (National Research Council: Washington, DC, 1993). The techniques discussed in the study are divided into 5 main groups: (*a*) low-temperature, low-pressure, liquid-phase detoxification (converting the agent to less toxic compounds); (*b*) low-temperature, low-pressure, liquid-phase oxidation processes, including biological oxidation (agent reacts with oxygen to form carbon dioxide, water and salts, in the event of complete oxidation, or mineralization); (*c*) moderate-temperature, high-pressure oxidation; (*d*) high-temperature, low-pressure pyrolysis (heat is used to destroy molecular bonds); and (*e*) high-temperature, low-pressure oxidation.

[42] One such technique, which originated from the treatment of nuclear industry waste, is based on an electrochemical process using nitric acid and silver at temperature of less than 100°C. Brown, P., 'Dounreay patents poisons destroyer', *The Guardian*, 30 Dec. 1992, p. 8.

and the focus of verification and compliance monitoring. The CWC is not a solution to the problems associated with the dumping of chemical weapons after World War II and related matters.

The future impact of old and abandoned chemical weapons on the budget of the OPCW and other verification activities could be enormous. There is a degree of uncertainty related to these weapons since states are not forced to declare information about them prior to entry into force of the CWC. In planning and preparing for entry into force the Provisional Technical Secretariat (PTS) was tasked to collect all available information to acquire as much data as possible.

There is also concern that old and abandoned chemical weapons may be present in some developing countries as a result of the military conflicts which took place there in the past 70 years. These countries could face technical difficulties if chemical weapons are discovered in the future, and the OPCW may need to provide advice and assistance to destroy such weapons.

Old and abandoned chemical weapons differ from stockpiled CW. Each such weapon is unique, and its destruction will cost more than that of a stockpiled chemical weapon. Old and abandoned CW may be leaking, unusable, recovered from former battlefields or unearthed with no evidence of their origin. Their handling, transport and destruction is difficult and stringent health and safety requirements must be met. A state which possesses old chemical weapons may not possess chemical weapons or have the technical capability to destroy chemical weapons. This problem is now being studied in several countries.

Judging from the information which some parties to the CWC have made available, the discovery of old or abandoned CW will remain an issue long after the CWC enters into force as discoveries continue to be made. It will probably never be possible to declare a state free of these weapons. However, old or abandoned chemical weapons rarely pose a threat to the CWC, and this is one reason why the regime for the declaration, verification and destruction of such weapons is different from that for CW produced after 1 January 1946.

Particular procedures apply to the declaration and destruction of old chemical weapons produced between 1925 and 1946. When the Technical Secretariat has determined that the OCW are abandoned chemical weapons measures will be taken to assist the abandoning State Party and the territorial State Party to work together to fulfil the requirements for destruction. If the abandoning State Party cannot be identified the territorial State Party may be able to obtain technical and logistical assistance from the Technical Secretariat to destroy the weapons.

The CWC framework for dealing with the destruction of old and abandoned chemical weapons includes procedures to identify the origin, status and date of production of such weapons and to evaluate their usability. These procedures are designed to ensure that the weapons are destroyed and that the destruction process is supervised and verified. Because some states which discover OCW will not have a destruction facility the States Parties to the CWC need to cooperate. Existing destruction capacities are limited, but the OPCW can help to coordinate common efforts. States will, however, remain individually responsible for all actions related to the discovery and destruction of OCW.

The problems related to old and abandoned chemical weapons are serious, and states should use the CWC as an opportunity to destroy these weapons, to develop co-operation and to exchange experience and technical know-how.

22. The legal problems associated with sites contaminated by chemical warfare agents in Germany

THOMAS KURZIDEM

I. Introduction

The general problem of contaminated sites and of sites contaminated by chemical warfare agents, in particular, has led to increased public interest and debate in recent years. After the reunification of Germany in 1990 and the subsequent reduction of military forces in Germany land was transferred from military to civil use, and it became apparent that there were environmental problems related to the former and current military use of land.

Contamination was caused by chemical weapons and related activities, such as CW production, stockpiling, use, disposal and destruction. For example, at Munster,[1] where troops were stationed, and at Spandau Citadel[2] major clean-up operations are under way or have been carried out. These and other similar projects have demonstrated that measures to clean up sites contaminated by warfare agents are extremely expensive, and in Germany this has created political tension between the federal government and the states (Länder). The central legal question relates to assigning responsibility and liability for the measures which need to be taken, simply put: who has to pay for what?[3]

[1] See chapter 12 in this volume.

[2] See chapter 5 in this volume.

[3] For an overview of the state of discussion see *inter alia* Becker, B., 'Zur intertemporalen Verwirklichung der Handlungsstörerhaftung bei Rüstungsaltlasten' [Establishing the interim liability of a person responsible in the case of sites contaminated by old armament wastes], *Deutsches Verwaltungsblatt*, vol. 106, no. 7 (1991), pp. 346–55; Becker, B., 'Zur öffentlich-rechtlichen Verantwortlichkeit für die Sanierung von Rüstungsaltlasten' [Responsibility under public law for the clean-up of sites contaminated by old armament wastes], ed. Hessisches Ministerium für Umwelt, Energie und Bundesangelegenheiten [Hessian Ministry for Environment, Energy and Federal Matters], *Expertengespräch Rüstungsaltlasten aus der Produktion und Verarbeitung von Sprengstoff* [Expert hearing on sites contaminated by old armament wastes as a result of the production and processing of explosives] (CPM GmbH: St Augustin, 1992), pp. C32/01–C32/28; Bielfeldt, C., 'Rechtliche Probleme der Sanierung von Rüstungsaltlasten' [Legal problems of cleaning up sites contaminated by old armament wastes], ed. W. Spyra *et al.*, *Untersuchungen von Rüstungsaltlasten* [Examination of sites contaminated by old armament wastes] (EF Verlag: Berlin, 1991), pp. 33–46; Brandt, E., 'Zur aktuellen Rüstungsaltlastendiskussion aus rechtlicher Sicht' [The current discussion of sites contaminated by old armament wastes from a legal point of view], eds K. J. Thomé-Kozmiensky *et al.*, *Management zur Sanierung von Rüstungsaltlasten* [Managing the clean-up of old armament wastes] (EF Verlag: Berlin, 1992), pp. 123–37; Brandt, E., 'Rüstungsaltlasten' [Sites contaminated by warfare agents], ed. E. Brandt, *Altlastenrecht* [Contaminated sites legislation], 3rd edn, (C. F. Müller Verlag: Heidelberg, 1993), pp. 187–220; Donner, H. and Fischer, C., 'Rechtsfragen der Sanierungs- und Finanzierungsverantwortung für "Rüstungsaltlasten"' [Legal questions concerning the responsibility for cleaning up and financing the cleanup of 'sites contaminated by old armament wastes'], *Natur und Recht*, vol. 12, no. 9 (1990), pp. 385–91; Knopp, L., 'Rüstungsaltlasten aus rechtlicher Sicht' [Sites contaminated by old armament wastes from a legal point of view], eds D. Martinetz and G. Rippen, *Handbuch Rüstungsaltlasten* [Old armament waste handbook] (Ecomed: Landsberg, 1996), pp. 53–70; Peine, F.-J., 'Rüstungsaltlasten' [Sites contaminated by old armament wastes], *Deutsches Verwaltungsblatt*, vol. 105, no. 14 (1990), pp. 733–40; Schlimm, J., 'Erfassung von Rüstungsaltlasten durch den Bund' [Registration of sites contaminated by old armament wastes by the Federation], eds R. Kompa and K.-P. Fehlau, *Altlasten und kontaminierte Böden '92* [Contaminated sites and soil '92] (Verlag TÜV Rheinland: Köln, 1993), pp. 195–207, Der Rat von Sachverständigen für Umweltfragen [Advisory Board on Environmental Matters], *Altlasten-Sondergutachten Dezember 1989* [Contaminated sites: special expertise December 1989] (Metzler-Poeschel: Stuttgart, 1990), sects 804–12, pp. 203–204; Der Rat von Sach-

II. Defining the problem

It is difficult to translate the terms related to old armament wastes from German to English. The German term 'Altlast' can be adequately translated as 'contaminated site',[4] but 'Rüstungsaltlasten' and 'Militärische Altlasten' are not easily translated. In addition, there are no legal definitions of these terms in German, and there is dispute about the definitions.

In a reply to an interpellation in the Parliament (Bundestag) in 1990,[5] the federal government suggested a definition of the term Rüstungsaltlasten, explicitly stating that the problem cannot be limited only to former military production facilities and storage sites. The proposed definition included any contamination of soil, water or air caused by chemicals from either conventional or chemical warfare agents. The federal government enumerated substances likely to cause such contamination and the sites suspected of being contaminated by conventional or chemical warfare agents. Former munition stocks, production, dismantling and testing sites and the sites of the temporary or final disposal of chemical warfare agents were also listed.

The definition was further refined and it has been widely accepted to include sites where the German Wehrmacht before and during World War II produced, tested, stockpiled or disposed of warfare agents, their precursors, by-products and intermediate and waste products, and which are currently a threat to public health and safety (i.e., to people and the environment including the soil and groundwater).[6] The definition limits the period when contamination occurred to activities that were carried out during German rearmament in the 1930s or during World War II. The definition also includes the post-World War II activities of the Allies in 1945–49,[7] such as ammunition dumping. However, contamination that occurred in World War I[8] or other environmental damage caused by military activities are not included, and the definition has therefore received much criticism.

The term 'Militärische Altlasten' applies to contamination caused by any kind of military activity after World War II.[9] This category is of particular

verständigen für Umweltfragen, *Altlasten II–Sondergutachten Februar 1995* [Contaminated sites II: special expertise February 1995] (Metzler-Poeschel: Stuttgart, 1995), sects 332-405, pp. 160–92.

[4] This term applies mainly to 'normal' contaminated civil sites.

[5] 'Antwort der Bundesregierung auf die Große Anfrage der Abgeordneten Frau Garbe, Frau Hensel, Frau Teubner, Frau Dr Vollmer und der Fraktion DIE GRÜNEN' [Reply of the Federal Government to the interpellation of the members of Parliament Mrs Garbe, Mrs Hensel, Mrs Teubner, Mrs Dr Vollmer and the Green Party], Bundestagsdrucksache (BT Drs.) 11/6972.

[6] Bundestagsdrucksache (BT Drs.) 11/4104; 'Antwort der Bundesregierung auf die Kleine Anfrage der Abgeordneten Vera Wollenberger und der Gruppe Bündnis 90/DIE GRÜNEN [Reply of the Federal Government to the question of the member of Parliament Vera Wollenberger and Group 90/the Green Party], Bundestagsdrucksache (BT Drs.) 12/331. See also *inter alia* Brandt, 'Zur aktuellen Rüstungsaltlastendiskussion aus rechtlicher Sicht' (note 3), p. 125; Schütz, H.-J., 'Verantwortlichkeit und Haftung für militärische Altlasten im deutschen öffentlichen Recht' [Responsibility and liability for contaminated military sites in German public law], Thomé-Kozmiensky *et al.* (note 3), p. 175.

[7] The period covered is limited to the time before the establishment of the two German states (23 May 1949 for the former Federal Republic of Germany and 6 Oct. 1949 for the former German Democratic Republic).

[8] In this context a major problem is determining when contamination has, in fact, occurred. In particular, former and current testing grounds and troop stationing areas have long been used by the military, often dating back to the 19th century. It is therefore often not possible to establish when or by whom the contamination was caused which currently threatens people and the environment.

[9] For an overview of the legal problems relating to sites which were contaminated by military use after World War II, see Schütz (note 6).

importance as regards the contamination caused by the German Armed Forces (Bundeswehr) and the foreign troops stationed in Germany after World War II and covers a broader scope of activities. Contamination by chemical warfare agents generally occurred before 1949,[10] and this chapter focuses on the legal problems created by the contamination which took place in that period.

As mentioned above, the accepted definitions pose substantial problems. In a 1989 expert report on the legal problems associated with contaminated sites[11] the Advisory Board on Environmental Matters pointed out that an adequate definition of the problem did not exist and that the problem would need to be addressed by federal legislation, including a commitment by the Federation to assume responsibility for a substantial part of the cost of decontamination. A second expert report by the Advisory Board in 1995 discussed the problem extensively,[12] highlighting the weaknesses and shortcomings of the accepted definition. The Advisory Board suggested that the term for contaminated military sites should include two subgroups: former warfare material production facilities, and installations formerly used by the military (e.g., testing grounds).[13] Because of the peculiarities of the contamination caused by conventional and chemical warfare agents it was also suggested that a further subcategory (militärchemische Altlasten) be introduced for military sites contaminated by chemicals. However, in the absence of a legal definition, in this chapter 'sites contaminated by old armament wastes' refer to Rüstungsaltlasten and 'contaminated military sites' to militärische Altlasten.

Until there is a clear legal definition the matter cannot be resolved, but for legal analysis of the problem the definition is only of importance for the analytical framework that it provides.

Remnants of warfare and other explosive devices

Remnants of warfare, such as unexploded munitions or other explosive devices from World War II that remain buried in Germany, are dealt with by a special administrative practice that developed after World War II. The deactivation, removal and final disposal or destruction of such remnants is the responsibility of the Munitions Disposal Services (Kampfmittelräumdienste), agencies of the states which are responsible for all activities relating to remnants of warfare. Most Munitions Disposal Services are special divisions of the police administration or of the responsible ministries of the states (normally, the Ministry of the Interior).[14] Under the General War Effects Act (Allgemeines Kriegsfolgengesetz), which enumerates the obligations of the Federation for damage caused by World War II, the federal government reimburses the costs for the Munitions Disposal Services to the states. However, the federal government stresses that this is an expense which it has assumed voluntarily[15] and that it has no

[10] Thus far there are no reports of contamination of the former CW storage sites in Germany that are maintained by the US Army. Contamination which may have been caused by the destruction of old and abandoned CW in the former GDR is discussed in chapter 13 in this volume.

[11] Der Rat von Sachverständigen für Umweltfragen, 1989 (note 3), sect. 994, p. 234.

[12] Der Rat von Sachverständigen für Umweltfragen, 1995 (note 3).

[13] Der Rat von Sachverständigen für Umweltfragen, 1995 (note 3), sects 336 and 337, p. 161.

[14] This is the case *inter alia* in Hessen (Hesse), Niedersachsen (Lower Saxony) and Nordrhein-Westfalen (North Rhine Westphalia).

[15] See *inter alia* Schlimm (note 3), p. 200. This practice is governed by administrative regulations of the Federal Minister of Finances.

legal obligation under the War Effects act to do so.[16] Since chemical weapons were not used in World War II, old chemical weapons do not represent a significant problem in this context.

III. German environmental law and sites contaminated by old armament wastes

There is no specific legislation on the problem of old armament wastes, and responsibility and liability for their decontamination and related measures must therefore be examined in terms of general environmental legislation that is applicable to contaminated civil sites. Because no specific or comprehensive legislation on contaminated sites exists it is necessary to examine a variety of environmental regulations that address the environmental threats likely to be caused by warfare agents. These include federal and state legislation on waste, soil and water and the police acts of the states. Since there is both federal and state legislation it is useful to describe briefly the division of power in this type of legislation and administration in Germany.[17]

The division of power relating to environmental protection

In Germany the division of legislative power is characterized by a high degree of centralization.[18] As a rule the legislative powers of the Federation are limited to those enumerated in the German Constitution (Basic Law). However, with regard to environmental protection the federal powers are extensive, including *inter alia* those on the disposal of waste (Article 74, paragraph 24 of the Constitution). In addition, the Federation has legislative powers to enact framework provisions, including *inter alia* on water management (Article 75 of the Constitution). The states may enact their own legislation in areas which are not covered by the federal legislative powers and additional concurrent and framework legislation,[19] but only to the extent that the Federation has not done so. In general, the Federation has used its legislative power extensively as regards environmental protection.

There are various types of such legislation. At the federal level there is, however, no specific legislation on contaminated sites, and other areas of federal legislation apply. At the federal level the Waste Avoidance and Waste Management Act (Abfallgesetz) and its successor the Closed Substance Cycle and Waste Management Act (Kreislaufwirtschaftsgesetz) address waste prevention and the disposal of wastes. The federal government also currently plans to use its legislative powers (which are disputed) to enact a new Soil Protection Act (Bodengesetz). With regard to water management, the Federation has enacted

[16] See *inter alia* Brandt, 'Rüstungsaltlasten', (note 3), pp. 189–90; and Schlimm (note 3), p. 201.

[17] For a discussion of this issue see, Bothe, M., 'Government and environment in Germany between division of powers, federalism and centralism', *Studi parlamentari e di politica constituzionale* [Parliamentary studies and constitutional politics], vol. 104, no 2 (1994), pp. 117–31.

[18] Bothe (note 17), p. 117.

[19] Concurrent legislation is the area of legislation where both the Federation and the states are competent to enact legislation. However, the states may only enact legislation to the extent the Federation has not used its powers to legislate or at least not to its full extent (e.g., by enacting framework legislation). Framework legislation is a type of legislation where the basic principles are established by federal legislation, but the details of its application are defined by state legislation.

framework legislation: namely, the 1957 Water Management Act (Wasserhaushaltsgesetz).

Legislation in the states varies considerably. Some states have enacted legislation that specifically addresses contaminated sites in the context of waste legislation, while others have done so under legislation that protects the soil. Some of these acts explicitly state that the legislation applies to the problem of sites contaminated by old armament wastes and contaminated military sites, while others do not mention these problems (thus not excluding application of the legislation) or contain regulations which explicitly exclude applying the legislation to them. Another area where the states have enacted relevant legislation is the regulations to protect water that complement respective federal legislation. Finally, the police acts of the states are another important type of law. Police legislation is the exclusive province of the states, except for the laws which relate to the Federal Border Police (Bundesgrenzschutz). These police acts are often the basis of administrative action in the event of imminent danger to public health and safety, and they are of major importance in environmental protection.

In contrast to legislative power, administrative power in Germany is highly decentralized. Article 83 of the Constitution gives the states the right to administer their own laws and also those of the Federation. The Federation may establish its own administration only if it is explicitly permitted to do so by the Constitution. The relevant agencies of the states therefore have primary responsibility for sites contaminated by old armament wastes. For such sites and for contaminated civil sites the states both administer the legislation and bear most of the financial responsibility. Generally, the states have primary responsibility for contaminated sites unless a legal or natural person can be identified as responsible and liable to make compensation. The states are therefore interested in clear and comprehensive regulation of the problem.

The following discussion assesses the extent to which existing or draft legislation provides sufficient instruments to handle the problem (e.g., ordering decontamination of a contaminated site). Some regulations may only provide tools that can be used to deal with immediate danger or to secure a contaminated site and may be inadequate as regards the complete clean-up of a contaminated area.

Legislation on waste and contaminated sites[20]

Federal waste legislation

In Germany there has been continuity in waste legislation since the early 1970s. At the federal level the 1994 Closed Substance Cycle and Waste Management Act (Kreislaufwirtschaftsgesetz)[21] entered into force in 1996, replacing the 1986 Federal Waste Avoidance and Waste Management Act (Abfallgesetz),[22]

[20] For a comprehensive review of German legislation related to contaminated sites, see Schmidt, C., 'National report on the legal regulation of the contaminated sites in Germany', ed. M. Prieur, *Sites contaminés–En droit comparé de l'environnement* [Contaminated sites: comparative environmental law] (Pulim: Limoges, 1995), pp. 213–27.

[21] *Bundesgesetzblatt* [Federal gazette] *1994*, I, p. 2705.

[22] *Bundesgesetzblatt* [Federal gazette] *1986*, I, p. 1410. The act was amended several times before being replaced by the Closed Substance Cycle and Waste Management Act.

which in turn replaced the 1972 Federal Waste Disposal Act. However, since the contamination in question occurred before, during or shortly after World War II, the application of this legislation poses major problems. The Constitution prohibits retroactive application of legislation, and thus the legislation passed in 1972 and later cannot be applied to installations or disposal sites which were abandoned or ceased to be used before 1972. In addition, the 1986 and 1994 acts explicitly state that they do not apply to the search for, recovery, transport, storage, treatment and destruction of warfare agents.[23] Under certain circumstances some elements of the existing legislation could be used to cover the cost of evaluating a contaminated site,[24] but federal waste legislation is not sufficient to solve the problem.[25]

The waste acts of the states

Because of the inadequacy of federal waste legislation, the waste legislation of the states is of major importance, and the majority of them have adopted regulations on contaminated sites in this context. The Hessian Waste Management, Waste Disposal and Contaminated Sites Act (Hessisches Abfallwirtschaftsgesetz),[26] for example, regulates the preconditions, extent and responsibility for the clean-up of contaminated sites. It provides a comprehensive screening of the entire state to ascertain whether or not contaminated sites exist. In a first step, a register of areas that are suspected to be contaminated will be established which includes all 'suspected contaminated sites' (Altlastenverdachtsflächen). According to Article 16, paragraph 1 of the act a 'suspected contaminated site' can be any site where waste is treated, stored or disposed. This general definition can be interpreted as applying to sites that are suspected of being contaminated by old armament wastes. A December 1992 amendment to the act explicitly provides for its application to 'lands which were contaminated by military activities'.[27] According to the act a suspected site will be subject to close examination which may lead to one of the following results: the suspicion may be dispelled; it may continue to exist, even though contamination cannot be sufficiently proven to exist; or the suspicion may be proven true.

The results of the investigation are presented in a formal administrative decision which forms the basis for remedial action. The appropriate agency may order that such action be taken by the owner of the land, the operator or former operator of an installation which caused the contamination or any other person or enterprise which caused the contamination. However, if the current owner acquired the land in good faith after the contamination occurred he is exempt from responsibility. The administrative agency has the discretionary power to allocate responsibility to various persons or legal entities and may apportion the cost of remedial action among them. If responsibility for remedial action cannot

[23] Article 2, para. 2 (7) of the Recycling Economy and Waste Management Act, and Article 1, para. 3 (8) the Federal Waste Avoidance and Waste Management Act of 1986.

[24] With regard to the Federal Waste Avoidance and Management Act, see Brandt, 'Zur aktuellen Rüstungsaltlastendiskussion aus rechtlicher Sicht' (note 3), p. 125.

[25] Knopp (note 3), p. 54.

[26] *Hessisches Gesetzes- und Verordnungsblatt, I* [Hessian gazette], p. 106, promulgated on 26 Feb. 1991 and later amended.

[27] Translation by the author. In addition to the Hessian Act only the First Act on Waste Management and Soil Protection of Saxony has a similar provision in Article 8, para. 3 (3) (see discussion below).

be assigned or those liable are unable to bear the cost, remedial action becomes the responsibility of a specific entity which is created for that purpose.

The Hessian act provides the necessary instruments for a satisfactory determination of responsibility for contamination if the current condition of a contaminated site can be ascertained. However, it is a sophisticated type of legislation, and similar regulations in other states may not be as comprehensive.[28]

Soil protection legislation

There is no federal legislation which is specifically designed for soil protection. In October 1993 a first draft of the Federal Soil Protection Act was presented.[29] It was revised and submitted to Parliament in September 1996.[30] However, it seems unlikely that the draft act will be passed in the near future. Of the states, only Baden-Württemberg and Saxony (Sachsen) have passed soil protection legislation.

The draft Federal Soil Protection Act

The draft Federal Soil Protection Act is intended to prevent environmental damage to the soil. It provides for the clean-up of existing environmental threats and requires *inter alia* preventive action against possible future soil contamination. However, Article 3, paragraph 10 of the draft explicitly excludes applicability of the act to sites contaminated by old armament wastes: 'This Act does not apply to the search for, recovery, transport, storage, treatment or destruction of warfare agents'.[31] If the Soil Protection Act is passed with Article 3 in its current form no progress will have been made on the problem. The draft act has therefore been severely criticized by the states.

Soil protection acts of the states

Only Baden-Württemberg and Saxony have passed legislation that specifically focuses on protection of the soil. The Saxon Waste Management and Soil Protection Act (Erstes Gesetz zur Abfallwirtschaft und zum Bodenschutz im Freistaat Sachsen) of 12 August 1991[32] is of particular interest. Its Article 8, paragraph 3 (4) is similar to the provision of the Hessian Waste Management, Waste Disposal and Contaminated Sites Act and makes the act applicable to sites that were contaminated by former military use. It enables the appropriate authority to order those responsible for environmental damage, or the current owner of the land, to take measures that include evaluating and securing the site; submitting a redevelopment plan; taking measures to remove, reduce or survey a contaminated site; taking action to avoid or reduce contamination and

[28] For a critical discussion of the legislation in Thüringen, see Kothe, P., 'Altlastenbehandlung in Thüringen als bundesdeutsches Modell' [Treatment of contaminated sites as a model for Germany], *Die öffentliche Verwaltung*, vol. 47, no. 17 (1994), pp. 716–28, p. 717 and ff.

[29] The draft was published by Ständigen Deputation des Deutschen Juristentages (ed.), *Vorschläge zur Regelung der Altlasten im Rahmen des Bodenschutzes* [Proposals concerning the regulation of contaminated sites in the context of soil protection], Diskussionsgrundlage zum 60, Deutschen Juristentag, Münster, 1994 (C. H. Beck: Munich, 1994).

[30] Bundesrat [Federal Chamber], Drucksache 702/96, 27 Sep. 1996.

[31] The formulation is similar to that of Article 2, para. 2 (7) of the Closed Substance Cycle and Waste Management Act.

[32] Gesetzes- und Verordnungsblatt des Freistaats Sachsen [Saxon gazette], 1991, p. 306.

to clean up a contaminated site; and restricting or prohibiting certain uses of the soil or the use of certain substances.[33] This regulation and the similar section of the Hessian Act are the most advanced such state legislation.

Water legislation

The disposal of armament products and ammunition wastes has often led to severe contamination of surface and groundwater and poses a threat to water quality in the future. Water legislation is therefore important, and this type of legislation has a long tradition compared to other types of legislation designed to protect the environment.

The Federal Water Management Act

The Federal Water Management Act of 1957 (Wasserhaushaltsgesetz)[34] defines the basic regulations on measures related to quantitative and qualitative water management. As with federal waste legislation this act cannot be applied retroactively. In addition, the instruments provided for in it are limited and cannot be used to order the securing or clean-up of a site.

Water legislation of the states

Most of the water acts of the states were enacted before 1960. Because these acts are tied to earlier legislation, such as the Prussian Water Act, and contain prohibitions that are identical to those contained in the earlier legislation, they may provide a basis for administrative action.[35] The Prussian Water Act was in force in most parts of Germany during the Third Reich and prohibited *inter alia* the discharge of harmful substances into the water (e.g., from the production of warfare material). Many activities which caused water pollution were therefore already prohibited at the time they occurred. However, the Prussian Water Act and its succeeding legislation are insufficient to do more than establish an obligation to take the necessary steps to protect public health and safety.[36] The Prussian Water Act did not provide for more restrictive measures such as requiring the clean-up of a contaminated site, and current water legislation is inadequate to deal with sites contaminated by old armament wastes.[37]

Police legislation

As mentioned above, in Germany the police are under the jurisdiction of the states, with the exception of the Federal Border Police. The police acts of the states have assumed major importance as regards enforcement of environmental standards

[33] Article 9, para. 1 (1)–(5).
[34] As promulgated on 23 Sep. 1986, *Bundesgesetzblatt* (Federal gazette) *1986*, I, p. 1529. For an English translation of the act see Umweltbundesamt (ed.), *A–Z of Environmental Protection in Germany* (Informationszentrum Umwelt: Düsseldorf, 1992), pp. 251–75.
[35] Becker (note 3), p. C32-11 and ff.
[36] Brandt, 'Rüstungsaltlasten' (note 3), pp. 201–202.
[37] Becker (note 3), p. C32-22.

Police acts of the states

The police acts of the states have an important function in the enforcement of environmental standards with regard to contaminated sites, in general, and sites contaminated by old armament wastes, in particular. Under the general clauses of these acts the police are empowered to take necessary measures to deal with potential danger. Such action may only be taken if the 'common good' protected by law, such as the life and health of people and the purity of the groundwater and soil, is affected and if the threat is imminent. In such a situation the police can compel those responsible to take appropriate measures to prevent harm. Police legislation does not establish a duty to clean up a contaminated site. However, because the burden of cost, which has thus far been borne by the states, is large an investigation should be made to determine if the legal instruments offered by police legislation can be applied so that at least part of the cost is assumed by some other entity than the states. German legislation is very complicated in this respect, but under certain circumstances, if legal responsibility can be proven, the federal government, a company which conducted business in World War II or its successor may be held liable for environmental damage. However, the most that could be achieved would be an administrative order to take measures to prevent imminent harm to the public health and safety.

Existing legislation is clearly insufficient. While federal legislation does not apply to this area, the regulations of the states, which vary in their usefulness, are important instruments and of some value. However, even if these regulations are applied, the problem of establishing responsibility and liability for the measures to be taken remains.

IV. Responsibility and liability for environmental damage

This section discusses the extent to which it is possible to establish responsibility and liability for environmental damage on the basis of existing legislation and suggests those who may be liable to make restitution. A summary of the organization of the German rearmament programme during the Third Reich is necessary in order to understand the current situation.

The legal construction of the production of war material in the Third Reich

In the Third Reich production of war material was organized under two different legal schemes: a private and a public law construction.[38] Under the former, private companies, such as I. G. Farben, Dynamit AG, Degussa and WASAG, built facilities for the production of war material and transferred ownership to the Verwertungsgesellschaft für Montanindustrie GmbH (Montan), which had been the sole property of the German Reich since 1934.[39] The cost of building

[38] See *inter alia* Bielfeldt (note 3), p. 34. For an overview of the organization for the production of chemical warfare agents and production facilities during the Third Reich, see Groehler, O., *Der lautlose Tod–Einsatz und Entwicklung deutscher Giftgase von 1914 bis 1945* [The silent death: use and production of German poison gas from 1914 to 1945] (Rowohlt: Reinbek, 1989).

[39] Peine (note 3), p. 735.

these facilities was paid by the High Command of the Wehrmacht. Through Montan the German Reich organized and controlled key aspects of the German rearmament programme. The facilities were leased to private companies which operated them (e.g., Verwertchemie, an affiliate of Dynamit AG). Most of the companies which were involved in the rearmament programme were liquidated after World War II, and it is difficult to identify who should be required to provide compensation for damage which resulted from their wartime activities.[40]

Under the public law construction the production of warfare agents was organized in munition depots, known as Munitionsanstalten or Munas, which were owned by the Reich. The Munas were used for the production, stockpiling and transfer of munitions and explosive devices and were supervised by a military commander.[41] These activities were controlled by the Reich.

Problems establishing responsibility and liability

The fundamental problem of establishing responsibility and liability is that the vast majority of the companies involved in the production of chemical warfare agents and in the general rearmament programme were liquidated after World War II. However, some of those involved can still be identified. The extent to which the Federal Republic has assumed responsibility for the actions of the Third Reich (e.g., the Munas) and for damage caused by Montan needs to be explored. In addition, it must be ascertained if the private companies that participated in the rearmament programme still exist, or if the companies that are successors to them are required to assume responsibility for former hazardous activities.

The potential liability of the Federation

The Federal Constitutional Court has ruled that the Federal Republic is identical to the German Reich to some extent and is thus responsible for actions attributable to the German Reich.[42] Furthermore, Article 120 of the Constitution makes the Federation responsible for all costs which are directly linked with effects caused by World War II based on the General War Effects Act. Except for the liabilities assumed under that legislation the federal government has rejected any attempt to assign full or even partial responsibility for the problem to the Federation, and it is unlikely to alter its view on the matter. It may be possible that the Federation could be forced to assume responsibility for the activities of the Munas, but damage caused by Montan is not covered by the War Effects Act, which does not allow private law claims to be made against the Federation.[43]

After World War II ownership of most of the facilities and lands used in the rearmament programme passed to the Federation. Some of these remained the property of the Federation, while others were sold or transferred to the states and communities or to private juridical and natural persons. Problems with

[40] A conciliation proceeding concerning Verwertchemie ended in 1952 without result. Peine (note 3), p. 735.

[41] Bielfeldt (note 3), p. 34.

[42] Bundesverfassungsgericht [Federal Constitutional Court], ruling of 31 July 1973, *BVerfGE* [Proceedings of the Court] vol. 36 (1973), p. 1.

[43] Becker (note 3), p. C32-23.

regard to environmental damage on these properties must also be considered in terms of whether they are owned by the Federation or by states, communities or private individuals. The Federation is responsible for the contaminated sites which it owns and must take measures to address threats to people and the environment. These measures must be taken in cooperation with the appropriate authorities of the states. The federal authority which is currently using the site or which is responsible for it must bear the cost. This is also true for property used by the German Armed Forces.

An additional problem exists. A significant degree of environmental damage was caused in World War II by the Allied Forces, including damage which resulted from the destruction of key military and industrial facilities. However, the Allies cannot be held responsible for their wartime actions since there were no legal restrictions prohibiting them.[44]

Responsibility and liability for environmental damage caused by the Munas

In principle, all of the activities of the Munas can be attributed to the German Reich and thus to the Federal Republic as well. However, even if it were possible under existing legislation for the states to establish that the Federation is responsible, it would be difficult to enforce such an obligation since the authorities of the states are not authorized to enforce an administrative act if it interferes with the powers reserved to the Federation.[45] This approach is therefore unfeasible unless the Federation commits itself to help meet the cost of environmental clean-up.

Responsibility and liability for environmental damage caused by Montan

Montan AG was responsible for numerous sites that were contaminated by the production of warfare agents and material. After World War II the ownership of Montan passed to the Federal Republic of Germany, and in 1951 it was renamed the Industrieverwertungsgesellschaft mbH (IVG). After World War II IVG sold much of its property to states, communities and private juridical and natural persons. The land was often used for development projects, housing and industry. In 1986 IVG was transformed into a private law corporation, the Industrieverwertungsgesellschaft AG (IVG AG), with the majority of shares held by the Federation. In January 1994 the Federation sold its shares in IVG AG, thereby transferring it into private ownership. IVG AG is identical with Montan AG, and since IVG AG still owns sites which are contaminated it is fully responsible and liable for the measures required by law—possibly including clean-up of the contaminated sites.

However, there are more important problems related to the contaminated sites which were sold by IVG/IVG AG after World War II. Lower Saxony (Niedersachsen), in particular, has numerous sites which were contaminated by Montan or its leaseholders, and it has taken a leading role in this respect. The central problem with regard to Montan and its leaseholding companies is the need to establish that Montan controlled the activities of these companies to such an extent that environmental damage caused by the operation of facilities can be

[44] Peine (note 3), p. 737.
[45] Bielfeldt (note 3), p. 41.

attributed to Montan. It can be argued that such an assumption is false and that these companies acted only as leaseholders for Montan, operating their facilities independently. However, the high degree of direct and indirect control demonstrated in contracts between Montan and the leaseholders[46] (e.g., provision of comprehensive information and the partial transfer of profits to Montan) suggests that these companies were subordinated to Montan.[47] IVG AG may thus be obliged to make compensation for environmental damage caused by Montan.

In 1990 Hesse reached an out-of-court settlement with IVG AG with regard to compensation for two contaminated sites, formerly owned by Montan, which were used in World War II for the production of TNT. IVG AG paid 20 million Deutschmarks (DM) to the Hessian Government and declared that this payment was not to be regarded as recognition of a legal obligation for compensation. The Hessian Government in turn declared that any additional claims, including those of any Hessian local authorities, were considered settled for all time. The agreement was widely criticized, especially by other states with sites contaminated by Montan, and it was argued that the agreement prejudiced their claims and diminished their prospects for a comprehensive settlement of existing and future claims against IVG AG.[48] There was also criticism that payment of 20 million DM was inadequate when compared to the estimated cost (200 million DM) for clean-up of only these two contaminated sites. Particularly because Hessian legislation in this area is extensive, it is questionable whether or not the agreement was politically wise.

Although IVG AG is identical with Montan privatization complicated the situation, and it remains to be seen whether or not IVG AG can be held liable in future cases. The policy of the federal government in this context has provoked criticism since the Federation has profited substantially from the activities of IVG AG for almost 40 years. It can be argued that at least some of these profits should be used to clean up sites contaminated by Montan, even in the absence of a legal obligation.

Responsibility and liability for environmental damage caused by I. G. Farben

After World War II the Allied Powers ordered the dissolution of the I. G. Farben cartel, which was *inter alia* responsible for the production of large quantities of chemical warfare agents during the Third Reich.[49] Law No. 9 of the Allied Control Council provided for the expropriation of I. G. Farben. In order to facilitate the final liquidation of I. G. Farben, the IG Farbenindustrie Aktiengesellschaft in Abwicklung (I. G. Farben Industries Joint Stock Company in Liquidation, or I. G. Farben i.A.) was founded with the sole aim of administering remaining property and managing the payment of compensation to persons and groups which had suffered damage from the activities of I. G. Farben. Almost 50 years after the end of World War II the final liquidation has not been completed, and only 30 million DM have been paid in compensation,

[46] Bielfeldt (note 3), p. 40.

[47] Bielfeldt (note 3), p. 40.

[48] 'Gorbatschow sorgt für Auftrieb' [Gorbachev provides new impetus], *Frankfurter Rundschau*, 31 Aug. 1994, p. 12.

[49] There are numerous references to the involvement of I. G. Farben in the production of chemical warfare agents in Groehler (note 38).

not including, among others, the claims of those who were forced to work for I. G. Farben during World War II. In addition, after German reunification I. G. Farben i.A. sought restitution for property in the former GDR which it claimed had been illegally expropriated.[50]

There have been no attempts made to establish the liability of I. G. Farben i.A. for the clean-up of sites which were contaminated as a result of activities related to the production of warfare agents.[51] However, two unsuccessful claims were made against I. G. Farben related to 'normally' contaminated sites, and one of these cases is of particular interest in this context.

In the 1930s I. G. Farben operated an enterprise in Mühlheim, a small town near Frankfurt, for the production of gas for domestic fuel use. When World War II began the facility was closed and the land became the property of the German Army, which used it as a troop stationing area. Shortly before the end of World War II the area was bombed by the Allied air forces, resulting in damage to the gasworks. After World War II the ground became federal property. In the 1970s the city of Mühlheim bought part of the land with the intent of building low-cost housing. During the planning stage it became obvious that the land was contaminated, and the relevant authority, the Chief Official of the Administrative District in Darmstadt, declared the site to be contaminated. The city of Mühlheim paid for the clean-up of the site. Since the contamination resulted mainly from the former gasworks, the Chief Official of the Administrative District issued an administrative act establishing the liability of I. G. Farben i.A.

However, because of a recent decision of the Hessian Administrative Court, the District Administration decided to annul the administrative act since it was not as specific as Hessian legislation required. I. G. Farben i.A. also argued that the contamination of the soil was not caused by the operation of the former gasworks but by the bombing of it and its subsequent destruction. I. G. Farben i.A. therefore rejected claims for compensation. The matter has not been settled. What is important in this context is that the annulment of the administrative act was partly the result of procedural difficulties and partly caused by the problem of establishing responsibility for the contamination. Thus, at least under Hessian legislation, the responsibility and liability of I. G. Farben i.A. for contaminated sites cannot be generally precluded.

The successors of I. G. Farben

After World War II major segments of the production sites which formerly belonged to I. G. Farben became the property of its successors Hoechst AG, Bayer AG and BASF, currently the leading producers of chemicals in Germany. However, these companies did not become the legal successors of

[50] 'Gorbatschow sorgt für Auftrieb' (note 48). So far all attempts to regain former property in the eastern states have been rejected by the courts since expropriation of I. G. Farben property took place during occupation under Soviet rule.

[51] This approach has never been sufficiently discussed. It is, however, addressed in Claus, F., 'Einführungsvortrag in Arbeitsgruppe 3: Sanierungsverantwortlichkeit, Finanzierung, Bürgerbeteiligung' [Introductory lecture to Working Group 3: responsibility for clean up, financing, public participation], ed. Hessisches Ministerium für Umwelt, Energie und Bundesangelegenheiten [Hessian Ministry for Environment, Energy and Federal Matters], *Expertengespräch Rüstungsaltlasten aus der Produktion und Verarbeitung von Sprengstoff* [Experts hearing on warfare agents contaminated sites caused by the production and processing of explosives] (note 3), p. C31-02.

I. G. Farben and thus cannot be held responsible or liable for contamination caused by it. In so far as these companies still possess or use former I. G. Farben facilities they may be held liable for remaining contamination in accordance with the relevant regulations. Since the problem of armament production and ammunition waste sites remains unresolved it is not surprising that the states, which bear most of the cost of decontamination, have made several attempts to find a solution to this problem.

V. Legislative initiatives of the states related to sites contaminated by old armament wastes and contaminated military sites

The states have made several attempts to achieve a financial solution to the problem of old armament production and ammunition waste sites. The most important of these was the Draft Act for the Financing and Clean up of Old Armament Waste Sites (Rüstungsaltlastenfinanzierungsgesetz). The states also attempted to address the problem during the consultations on the implementing legislation of the Chemical Weapons Convention and the Registerverfahrenvereinfachungsgesetz (Act for the Expedition of Register Procedures). In regard to the latter, the eastern states focused primarily on the problem of the contaminated military sites which had been used by former Soviet troops after World War II.

The Draft Act for the Financing and Clean up of Old Armament Waste Sites

The Draft Act for the Financing and Clean up of Old Armament Waste Sites[52] was initiated by Lower Saxony and co-sponsored by Bremen and Rheinland-Pfalz when the draft was submitted to the Federal Chamber (Bundesrat) in April 1992. In the summer of 1992 the Federal Chamber submitted the draft to Parliament,[53] where it was rejected in July 1993.[54]

The states argued that, in the course of German reunification, the problem of old armament production and ammunition wastes had acquired a new dimension which required a comprehensive legislative solution. The purpose of the draft act was to establish a clear and comprehensive regulation with regard to the financing of measures to protect people and the environment from the harmful effects of sites contaminated by old armament wastes. The scope of the draft was broad, and it defined *inter alia* all sites and facilities which had been used for the development, testing, production, storage, use or destruction of warfare agents before 8 May 1945 as 'Rüstungsaltlast',[55] provided that such sites or facilities had caused environmental damage or constituted an imminent danger to people and the environment. Article 3 of the draft act provided for a comprehensive set of protective measures including: 'the search for, investiga-

[52] For an introduction, see Nöthel, S., 'Niedersächsische Bundesratsinitiative zum Erlaß eines Gesetzes über die Finanzierung der Sanierung von Rüstungsaltlasten in der Bundesrepublik Deutschland' [Federal Chamber initiative of Lower Saxony concerning the financing and clean-up of sites contaminated by old armament wastes in the Federal Republic of Germany], Thomé-Kozmiensky *et al.* (note 3) pp. 203–20.

[53] *Deutscher Bundestag* [Official journal], Drucksache 12/3257, 15 Sep. 1992.

[54] *Deutscher Bundestag* [Official journal], 2. Beratung [2nd debate], 2 July 1993, 169. Sitzung [169th session].

[55] This is the date on which Germany unconditionally surrendered and World War II ended.

tion, registration and evaluation of contaminated sites; surveillance of contaminated sites; measures for clearance and removal; measures to prevent harm and lessen danger; and measures to clean up and rehabilitate contaminated sites'.[56]

The draft act also provided for the establishment of a register of sites contaminated by old armament wastes,[57] but the most controversial regulations related to cost. Article 7 stated that the federal government was liable to pay the cost of all measures taken under the act. The draft further stipulated that if possible the owner of a contaminated site or a person responsible for the damage should be held liable to make compensation. Owing to the tight federal budget and the cost implications of the draft act it was not surprising that the federal government opposed the draft when it was discussed in Parliament.[58] As noted above, it was later rejected.

In late 1996 the states took the initiative to a new draft act, and it will be submitted in 1997 for legislative approval. Discussions between the ministries of the states are currently under way.

Legislation on ratification and implementation of the CWC

During the consultations in the Federal Chamber on the CWC Ratification Act[59] and the Act on Implementing the CWC[60] the eastern states, in particular, were concerned about the possible cost implications of the CWC with regard to old and abandoned chemical weapons. In an explanatory note to the draft act the federal government stated that ratifying the CWC and implementing it at the national level would place no financial burden on the states. The eastern states expressed concern about contamination by the former Soviet Army and contamination resulting from other former uses. It was argued, for example, that the cost of securing contaminated sites and the interim storage of contaminated soil would be the responsibility of the states and that the assurance of the federal government was therefore insufficient. Recent information indicates that these fears were well-founded. The Act on Implementing the CWC was initially blocked during discussions in the Federal Chamber and was referred to the Mediation Committee. Ultimately, the federal government and the states agreed that the Federation will bear all financial cost relating to implementation of the CWC.

The Act for the Expedition of Register Procedures

Under German law any transfer of real property must be registered in a real estate register (Grundbuch). One of the major problems resulting from German reunification is the uncertainty about ownership of real property that was expropriated after World War II on the territory of the former GDR. The problem is complicated by the fact that registration procedures are time consuming. In order to address these uncertainties and expedite the registration procedure the

[56] Article 3 para. 1–6 (translation by the author).

[57] Article 5.

[58] *Deutscher Bundestag* [Official journal], 1. Beratung [debate] 5 Nov. 1992, 117. Sitzung [117th session].

[59] Gesetz zum Chemiewaffenübereinkommen [Act concerning the Chemical Weapons Convention], *Bundesgesetzblatt* [Federal gazette] *1994*, II, p. 806.

[60] Ausführungsgesetz zum Chemiewaffenübereinkommen [Act on the implementation of the Chemical Weapons Convention], *Bundesgesetzblatt* [Federal gazette] *1994*, II, p. 1954.

Act for the Expedition of Register Procedures (Registerverfahrensbeschleunigungsgesetz) was enacted in December 1993.[61] Amending the Ratification Act for the Withdrawal Treaty between Germany and the Soviet Union,[62] Article 17, paragraph 1 of the act provides a procedure to expedite the transfer of ownership of former military lands from the Federation to the states. Under the provision the new states may apply for transfer of ownership. During discussion of the act in the Federal Chamber the states proposed amending this section with regard to financing the clean-up of contaminated sites so that the cost of research on and clean-up of the contamination caused by warfare agents would be borne by the Federation.[63] However, the amendment was rejected, and Article 17 entered into force unchanged.

VI. Military forces and German environmental law

There are three different categories of military forces in Germany which have contributed to the problem of contamination: the German Armed Forces, the Allied and NATO forces, and the former Soviet forces. While German environmental legislation can be directly applied to the German Armed Forces, the applicability of German environmental legislation to foreign forces stationed on German territory is based on international treaties.[64]

The German Armed Forces

Neither the Federal Republic nor the GDR produced or acquired chemical weapons after World War II, and thus there should be no problem with regard to CW produced or acquired after 1945. However, both countries were involved in the destruction of World War II chemical weapons, as is the reunified Germany. The clean-up and destruction programme at Munster is run by the German Armed Forces, which are not exempt from federal or state environmental legislation.[65] Land that was formerly used by the German Armed Forces and which is now in civil use was discussed above, but the situation is somewhat different for land which the German Armed Forces continue to use. If the German Armed Forces cause environmental damage in violation of existing legislation they are responsible for the damage and obliged to clean up any sites which they have contaminated. However, the states do not have authority over the military since responsibility for such sites rest exclusively with the Ministry of Defence.[66]

There are specific problems related to military sites which have been released for civil use or which will be sold in the future.[67] Responsibility for these sites

[61] *Bundesgesetzblatt* [Federal gazette] *1993*, I, p.2182.

[62] *Bundesgesetzblatt* [Federal gazette] *1991*, II, p. 258.

[63] Bundesrat [Federal Chamber], Drucksache 360/93, 9 July 1993.

[64] For an overview, see Bothe, M., 'Völkerrechtliche Situation bei Rüstungsaltlasten' [International law and sites contaminated by old armament wastes], Thomé-Kozmiensky *et al.* (note 3) pp. 139–53.

[65] See *inter alia* Führen, K.-H., 'Behandlung militärischer Altlasten durch das Bodenrecht' [Soil legislation on the treatment of sites contaminated by warfare agents], Thomé-Kozmiensky *et al.* (note 3), p. 158. See also chapter 12 in this volume for a discussion of the Munster facility.

[66] Führen (note 65), p. 162.

[67] For an overview, see Fries, H.-J., 'Rechtliche Grundlagen der Sanierung und bisherge Maßnahmen zur Erfassung und Sanierung von Altlasten im Bereich der Bundeswehr' [Legal basis for the clean-up and

is generally transferred from the Ministry of Defence to the Federal Properties Agency (Bundesvermögensverwaltung), a special body of the Ministry of Finance, which handles the selling of such sites. Before the sites can be sold (e.g., to the states or communities) any substantial threats to people and the environment as the result of contamination by warfare agents must be eliminated. However, this does not mean that the Federation is obliged to clean up the sites. Since civil use of the sites often requires that additional decontamination measures be taken responsibility for the cost must be determined. If the contamination is severe the federal government has agreed to assume the cost of additional clean-up measures, but only to an amount equivalent to the profit it would make from selling such a site. Since the clean-up of contaminated sites is expensive in most cases the selling price will not cover the cost of clean-up, and substantial cost will need to be borne by the future owners (e.g., the state or local authorities).

A solution to the problem could be the creation of a 'superfund' into which the profits from selling former military lands would be transferred (including profits from the selling of former military lands that were not contaminated). The superfund would be used for the sole purpose of financing the clean-up of such contaminated sites. Although even a superfund probably would not cover all cost it would be progressive step to fair burden-sharing between the states and the Federation. However, because of the tight federal budget and general federal policy in this respect it seems likely that the current approach will remain unchanged.

NATO forces in Germany

US forces stockpiled chemical weapons on German territory only after World War II. The weapons were stockpiled for a long period near the small town of Clausen until their removal in 1990 to Johnston Atoll in the Pacific south-west of Hawaii. There is no indication that any damage to the environment occurred from the stockpiling or transport of these weapons, although it cannot be totally precluded. Should contamination be found it is questionable whether or not German environmental law can be applied to such a case and if so to what extent.

The NATO Status of Forces Agreement (NATO SOFA) governs the relationship between NATO forces stationed on German territory and the German government and its citizens.[68] Troops stationed on foreign territory must respect the law of the host country, and this is emphasized by Article II of NATO SOFA: 'It is the duty of a force and its civil component and the members thereof as well as their dependants to respect the law of the receiving State'. Although this formulation is vague it can be argued that it includes all German criminal, civil and administrative law.[69] The Allied Forces in Germany interpreted the term 'respect' rather loosely, but this does not alter the fact that the relevant German

measures taken so far concerning the registration and clean-up of contaminated sites under the responsibility of the German Army], Thomé-Kozmiensky *et al.* (note 3), pp. 113–21.

[68] *Bundesgesetzblatt* [Federal gazette] *1961*, II, p. 1190.

[69] K. J. Walker, 'A green peace: the closure of military installations and the environmental liability of the United States Government', *Georgetown International Environmental Law Review* no. 3 (1990), p. 439.

environmental legislation applies to environmental damage caused by NATO forces. Article 53 paragraph of the Supplementing Agreement to NATO SOFA states: 'Within accommodation made available for its exclusive use, a force or a civil component may take all the measures necessary for the satisfactory fulfilment of its defence responsibilities. Within such accommodation, the force may apply its own regulations in the fields of public safety and order where such regulations prescribe standards equal to or higher than those prescribed in German law'.[70]

This formulation ensures that the minimum standard is the application of German laws. The reference to regulations on 'public safety and order' further implies that German environmental law (federal or state) is applicable since that type of law definitely falls in this category. It should be stressed that the reference to German law is of dynamic character,[71] which means that it is always the legislation currently in force which must be applied. Theoretically, NATO forces therefore could be obliged under German environmental legislation to take measures to prevent environmental damage or even to clean up a contaminated site, including assuming the cost of such measures, unless otherwise agreed.

However, German legislation cannot simply be enforced on foreign forces by exercising jurisdiction because of the principle of sovereignty. The effective application of German environmental standards depends to a large extent on the degree of cooperation between the parties. If agreement cannot be reached conflict might be settled by the dispute settlement procedures in the agreements outlined above or in relevant international treaties. However, for political reasons it appears unlikely that this would occur, and resolution of such a conflict would most likely be achieved by diplomatic means.

In order to address such situations more adequately Germany initiated negotiations with its NATO partners to amend NATO SOFA and clarify existing uncertainties. These efforts led to conclusion of the Agreement to Amend the Agreement of 3 August 1959[72] (Supplementary Agreement to NATO SOFA) which was signed on 18 March 1993 but which has yet to enter into force. It *inter alia* provides that accommodation made available to NATO forces must be used in conformity with relevant German legislation. It further provides that in the case of imminent danger to public health and safety the competent German authorities are authorized to enforce the appropriate administrative act to prevent harm. Once the Supplementary Agreement enters into force the above-mentioned uncertainties related to the applicability of German environmental regulations will be resolved.

The forces of the former Soviet Union

The withdrawal of the troops of the former Soviet Union was accomplished by the end of August 1994. The relationship between the former Soviet troops and the reunited Germany was governed by the Agreement on the Withdrawal of Troops (Truppenabzugsvertrag) of 12 October 1990. Article 2, paragraph 5 of it

[70] *Bundesgesetzblatt* (Federal gazette] *1961*, II, p. 1183.
[71] Bothe, (note 64), p. 145.
[72] The text of the agreement is reproduced in *Deutscher Bundestag* [Official journal], BT-Drs. 12/6477.

states: 'The Soviet forces, its members as well as their dependants respect the sovereignty of the Federal Republic of Germany and German law'.

Regarding the use of accommodation by the former Soviet troops the provisions of Article 8 further underline the duty to respect German law, explicitly referring to legislation relating to public health, public safety and order, and environmental protection. Article 8 goes farther than the respective regulations of NATO SOFA in the version currently in force. There can be no doubt that the former Soviet troops could have been held liable for environmental damage caused by them, including clean-up of contaminated sites. The Agreement on the Withdrawal of Troops established a complex system for establishing compensation payments. However, according to a joint declaration made in Moscow by German Chancellor Helmut Kohl and Russian President Boris Yeltsin on 16 December 1992 both countries agreed to refrain from making financial claims on each other that were related to the transfer of military sites to German authority.

The eastern states subsequently claimed that this fundamental change to the agreement caused them substantial loss. They argued that the current standard (i.e., elimination of threats to people and the environment and contributing to the cost of clean-up measures only in an amount equivalent to the profit made by the Federation in the sale of a site) gave them less compensation than they would have received under the Withdrawal Agreement. In addition, they claimed that such a fundamental change of an international agreement, which was subject to approval by Parliament, would require the involvement of the Parliament, which was not the case.

VII. Conclusions

The German Government has failed to present an adequate solution to the problem of old chemical munitions and toxic armament wastes. Although there are reasons why the federal government ought not to assume full responsibility for these problems, the states argue convincingly that they are unable to deal with the problem alone. Neither the Federation nor the states can bear sole responsibility for the cost. It is high time for constructive dialogue which aims to achieve a fair sharing of the burden.

Part VIII. Final discussion and recommendations

23. Conclusions

THOMAS STOCK and KARLHEINZ LOHS

I. Introduction

The issue of old chemical munitions and toxic armament wastes did not arouse much public debate until recently, but it is now a topic which is receiving increased attention. The lack of interest was in part because during the cold war any activities of a military nature were regarded as necessary because of the East–West confrontation.

Environmental groups expressed concern about old chemical munitions and toxic armament wastes, but governmental authorities showed little interest. The division of Europe after World War II and the stationing of troops in Germany and elsewhere were taken for granted. Environmental damage from military activities was often felt to be a 'necessary by-product', and the toxic wastes which resulted from military training and service activities were regarded as unavoidable. With the end of the cold war and the withdrawal of foreign troops from Europe the opportunity arose to rethink the problem and to explore new ways to cope with the toxic armament wastes which result from military activities. The recovery of military areas is an expensive undertaking, and careful planning and assessment of the extent of the damage is essential before beginning renewal and redevelopment activities.

Past efforts related to chemical weapon disarmament focused on concluding the Chemical Weapons Convention. The major goal of the CWC is to eliminate all chemical weapons and to ensure that no new chemical weapons are produced. The conclusion of the CWC and its entry into force on 29 April 1997 are significant chemical disarmament measures.

Old and abandoned chemical weapons were first addressed in the latter part of the CWC negotiations at the Conference on Disarmament since it was understood that politically sensitive questions were involved that might undermine the primary goal—chemical disarmament. The final stage of the negotiations attempted to provide realistic solutions to a number of problems, including: (*a*) how to cope with the chemical weapons produced by Germany in the 1930s and 1940s which were captured by the Allies after World War II and dumped at sea; (*b*) how to determine what should be done with the dumped weapons; and (*c*) how to assign responsibility for clean-up activities and the funding of such efforts. The matter was complicated by the fact that states were often not eager to be reminded of their past dumping activities and the need to assume responsibility for them. Nevertheless, the dumped munitions exist and must be dealt with, even if the issue is morally and politically sensitive.

Many of the issues related to old chemical munitions and toxic armament wastes are complicated both to analyse and explain. In order to understand the magnitude and complexity of the problem it is necessary to study its history.

II. Major findings of the study

In the following summary of the study nine theses are presented and discussed. The recommendations made by the study are then briefly outlined.

Thesis 1. There is uncertainty about the type and quantity of existing old and abandoned chemical munitions.

In Europe and elsewhere there is lack of knowledge about the type and quantity of old and abandoned CW munitions which are present in various areas. More research on chemical warfare activities in World War I and World War II is needed.

The use of chemical weapons in World War I encouraged many Western countries to conduct research and development activities on the possible use of chemical weapons. Chemical weapons were tested; CW stockpiles were built up; and soldiers were trained to use chemical weapons. The history of such programmes is generally well documented in the countries (such as Canada, France, Germany, Russia, the UK and the USA) which had chemical warfare programmes. However, the general public is often unaware of these activities.

Entry into force of the CWC makes it essential that countries research their past CW activities, including production, so that they can prepare the necessary declaration on old and abandoned chemical weapons which the CWC requires. However, there may be reluctance to provide information on past chemical warfare programmes and CW disposal activities if it is not already generally known. Some countries, such as Russia, appear unwilling to disclose information about past activities. Even if countries are willing to open their historical records, the necessary data may no longer be available.

States, especially those in the developing world, which have been involved in military conflict in the past 50 years must recognize that there are probably remnants of munitions, including chemical munitions, on their territories.

Conclusion

States should give the problem of old and abandoned chemical weapons serious attention. States which are parties to the CWC must scrutinize relevant historical records and prepare an inventory of sites where old or abandoned chemical weapons may be located. The fundamental political changes which have occurred in Europe and the growing readiness to address environmental problems should encourage increased openness.

Thesis 2. The remaining stockpiles of old chemical munitions in Europe and elsewhere present a threat to people and the environment.

Although they may be more than 50 years old, some CW agents and munitions may still be hazardous even if they have been buried or dumped at sea. Much research has been done on these weapons, and some of them have been destroyed. However, it remains difficult to predict how these munitions will act outside the laboratory since environmental factors may have caused them to deteriorate. Each munition therefore must be treated as unique.

The status of old CW munitions and agents can be determined by studying past handling of old chemical weapons, including destruction efforts; acquiring data on past CW programmes and production activities, including information about the period in which such activities occurred; and by using modern analytical instruments and techniques. The review of the literature which was undertaken for this study[1] made clear that this is a topic which has been studied extensively. Nevertheless more data are needed on the deterioration and degradation of chemical agents and CW munitions in soil and water since laboratory findings are not always directly comparable to conditions in the field.

Conclusion

There must be more research on the long-term behaviour, stability and degradation of chemical warfare agents and munitions. In addition, more study is needed of the environmental damage caused by chemical warfare agents and by both conventional and CW munitions. There is a need for greater international coordination of research activities. The CWC could serve as an umbrella for such activities, but European and UN organizations, such as the United Nations Environment Programme, should also be involved.

Thesis 3. The consequences of dumping old chemical munitions at sea are of growing concern.

Chemical munitions were dumped in the Baltic Sea, the North Sea and other ocean waters both before and after World War II. Data about the amounts which were dumped are uncertain in part because of the passage of time and the lack of accurate records, and also because some of the countries involved in dumping operations are unwilling to disclose the information which they possess.

The states which border the Baltic Sea established the Ad Hoc Working Group on Dumped Chemical Munition of the Baltic Marine Environment Protection Commission of the Helsinki Commission (HELCOM CHEMU) to coordinate information from national reports about past dumping operations in the Baltic. The data in these reports have made it possible to estimate more accurately the quantity of CW munitions which were dumped in the Baltic Sea after World War II. Russia, however, has yet to present information about suspected Soviet dumping activities, which are believed to have occurred even later.

The working group recommended further action based on scientific investigation of the long-term behaviour and possible effect on the environment of the dumped chemicals. Cleaning (by raising and destroying the munitions) is not possible because of the extent of the problem. Removing large quantities of chemical munitions from the seabed would involve the following steps: recovery, transport, storage and destruction. This would be expensive and technically difficult, if not impossible. On the other hand, the munitions will degrade gradually as the shells deteriorate and their contents come in contact with sea water.

[1] Research for this study led to the creation of a SIPRI database on old chemical munitions and warfare agents and toxic armament wastes.

An additional danger is posed by the fact that CW munitions were often dumped together with conventional munitions, which could detonate and injure divers during a recovery operation.

Further study is needed of how dumped munitions affect the sea, including their effect on water quality, sediment and marine organisms. The working group concluded that the impact of the dumped munitions was not as great as had first been expected. The most important measure which can be taken is to ensure that no additional toxic contaminants are added to the sea by further dumping. Chemical weapons which were dumped at sea in the past are not affected by the CWC, which does not cover chemical weapons dumped at sea before 1985.

Conclusion

Chemical weapons were dumped at sea in the past, and there is growing concern about how these activities may have affected the marine environment. A distinction must be made between true threats to the environment and understandable public concern. International cooperation will be required to deal with the problem, and costs should be shared. The areas which present the greatest hazard need to be determined, and international organizations should be involved in clean-up efforts. The initial action should be a comprehensive mapping of the primary dumping areas throughout the world.

Thesis 4. The existing capability to destroy old chemical munitions is inadequate.

The processes involved in the dismantling and destruction of old chemical weapons are well known, but the technical capacity to destroy them is limited. The experience of some European states has shown that it is possible to dismantle, store and destroy old chemical munitions. In some countries dismantled munitions and large quantities of old agents have accumulated and await destruction because of inadequate destruction capacity. Environmental concerns have also led to delays.

Old and abandoned chemical weapons are often unique or difficult to handle, and there is a need to study unconventional alternative destruction technology and its application. Destruction methods which can be used on-site (e.g., chemical detoxification) may be particularly valuable. Some countries which possess old chemical munitions are unable to maintain a facility to destroy them. New types of destruction (such as chemical and biological degradation, plasma arc techniques, and the like) and the development of portable units for on-site destruction may be possible to use in such instances.

Determination of the most reliable and appropriate destruction technique must be made on-site by experts and by analysis of the impact that the selected technique will have on the environment. Each discovery of old chemical munitions or toxic agent wastes is unique, and the choice to destroy the munitions on-site or to remove them to a central destruction site must be made on a case-by-case basis.

Conclusion

The destruction of old and abandoned chemical weapons and toxic armament wastes provides an opportunity to examine new and unconventional destruction techniques. The munitions in question must be handled by experts. Existing destruction capacities are limited. Under the CWC it may be possible to establish bilateral agreements to share information about destruction technology.

Thesis 5. International cooperation is needed to solve the problem of old and abandoned chemical munitions.

Discovering, dismantling, transporting and destroying old chemical munitions require experience and expertise, and some states already possess that knowledge. It does not matter whether experts have received their training from the military or private industry, but most dismantling and destruction activities have been conducted by military specialists. The skills and knowledge needed to disarm chemical weapons are complex, and the individuals who are involved in these activities must continue to train to maintain their expertise.

Old and abandoned chemical weapons may be more than 50 years old, and knowledge of their chemical fillings, explosive devices, shell designs or special components is not always available or well documented. The information and data which some countries possess should be made available to more countries. The OPCW could perhaps facilitate such cooperation between states.

The CWC provides a unique opportunity to develop cooperation between states which possess old chemical weapons. While the OPCW will serve as an umbrella organization, states will continue to be responsible for the discovery and destruction of old chemical weapons on their territories. Abandoned chemical weapons may create political tension between states, and it is recommended that the OPCW develop clear guidelines and recommendations to deal with such problems.

Conclusion

It is essential that states share information about the technical aspects of the dismantling and destruction of old and abandoned chemical munitions. The various bodies of the CWC can foster cooperation and assist in the development of mechanisms for the exchange of data and information.

Thesis 6. The CWC provides a framework for dealing with old and abandoned chemical munitions.

The CWC mandates the OPCW to coordinate efforts to solve the problems related to old and abandoned chemical munitions. The OPCW will have access to all available data about old and abandoned chemical weapons now that the Convention has entered into force and after its parties have made their first declaration regarding old weapons. Currently, the extent of the problem is unknown since only a few countries have disclosed information about the existence of old or abandoned chemical weapons on their territories.[2] The PrepCom requested that signatory states provide information about these weapons in advance of entry into force of the CWC, but most states did not comply with the request.[3]

Conclusion

The national and international institutions under the CWC can provide a framework to encourage and assist in the dismantling and destruction of old and abandoned chemical weapons.

Thesis 7. The CWC assigns the National Authority a special responsibility related to old and abandoned chemical munitions.

In most countries the military is responsible for dealing with old chemical munitions, including their storage and destruction. The national implementation obligation of the CWC requires a state's National Authority to assume responsibility for the state's declaration obligation. A party to the CWC must implement the necessary legal and administrative measures to ensure that its National Authority can fulfil this responsibility. The CWC contains provisions which cover chemical weapons produced before 1925. These old weapons are treated as toxic wastes, and the responsibility for their disposal or destruction remains entirely with the state concerned. National legislation must ensure that these toxic wastes are destroyed in an environmentally safe manner. There is an additional obligation to report to the OPCW.

Conclusion

States are required to review and modify their national legislation (including environmental laws) as needed to provide for the destruction of toxic wastes, in general, and old chemical weapons or armament wastes, in particular.

Thesis 8. The withdrawal of forces from former military bases in Europe necessitates the removal of toxic armament wastes and provides an opportunity to develop new decontamination techniques.

The withdrawal of military forces from European bases focused attention on the contamination of these areas by past training and service activities. An opportunity now exists to study the pattern of environmental damage, the possibility to redevelop such sites and new approaches to decontamination. Toxic armament wastes are remnants of a variety of past military activities. Such wastes include: chemical substances and precursors used in the production of munitions; oil, petrol, flammable fuels and chemicals used in military vehicles; and wastes generated by the dismantling of weapons. The decontamination agents used in the dismantling process have often had a negative effect on the environment.

[2] Germany and China have done so. This study presents information which indicates that Belgium possesses stocks of old chemical munitions from World War I.

[3] Note by the Executive Secretary: request for information on old chemical weapons, PrepCom document PC-V/B/6, 7 Dec. 1993.

Clean-up activities at former CW production and storage sites can provide valuable information about how best to investigate such sites and to develop a plan of action for their redevelopment. If redevelopment activities are inadequate significant harm to people and the environment could result.

Conclusion

The CWC does not cover the clean-up of former armament production sites. Such activities could be encouraged by mutual cooperation and the exchange of data, and past experience in dealing with toxic armament wastes should be more openly shared. There is a need for international cooperation, exchange of information and expertise, and more active involvement by organizations such as the European Union.

Thesis 9. States must review their legislation on toxic and armament wastes.

In many countries the environmental legislation is inadequate to deal with the problems of cleaning up and redeveloping areas which are contaminated by toxic wastes. It is estimated that only 5–8 per cent of all toxic waste problems result from old and abandoned armaments, but the contamination is often extremely hazardous. A large financial and technological investment may be needed to deal with the toxic waste problem. International cooperation and the exchange of information on decontamination are advisable because financial resources are limited.

Conclusion

States should review and amend their national legislation as needed in order to deal adequately with toxic and armament wastes.

SIPRI publications on CBW

• The Problem of Chemical and Biological Warfare, 6 volumes

1. The Rise of CB Weapons. 1971. ISBN 91–85114–10–3.
2. CB Weapons Today. 1973. 91–85114–16–2.
3. CBW and the Law of War. 1974. 91–85114–17–0.
4. CB Disarmament Negotiations, 1920–1970. 1971. 91–85114–11–1.
5. The Prevention of CBW. 1971. 91–85114–13–8.
6. Technical Aspects of Early Warning and Verification. 1975. 91–85114–18–9.

• Chapters in the SIPRI Yearbooks

SIPRI Yearbook of World Armaments and Disarmament, 1968/69. 1969. 91–85114–03–0.
pp. 112–34. Development in chemical and biological warfare.

SIPRI Yearbook of World Armaments and Disarmament, 1969/70. 1970. 91–85114–07–3.
pp. 185–206. The CBW debate and other disarmament measures.

World Armaments and Disarmament: SIPRI Yearbook 1972. 1972. 91–85114–12–X.
pp. 501–22. Chemical and biological disarmament.

World Armaments and Disarmament: SIPRI Yearbook 1973. 1973. 91–85114–19–7.
pp. 383–91. Chemical disarmament.

World Armaments and Disarmament: SIPRI Yearbook 1974. 1974. 0–262–19129–6.
pp. 370–84. Chemical disarmament.

World Armaments and Disarmament: SIPRI Yearbook 1975. 1975. 0–262–19140–7.
pp. 426–32. Chemical disarmament.

World Armaments and Disarmament: SIPRI Yearbook 1977. 1977. 91–22–00116–6.
pp. 86–102. Dioxin: a potential chemical-warfare agent; and
pp. 364–67. Prohibition of biological and chemical weapons.

World Armaments and Disarmament: SIPRI Yearbook 1978. 1978. 0–85066–134–X.
pp. 360–76. The destruction of chemical warfare agents.

World Armaments and Disarmament: SIPRI Yearbook 1979. 1979. 0–85066–181–1.
pp. 470–89. Stockpiles of chemical weapons and their destruction.

World Armaments and Disarmament: SIPRI Yearbook 1980. 1980. 0–85066–201–X.
pp. 365–79. Chemical disarmament.

World Armaments and Disarmament: SIPRI Yearbook 1982. 1982. 0–85066–230–3.
pp. 317–61. The changing status of chemical and biological warfare: recent technical, military and political developments; and
pp. 456–57. Chemical weapons.

World Armaments and Disarmament: SIPRI Yearbook 1983. 1983. 0–85066–247–8.
pp. 391–426. Chemical and biological warfare: developments in 1982; and
pp. 563–67. Chemical disarmament.

World Armaments and Disarmament: SIPRI Yearbook 1984. 1984. 0–85066–263–X.
pp. 319–49. Chemical and biological warfare: developments in 1983; and
pp. 421–54. Implications of genetic engineering for chemical and biological warfare.

World Armaments and Disarmament: SIPRI Yearbook 1985. 1985. 0–85066–297–4.
pp. 159–205. Chemical and biological warfare: developments in 1984; and
pp. 206–19. An analysis of the reports of Iraqi chemical warfare against Iran, 1980–84.

World Armaments and Disarmament: SIPRI Yearbook 1986. 1986. 0–19–829100–0.
pp. 159–79. Chemical and biological warfare: developments in 1985.

SIPRI Yearbook 1987: World Armaments and Disarmament. 1987. 0–19–829114–0.
pp. 97–115. Chemical and biological warfare: developments in 1986.

SIPRI Yearbook 1988: World Armaments and Disarmament. 1988. 0–19–829126–4.
pp. 101–25. Chemical and biological warfare: developments in 1987.

SIPRI Yearbook 1989: World Armaments and Disarmament. 1989. 0–19–827751–2.
pp. 99–130. Chemical and biological warfare: developments in 1988.

SIPRI Yearbook 1990: World Armaments and Disarmament. 1990. 0–19–827862–4.
pp. 107–40. Chemical and biological warfare: developments in 1989; and
pp. 521–44. Multilateral and bilateral talks on chemical and biological weapons.

SIPRI Yearbook 1991: World Armaments and Disarmament. 1991. 0–19–829145–0.
pp. 85–112. Chemical and biological warfare: developments in 1990; and
pp. 513–39. Multilateral and bilateral talks on chemical and biological weapons.

SIPRI Yearbook 1992: World Armaments and Disarmament. 1992. 0–19–829159–0.
pp. 147–86. Chemical and biological warfare and arms control developments in 1991; and
pp. 509–30. The United Nations Special Commission on Iraq.

SIPRI Yearbook 1993: World Armaments and Disarmament. 1993. 0–19–829166–3.
pp. 259–92. Chemical and biological weapons: developments and proliferation;
pp. 293–305. Benefits and threats of developments in biotechnology and genetic engineering;
pp. 691–703. The United Nations Special Commission on Iraq: activities in 1992;
pp. 705–34. The Chemical Weapons Convention: the success of chemical disarmament negotiations; and
pp. 735–56. The Convention on the Prohibition of the Development, Production, Stockpiling and Use of Chemical Weapons and on their Destruction.

SIPRI Yearbook 1994. 1994. 0–19–829182–5.
pp. 315–42. Chemical weapon developments;
pp. 685–711. The Chemical Weapons Convention: institutionalization and preparation for entry into force;
pp. 713–38. Biological weapon and arms control developments; and
pp. 739–58. UNSCOM: activities in 1993.

SIPRI Yearbook 1995: Armaments, Disarmament and International Security. 1995. 0–19–829193–0.
pp. 337–57. Chemical and biological weapons: developments and destruction;
pp. 597–633. Multilateral weapon-related export control measures; and
pp. 725–60. Chemical and biological arms control.

SIPRI Yearbook 1996: Armaments, Disarmament and International Security. 1996. 0–19–829202–3.
pp. 537–59. Multilateral military-related export control measures; and
pp. 661–708. Chemical and biological weapon developments and arms control.

SIPRI Yearbook 1997: Armaments, Disarmament and International Security. 1997. 0–19–829312–7.
pp. 345–63. Multilateral military-related export control measures; and
pp. 437–68. Chemical and biological weapon developments and arms control.

• Other SIPRI publications

Chemical Disarmament: Some Problems of Verification. 1973. 91–85114–20–0.

The Effects of Developments in the Biological and Chemical Sciences on CW Disarmament Negotiations. SIPRI research report 13, SIPRI, 1974.

Delayed Toxic Effects of Chemical Warfare Agents. 1975. 91–85114–29–4.

Chemical Disarmament: New Weapons for Old. 1975. 91–85114–27–8.

Medical Protection against Chemical-Warfare Agents. 1976. 91–22000–44–5.

Ecological Consequences of the Second Indochina War. 1976. 91–22000–62–3, pp. 24–45, 53–55.

Weapons of Mass Destruction and the Environment. 1977. 0–85066–132–3, pp. 31–48.

The Fight against Infectious Diseases: A Role for Applied Microbiology in Military Redeployment. SIPRI, 1979, mimeo. 91–85114–26–X.

Chemical Weapons: Destruction and Conversion. 1980. 0–85066–199–4.

Herbicides in War. 1984. 0–85066–265–6.

Biological and Toxin Weapons Today. 1986. 0–19–829108–6.

SIPRI publications may be obtained through all main bookshops. In case of difficulty, please contact the publishers or SIPRI.

SIPRI Chemical & Biological Warfare Studies

No. 1 Effects of Chemical Warfare: A selective review and bibliography of British state papers, by Andy Thomas

No. 2 Chemical Warfare Arms Control: A framework for considering policy alternatives, by Julian Perry Robinson

No. 3 The Detoxification and Natural Degradation of Chemical Warfare Agents, by Ralf Trapp

No. 4 The Chemical Industry and the Projected Chemical Weapons Convention: Proceedings of a SIPRI/Pugwash Conference, volume I

No. 5 The Chemical Industry and the Projected Chemical Weapons Convention: Proceedings of a SIPRI/Pugwash Conference, volume II

No. 6 Chemical and Biological Warfare Developments: 1985, by Julian Perry Robinson

No. 7 Chemical Weapon Free Zones?, edited by Ralf Trapp

No. 8 International Organization for Chemical Disarmament, by Nicholas A. Sims

No. 9 Non-Production by Industry of Chemical-Warfare Agents: Technical Verification under a Chemical Weapons Convention, edited by S. J. Lundin

No. 10 Strengthening the Biological Weapons Convention by Confidence-Building Measures, edited by Erhard Geissler

No. 11 National Implementation of the Future Chemical Weapons Convention, edited by Thomas Stock and Ronald Sutherland

No. 12 Views on Possible Verification Measures for the Biological Weapons Convention, edited by S. J. Lundin

No. 13 Verification of Dual-use Chemicals under the Chemical Weapons Convention: The Case of Thiodiglycol, edited by S. J. Lundin

No. 14 Verification under the Chemical Weapons Convention: On-Site Inspection in Chemical Industry Facilities, by Ralf Trapp

No. 15 Control of Dual-Threat Agents: The Vaccines for Peace Programme, edited by Erhard Geissler and John P. Woodall

No. 16 The Challenge of Old Chemical Munitions and Toxic Armament Wastes, edited by Thomas Stock and Karlheinz Lohs

The manufacturer's authorised representative in the EU for product safety is Oxford University Press España S.A. of El Parque Empresarial San Fernando de Henares, Avenida de Castilla, 2 - 28830 Madrid (www.oup.es/en or product.safety@oup.com). OUP España S.A. also acts as importer into Spain of products made by the manufacturer.
Printed and bound by CPI Group (UK) Ltd, Croydon, CR0 4YY
12/09/2026
02215894-0005